SPARK

SPARK

THE LIFE OF ELECTRICITY AND THE ELECTRICITY OF LIFE

TIMOTHY J. JORGENSEN

PRINCETON UNIVERSITY PRESS
PRINCETON & OXFORD

Published by Princeton University Press
41 William Street, Princeton, New Jersey 08540
6 Oxford Street, Woodstock, Oxfordshire OX20 1TR

press.princeton.edu

Library of Congress Cataloging-in-Publication Data

Names: Jorgensen, Timothy J., author.
Title: Spark : the life of electricity and the electricity of life / Timothy J. Jorgensen, Princeton University Press.
Description: Princeton : Princeton University Press, [2021] | Includes bibliographical references and index.
Identifiers: LCCN 2021026499 (print) | LCCN 2021026500 (ebook) | ISBN 9780691197838 (hardback) | ISBN 9780691232652 (ebook)
Subjects: LCSH: Electricity. | BISAC: SCIENCE / Physics / Electricity | SCIENCE / History
Classification: LCC QC527 .J66 2021 (print) | LCC QC527 (ebook) | DDC 537—dc23
LC record available at https://lccn.loc.gov/2021026499
LC ebook record available at https://lccn.loc.gov/2021026500

British Library Cataloging-in-Publication Data is available

Editorial: Ingrid Gnerlich, Maria Garcia, and Whitney Rauenhorst
Production Editorial: Natalie Baan
Production: Danielle Amatucci
Publicity: Kate Farquhar-Thomson and Sara Henning-Stout
Copyeditor: Jennifer McClain

Jacket art and design by Sukutangan

This book has been composed in Arno

Printed on acid-free paper. ∞

Printed in the United States of America

10 9 8 7 6 5 4 3 2 1

For my vivacious, bright, and beautiful wife, Helen. She gets the credit for first proposing I write a book about electricity from a biological perspective, a creative idea for which I am very grateful. My life would be severely lacking without her unwavering love and devotion.

(What she sees in me I have no idea.)

CONTENTS

PREFACE

CURRENT EVENTS

> Electricity is often called wonderful, beautiful; but it is so only in common with the other forces of nature.
>
> —MICHAEL FARADAY

We live in an electrical society. Most of our daily activities are powered, illuminated, and otherwise enhanced by electricity. If you've ever experienced a prolonged electrical power outage, you know what I mean. Yet, some people have curious ideas about what electricity actually is. They think of electricity as an alien physical force, outside of their bodies, that's confined within their electronic devices and channeled from appliance to appliance through wires. They don't appreciate that electricity is also a biological force, essential to the life of all animals that have a nervous system, and even those that don't.

Many people also are unaware that electricity is really the very foundation of life. It's the spark that brought the first primitive life-forms into existence and started them down the evolutionary path leading to today's complex species with sophisticated internal electrical systems. Life is nothing if not electrical. You might as well call it "eLife."

This is a book about electricity from a biological perspective. That may seem an unusual approach, but I think it's the best way to tell the story of electricity. That's because many attributes of electricity were first discovered through the study of electricity's effects on bodily senses and muscle tissues. And, amazingly, almost everything we know about our nervous system comes from electrically probing nerves. So it might not be too surprising that electrical science and neurological science share a joint story.

Discoveries in the electrical and neurological fields leapfrogged each other over the centuries, as developments in one field enabled advancements in the other. The term *electricity* was first used in the late sixteenth century by the English scientist William Gilbert to describe his investigations of electricity's physical powers. And the term *neurology* was first used shortly thereafter by the English physician Thomas Willis to describe his studies of human reflexes and paralysis. Such neurological studies progressed from mere descriptions of neural anatomy to functional investigations of how the nervous system works, using targeted electrical stimulation as a probe.

But it's not just nerves that respond to electrical signals. Muscle cells, including heart muscle, rely on electrical stimuli to control their mechanical functions. And even nonmuscle tissues send and receive electrical signals through the nervous system, as they report their status to, and receive their instructions from, the brain. In fact, virtually all of our bodily functions are electrically monitored and controlled.

It's because electrical regulation of cells, tissues, and organs is so vital to survival that disruption of the body's "electronics" can be so damaging, and even lethal. With electricity all around us, we need to better know when, why, and how electricity kills. Fortunately, we have learned much about the lethal effects of electricity, especially when it's delivered in large, abrupt doses, sometimes call *jolts*. This information is important for the prevention and treatment of severe electrical shocks.

Since its discovery, electricity has repeatedly been employed to treat various diseases. Hampered by poor understanding of human physiology, ignorance of the mechanisms of diseases, and misunderstanding of the role of electricity in the body, the track record of early electrical therapies was spotty, to say the least. Hair removal by electricity (hair electrolysis) is perhaps one of the few early electrical medical procedures that persists, little changed, to the present day.

After its early heyday, in the late nineteenth century, electrical therapy fell out of fashion and even became taboo within the medical community. But as the electrical and neurological sciences have matured, there has been a resurgence of interest in using electricity to treat diseases. This is particularly true for treatments of neurological disorders, such as electroconvulsive therapy for depression and deep brain stimulation for Parkinson's disease, but is also true for many other illnesses. The use of electricity for medical purposes ranges from the mental control of prosthetic limbs by amputees to the restoration of vision to the blind. In the future, electricity may even replace many drugs as first-line therapies for treating a variety of diseases. As such, it might be possible to avoid the adverse side effects of those drugs, which can often be severe. But we are also learning that electricity's use to treat anxiety can, unfortunately, sometimes be nearly as addictive as drugs.

This book is *not a science textbook* on electricity or electrophysiology. If you are an electrical engineer or an electrophysiologist, this book will not satisfy your specialized technological needs. Rather, it is my hope that the book will be accessible to the widest possible audience. I have, therefore, avoided technical jargon, and specifically focused on the stories about electricity that are important to understanding electricity's relationship to biology and health.

This is also *not a history of science book*. Although some of the stories I relate are taken from history, and everything in the book is historically accurate, I have made no attempt to compile a complete history of electricity. Neither this book, nor any other single-volume work, can do justice to electricity's long and rich history. Rather, I have selectively chosen stories that illustrate the scientific points I want to focus on regarding the intersection of electricity and biology. I also plead guilty of practicing *presentism*—the tendency to interpret past events in terms of modern concepts, values, and understandings. Presentism is a cardinal sin for historians of science. But I am not a historian and this is not a history book. I actually find presentism a useful tool to explain how it is that the science got us from where we were then to where we are now, which is one of my main objectives for this book.

Which brings us to the logical question: If not a science textbook or a history of science book, what, then, is this book? I would describe it as a book of explanatory science, written in narrative style that seeks to educate and entertain. I hope you will agree that it fulfills both goals.

My intended audience is inquisitive people interested in learning what electricity is and how it works, particularly with regard to how it affects people's lives. I hope that readers of this book will learn a lot of new information about electricity. But I also suspect that many people will find that what they thought they knew about electricity is wrong, or at least incomplete.

For example, did you know that Benjamin Franklin almost killed himself while performing an electrical experiment with a turkey? (I'll bet you thought I was going to say a kite.) Do you know why people and animals aren't killed by touching electric fences? Did you know that while the measured lengths of lightning bolts keep going up worldwide, with some bolts now hundreds of miles long, the annual lightning death rate in the United States keeps going down? Do you know why people in Denmark are virtually never electrocuted at home? Do you know how amputees with state-of-the-art prosthetic limbs can control the limb's movement just by thinking about it? Did you know that electricity can help totally blind people to see and deaf people to hear? Did you know that shock therapy for mental disorders has nothing to do with electricity? If you don't know the answers to these questions, or even if you do, I think you'll enjoy taking a fresh look at these electricity topics from a somewhat different perspective—a biological one.

You will get the most out of this book if you appreciate how it is styled and organized. The book is written as a work of nonfiction, told as an unfolding story of exploration and discovery. It is meant to be read as you would read a novel—from beginning to end, without skipping chapters or reading them out of sequence. Each chapter builds on the previous one and sets the stage for the next. Reading chapters out of order will result in a confusing and garbled storyline. Also, you will find no graphs or tables to distract you, and the mathematics is kept to a minimum. Rather, this book conveys all technical material exclusively within the context of the story.

Each chapter in the early part of the book is a self-contained story focused on a particular topic related to electricity or biology, typically neurobiology. The electricity and biology chapters alternate back and forth to support the main thesis of the book: that the sciences of electricity and neurobiology progressed together—hand in hand, so to speak—with advances in one field dependent upon advances in the other.

The theme of the early chapters is the premature use of electrical therapies (or *electrotherapies*) by physicians, based on a fundamental misunderstanding of the underlying sciences of both electricity and human physiology. Many of these treatments were painful, and virtually none was effective. These early electrotherapy stories are told anecdotally and serve to underscore the level of ignorance about electricity and human physiology that was prevalent in medicine at the time.

In the latter part of the book, the stories about electrical physics begin to wane, as I have already presented most of the information needed to understand the fundamental electrical nature of neurobiology. Still, all the stories about neurobiology and electricity have not yet been told. At this point, the neurobiology chapters begin to alternate with medical chapters. The theme of the latter part of the book is how recent insight into the mechanisms underlying electrical and neurological interactions is leading to the development of new electrical treatments for various human maladies. Many of these novel electrical treatments, unlike their predecessors, are currently being scientifically validated as safe and effective. After a hiatus of more than a century, the use of electricity to treat human disease is returning in a big way, and it seems to have a bright future across many medical specialties.

The text of the book has ample endnotes. The purpose of the endnotes is twofold—to provide citations and sources for the statements in the narrative, and to supply expanded explanations of something mentioned in the text for readers who have a particular interest in that topic. The narrative does not rely on any information in the endnotes, so they need not be read, but a deeper or more thorough understanding of the topic can be obtained by reading the endnotes.

The book has no glossary. Just as the best understanding of a word is usually achieved by seeing the word used in context rather than by looking up its definition in a dictionary, such is often the case with technical terminology. All technical words, or common words that are used in a specifically technical way, are italicized at their first appearance in the narrative and defined within the context of the narrative. These words also appear in the book's comprehensive index, so that the reader can easily find the word again if its meaning has been forgotten.

And there is one last thing you should know before you start reading. Although I believe this book is appropriate for anyone, there are two chapters some might find objectionable. The chapter on early electrotherapies (chapter 4) has some discussion of sexual matters, and the chapter on electrocution (chapter 6) contains a good bit of violence. So I've written those chapters in such a way that they can be skipped without a significant loss of continuity to the book's narrative. But I hope you don't skip them, as they convey important information regarding health and safety issues, which is one of the reasons I wrote this book in the first place. In any event, you have now been warned; proceed accordingly.

Finally, I assure you I have no connection, financial or otherwise, with any of the companies or products mentioned in this book.

Beyond just informative and enjoyable, I'd like the scientific story of this book to be useful. I want it to enable readers to enhance their everyday interactions with electricity in many practical ways. Electricity is a powerful force of nature that everyone deals with to one extent or another. If we all become a little more knowledgeable about it, our society can use electricity's many powers more wisely . . . for the betterment and safety of all.

I hope you enjoy reading the story of electricity, and come to see sparks in an entirely new light.

SPARK

1

SPARKS WILL FLY

STATIC ELECTRICITY

We see spiders, flies, or ants entombed and preserved forever in amber, a more than royal tomb.

—SIR FRANCIS BACON

Copenhagen, Denmark, is an eye-catching city with bold architecture complementing natural beauty. But if you perform an internet search for images of Copenhagen, most of the hits will show colorful photographs taken from just a single vantage point: the dockside view at Nyhavn harbor. At Nyhavn, a seventeenth-century waterfront docking area and now popular entertainment district, many vividly colored buildings and boats line both sides of its narrow waterway. On a sunny day with a blue sky, it is unquestionably the most photogenic location in a city full of photogenic locations. Yet, the most beautiful things in Nyhavn cannot be seen in any of these photographs.

Nyhavn's stunning treasures are housed in a small, white building on the extreme far left, out of frame for most pictures of the harbor. The treasures are literally jewels, and they are on display in the House of Amber, a combination jewelry store and amber museum.[1] The three-story building itself is historic, dating to 1606, but its contents are prehistoric, with some amber pieces estimated to be 30–50 million years old. As I enter the House of Amber, a young professional woman, smartly dressed, introduces herself as Birgitte Niclasen and welcomes me. She works in the House of Amber's human resources and special projects section, and she offers to give me a tour of the museum's amber collection.

Niclasen knows a lot about amber, the national gemstone of Denmark. She explains that amber is fossilized sap from prehistoric pine trees and is found in various parts of the world, but particularly in the Baltic Sea region because of the vast prehistoric conifer forests that existed in this area. Amber, with a density only slightly heavier than water, is remarkably light. As such, it easily was washed into prehistoric rivers, where it was carried out to sea. For this reason, amber often can be found on beaches in areas that were once prehistoric river estuaries. In fact, much of the world's amber stockpiles simply were picked up from beaches by professional or amateur collectors. The collectors know which stretches of shoreline are most likely to yield amber, and they visit them after every storm in the hope that the storm will have dislodged more amber from the sea floor and washed new deposits onto the beach.

Amber is famous for having *inclusions,* which are imbedded small objects—often insects—frozen in time. The insects were trapped in sticky tree sap that ultimately solidified and fossilized to become amber.[2] High-quality inclusions increase the amber's ornamental value and frequently end up in some of the finest jewelry. Niclasen shows me the museum's best examples of amber pieces with insect inclusions. She notes that such inclusions are typically not pristine; rather, the insects are often damaged or have misshapen appendages. Perfect insect inclusions are rare, making them tremendously valuable for jewelry. Jewelry with inclusions of animals larger than the size of insects is extremely rare, and most such pieces are fakes. The fakes are modern-day animals simply imbedded in amber-colored synthetic resins. Niclasen tells the story of a woman visitor to the museum who showed off a pendent that appeared to have a perfect inclusion of a small snake. The woman was so proud of her "amber" pendant that Niclasen didn't have the heart to tell her it was undoubtedly a fake. Unfortunately, the woman probably

FIG 1.1. Amber. Although valued since antiquity as a gemstone, amber is actually fossilized tree resin that is millions of years old. Different types of amber vary in appearance, but Baltic amber (shown here), the most common variety, is typically clear or translucent. When polished, impurities called inclusions can sometimes be seen. The inclusions are often insects (such as the ant shown here) that were trapped in the thick liquid resin before it solidified. (Photo © Anders L. Damgaard; www.amber-inclusions.dk)

had paid a lot of money for her treasure, but her naïveté is understandable. Hawkers of fake amber have deceived even financially shrewd people. Most famously, J. P. Morgan, a wealthy nineteenth-century American financier and banker, once paid $100,000 ($2,900,000 in 2020 dollars) for an amber collection whose most precious piece had an inclusion of a frog. When he donated the collection to the American

Museum of Natural History in New York, the curators quickly recognized that the prized amber piece with the frog inclusion was actually a fake.[3]

Fortunately, there are several chemical means to identify fake amber from the real stuff. Unfortunately, most of them require a small sample of the amber for analysis and thus are modestly destructive to the piece in question.[4] There is a nondestructive method, though. Niclasen remarks, "The easiest way to identify a fake is simply to rub the suspect amber with a woolen cloth and see if the rubbed amber can move small dust particles." If it is real amber, you can electrify it by rubbing it in this way. That is, the rubbing will cause the amber to take on static electric properties: it will attract small particles, and it may even produce an electrical spark you can feel. In fact, amber's electrical properties make it unique among gemstones and enhance its appeal.[5]

All amber is beautiful, and people have likely been adorning themselves with it for as long as there have been people. Archaeologists have found amber pendants dating from 12,000 BC, which was just at the tail end of the last Ice Age.[6] Yet, the appeal of amber jewelry to primitive peoples went well beyond simply its beautiful appearance. Amber was also known to be magical. People who wore amber jewelry would often feel it give off shocks and could see it attract particles of soil or small seeds after being rubbed with wool.[7] No wonder all sorts of mystical qualities were attributed to amber!

Humans have a propensity to immediately apply all new discoveries of natural phenomena toward the treatment of their ailments. We will see this recurring pattern time and again throughout the book. Amber was no different.

The first recorded medical treatments with amber come from the ancient Greeks, but such practices probably started much earlier. The Greeks sought to utilize the attractive forces of amber to fight disease. They massaged the bodies of sick people, rubbing pain-afflicted areas with amber stones in the hope that amber's attractive forces would pull the pain out of their bodies. Later, the Romans employed amber as both a disease treatment and a disease preventative. Pliny the Elder (23–79 AD), the great Roman naturalist and naval commander who was killed at Pompeii by the eruption of Mount Vesuvius, recommended wearing amber around the neck to ward off throat diseases and mental disorders.

Powdered amber later found its way into various medicines and potions, and many people claimed to have benefited from its healing powers. As late as the sixteenth century, the famous Swiss physician Paracelsus recommended amber as "a noble medicine for the head, stomach, intestines and other sinews complaints."[8] Whether anything more than a placebo effect was at play is quite doubtful. But even a placebo can have clinical value.[9]

Apart from the attractive forces coming from rubbed amber, the stone also could produce sparks. Both the attractive forces and the sparks required the amber to be rubbed, suggesting the two phenomena were somehow related and produced in the same way. But unlike its attractive forces, amber's sparks could be felt and seen. And sometimes the feeling was strong enough to be quite unpleasant. What was the nature of these sparks and were they good or bad for the body? Was the pain level of the sparks a consequence of the attractive forces? The sparks raised many questions.

Before going any further, we should stop here and define exactly what we mean by a *spark*. People are sometimes sloppy with how they use the term.

The modern English word *spark* is thought to come from the Old English word *spearca,* meaning "a glowing or fiery particle thrown off."

The origin of the word is probably extremely old, since the production of heat sparks is one of the most evident consequences of burning wood or smacking two stones together—things that even cave people did regularly. Because sparks can burn skin and eyes, people were careful when dealing with this menacing phenomenon. So it seems likely there would have been a word for such a ubiquitous hazard even in very primitive lexicons.

But the use of the term *spark* in connection with electrical events came much later, when people started to recognize that the shocks they sometimes felt when they touched objects, like amber, were typically accompanied by very faint flashes of light at the exact point of contact. Since they looked similar to the sparks from the fires and stones, and they were also similarly painful to the touch, they were given the same name. But such sparks are, in fact, quite different.

The sparks from fires and stones are really just flying fiery particles produced by rapid burning or intense friction. In contrast, the sparks that produce the shocking sensation from amber are due, as we shall soon see, to electricity rather than heat. And the sensation they produce isn't a burn but a direct stimulation of the body's nerves by the electricity. So the fact that electrical sparks could interact with the body and produce a physical sensation was appreciated at the very moment of their discovery. It is the electrical spark that concerns us here.

As time progressed, it gradually became apparent that other substances behaved similarly to amber. If you rubbed them, you could likewise produce attractive forces and sparks. As such, these dual properties that could be produced by rubbing different materials together were called "electrical" properties, derived from the Latin word *electricus,* meaning "amber-like."

The gradual realization that different combinations of paired materials could be rubbed together to produce electrical properties, and the appreciation that there were patterns, orders, and rules as to which material of the rubbed pair would become electrified was a breakthrough.

It opened the door for the systematic study of electrical properties, and moved electricity from the dominion of mysticism into the realm of science.

Most people today will confidently tell you electricity is a current of electrons, and they may even volunteer that an electron has a negative charge. Beyond that, if you challenge them to define what they mean by "electron" and "charge," they probably will start stammering because these are extremely difficult concepts to grasp, even for physicists. If you press them on what they mean by "current," however, their spirits will brighten, because everyone thinks they know what a current is: it is the pattern of flow of a fluid (like water) or a gas (like air).

Since electrons flow in currents, exploring electricity as you might investigate currents of water could result in some very important and highly comprehensible insight. This is exactly what the first scientific investigators of electricity did. And they did so with no knowledge of electrons. They just imagined electric currents to be equivalent to water currents, and they simply thought electricity to be a current of some invisible fluid. For the most part, this approach worked and people learned much by thinking of electricity as a current of an invisible fluid.

Of course, electricity isn't really an invisible fluid, so if you take this model too literally and start trying to find the invisible fluid, you will be doomed to failure. Michael Faraday, the brilliant nineteenth-century scientist who would transform our understanding of electricity, knew this well.[10] In 1821, he admonished his colleagues: "Those who consider electricity as a fluid, or as two fluids, conceive that a current or currents of electricity are passing through the wire during the whole time [that it remains connected to a battery]. . . . There are many arguments in favor of the materiality of electricity, and but few against it; but still it is only a supposition; and it will be well to remember . . . that we have no proof of the materiality of electricity, or of the existence of any [fluid flow] through the wire."[11]

Notwithstanding Faraday's admonition, "electrical fluid" is a good place to start if we want to understand electrical principles because there are many parallels between electric currents and water currents. So that's what we'll do here; we'll begin our exploration of electricity by focusing on its behavior as a current, and we'll worry about the exact physical nature of the imaginary "fluid" that comprises electrical current later.

Of course, almost everyone knows the famous eighteenth-century American printer, inventor, colonial statesman, and scientist, Benjamin Franklin, is very important to the history of electricity, and that he once flew a kite in a lightning storm to demonstrate such storms were electrical. But his accomplishments as an electricity scientist wouldn't have been diminished even if he had never flown his kite. The kite stunt turned out to be a double-edged sword for Franklin's image. Rather than being a scientific discovery, it was really just a demonstration of electrical principles he already understood better than anyone else.[12] Franklin proved his electrical knowledge by dramatically tempting fate with a kite, and he received outsized attention for it. The kite makes an interesting story, and I promise we'll discuss it in some detail. For now, however, let's focus on Franklin's experiments that showed electricity to behave like a fluid. That was his true insight.

In Franklin's day, the only technique to produce electricity was to rub materials together, as we have just discussed. Producing electricity in this way leaves the electricity localized on the surface of the object being rubbed. Since this type of electricity remains localized and isn't moving in currents, we call it *static*, meaning unchanging or fixed in position. But if static electricity is provided a route to go elsewhere, it will cease being static and start to flow, thereby producing a current, like water being syphoned from a bowl. In the bowl, the water is static, but if provided a route through the syphon, the water flows in a downhill current, with gravity providing the required force. Once we have such a current, we have something we can measure . . . and study. The flow of the water allows us to deduce the properties of water, such as its weight, its density, its resistance, its speed, its pressure, etc. Similarly, through the study of electrical current, we can begin to understand electricity.

There is a limit to how much electricity one can generate by rubbing amber by hand, so people sought alternatives. Common glass can substitute for amber when it is rubbed with silk. The major appeal of glass was that it was relatively cheap and could readily be formed into any desired shape and size. This led to the production of static electricity–producing machines based on rotating a glass rod, cylinder, or sphere against a piece of silk. The rotation usually was achieved with a simple hand crank.[13] It was these static electricity machines that first caught Franklin's attention.

Franklin reported he had first become interested in electricity when he attended a demonstration in Philadelphia given by a man whom he referred to simply as "Dr. Spence." We aren't exactly sure who Dr. Spence was, but he was most likely Archibald Spencer. A man by such a name had run an advertisement that appeared in Franklin's *Pennsylvania Gazette* in April of 1744. The announcement promoted a series of lectures to be given by Archibald Spencer on the topic of "Experimental Philosophy," and it instructed interested parties to register for the lecture series and pick up their *Catalogue of Experiments* at the Philadelphia Post Office. (Spencer apparently was moving from city to city with his electrical demonstrations, because a similar advertisement promoting demonstrations in Boston had appeared in a newspaper there a few months earlier.) Since Franklin owned the newspaper that published the advertisement, and he was, at the time, postmaster of Philadelphia, it seems highly likely that Franklin was aware of Spencer's lecture series and decided to attend.

Apparently, Spencer's Experimental Philosophy lectures contained a number of static electricity demonstrations to entertain his audience, one of which was called the Flying Boy demonstration.[14] This particular demonstration had become the main attraction for the many traveling electricity performances that were captivating upper-crust society at the time. People found the demonstration noteworthy because it showed that electrical phenomena could be produced apart from a machine. Stephen Gray, a distinguished British scientist, debuted the Flying Boy in 1729.[15] The demonstration has been described in detail by him and others, so we know exactly how it worked.

"Flying Boy" really was a misnomer. No boy ever flew. Rather, the demonstration involved a boy being suspended by silk ribbons from the ceiling, facedown in a prone position—a simulated "flying" posture. Then a glass static electricity generator was used to electrify the boy's bare feet. An open book was then put on the floor beneath the boy's face, and he was asked to turn the page. The boy waved his finger across the book and the page turned without his touching it! That was what astounded the audience.[16] How did the boy turn the page without touching it?

Of course, the page could have been turned without any touching simply by rubbing a glass rod to electrify it and give it attractive forces, and then waving the rod itself across the book. So the page turning alone wasn't any cause for excitement. The interesting part was that the electricity generated with the static electricity generator was being transferred *to the boy's body*, such that his whole body had become electrified, just as if his body were an electrified glass rod. And if the page turning didn't convince an audience member about the boy truly being electrified, one could just touch the boy's body and receive the same type of shock that resulted from touching an electrified glass rod.[17]

It turns out the "flying boy" part of the performance wasn't merely theatrics; it was essential to the demonstration's success. The boy needed to be electrically isolated in order for his body to become electrified. If there had been a way for the electricity to move from the boy to the ground, it would quickly enter the ground and dissipate. But air and silk are poor conductors of electricity. (*Conductors* are materials

FIG. 1.2. The Flying Boy. The ability of the human body to conduct electricity and hold a static charge was first demonstrated by British scientist Stephen Gray using a boy suspended in a prone ("flying") position by means of electrically insulating silk ribbons. The boy, whose body was isolated from the ground, could be electrically charged by transferring electricity to him through a glass rod that had been precharged with static electricity. Once charged, the boy was able to turn book pages without touching them and attract small pieces of paper or feathers to his fingers, due to the attractive forces of his electrified body. Although the demonstration was originally done purely as part of Gray's scientific investigations, modified versions of the Flying Boy soon became the main attraction at traveling static electricity shows that toured the American colonies in the mid-1700s to entertain the general public, and through which Benjamin Franklin first became acquainted with the mysterious properties of electricity. This historical image depicts the Flying Boy demonstration as originally performed by Gray around 1730. (Science Source)

through which electricity can pass, while *insulators* are materials through which it doesn't easily flow.) By suspending the boy in the air with silk, the electricity in the boy's body had nowhere to go, so it remained stationary (static) in the boy's body. Why a boy and not a girl? No reason; girls work just as well. Why a boy and not a man? Probably because men are too heavy to easily suspend with silk ribbons, and boys worked for less pay.

Traveling static electricity performances like Spencer's were typically offered only during the colder months of the year. Not because there were fewer other amusements during the winter, and not because the performers wanted to have their summers off. Rather, the static electricity performances didn't happen in the summer because the electrical demonstrations didn't work during the summer, for reasons nobody knew. We now know the reason is because the warm air of summer is typically very humid, and moist air is the enemy of static electricity. It is difficult to electrify an object when the air is humid, and once electrified, the object tends to lose its electricity quickly because water in the air bleeds the electricity away. Summer months, being humid, thwart all attempts to play with (or study) static electricity. So, April of 1744 must have been very near the closing of Spence's electricity demonstration season. Franklin, therefore, would have seen one of Spencer's last shows of the season. Spencer wouldn't be returning to his show circuit until fall.

Franklin was astounded by Spencer's electricity show, and he became obsessed with learning more about electricity. He had several friends and colleagues with whom he regularly corresponded about scientific matters. He soon started asking them what they knew about these electricity performances, and he began speculating as to exactly how electricity worked.

What he found was that many people had been studying such electrical marvels for a long time, but a universal theory that explained the various electrical effects still seemed a long way off.[18] One reason was that there were technical obstacles to studying electricity. In addition to the humidity problem, there was no easy way to measure electricity. It also was hard to produce large amounts by just rubbing objects together, and it remained localized for only a few minutes; that is, it couldn't be stored. These crude static electricity generators also produced highly variable results, and measuring the amount of electricity by gauging the pain intensity from an electrical shock wasn't reliable either.

But timing is everything. Franklin had the good fortune to become interested in electricity at just the perfect time. While he was watching Spencer's electricity demonstrations in Philadelphia, Pieter van Musschenbroek, working in Leyden, Germany, was in the process of conquering the electricity storage problem.[19] Musschenbroek invented a very simple device—something anyone could make—that was capable of storing very large amounts of electricity in a simple glass jar, just as one might store water. This discovery revolutionized electricity research, because it allowed the relatively modest amounts of electricity produced by the hand-operated electrostatic generators to be collected and stored in jars for an extended period of time. Need a little electricity? Just take a jar of it off the shelf. Need a lot of electricity? Gather a bunch of jars and connect them together.[20]

With the arrival of what was commonly known as the *Leyden jar*—Franklin often called it "Musschenbroek's wonderful bottle"—electricity demonstrations took a completely new turn. They went from electrifying a single boy to electrifying a chain of men!

In April 1747, just three years after Franklin had seen Spencer's fairly tame electricity demonstration in Philadelphia, Jean-Antoine Nollet, a French cleric and scientist, performed his own electricity demonstration for King Louis XV of France.[21] He had 180 men from the king's Royal Guard stand in line holding hands. He then had the soldier at one end of the line use his free hand to touch the top of a fully electrified Leyden jar. Instantly, all 180 men in line reeled from the strong shock they felt. The king was impressed.

If a series of 180 men could be shocked, how many more? Nollet staged an even larger spectacle. As his day job was running a monastery, he was able to recruit Carthusian monks to make an even longer hand-to-hand chain. Since monks were more common then than they are now, Nollet was able to gather a group of 700. Results were the same; all 700 monks received a jolting shock. No, that Flying Boy demonstration wasn't going to impress anyone anymore.

Nollet offered a scientific explanation for his results. He proposed that electrical transmissions between people were achieved by currents of two competing fluids—he called them "affluent" and "effluent" currents—that moved among human bodies by means of tiny body pores, too small to be seen. This two-fluid electrical theory gained some traction among scientists, because two different electrical fluids might somehow explain why pairs of electrified objects sometimes attracted each other and sometimes repelled each other. The explanation for the alternative repulsive and attractive properties of electrified objects might be explained if currents from two different types of fluids were electrifying them. However, other scientists had a problem with the theory because all the men in the chain seemed to be shocked simultaneously, and it was difficult to imagine fluids flowing through body pores could move so quickly. Nevertheless, the two-fluid theory would persist until replaced by Franklin's alternative one-fluid theory, which was appealing because it not only accounted for the movement of electricity between human bodies but also provided a working model of how the Leyden jar worked.

Imagine a wide-mouth glass jar, like a mayonnaise jar, with a plastic lid. Cover the outside of the glass jar with aluminum foil and line the inside with aluminum foil, making sure the outside foil has no physical contact with the inside foil. Put the lid back on the jar, punch a hole in the lid, and run a metal wire from the inside foil out through the hole, protruding just a few inches. Now, either place the jar on the ground or run another wire from the outside foil to the ground.[22] You have just made yourself a Leyden jar, ready to be electrified.[23]

To fill the jar with electricity, rub a glass rod with silk to electrify the rod, and then touch the wire protruding from the top of the jar. The electricity from the glass rod will flow into the aluminum foil lining the inside of the jar. Repeat. With each repetition, more and more electricity goes into the jar. As the jar becomes filled with electricity, electrical "pressure" builds up in the jar, just as pressure would build up if we were forcing air into the jar. The inside electricity would like to flow to the outside foil, escape to the ground, and thus relieve the pressure, but the glass is blocking its flow, so that cannot happen. The electricity just sits there, statically, waiting until an opportunity for escape presents itself. And that's exactly what happens if a person should touch the wire on the jar's top. The electricity escapes by going into the person's body, down through his feet, and into the ground. He briefly feels the shock of the passing electricity. Then it's gone, and the jar has lost all of its stored electricity.

Now suppose the man who touches the wire is standing on an insulator (or wearing nonconducting shoes), which blocks the electricity's route to the ground. In that case, we have the equivalent of the Flying Boy. The electricity enters the man's body, but it has nowhere to go, so it stays statically in the man's body, waiting to escape his body by another route. Touch the man, and it goes into you. Make a chain of men holding hands, and it will go through all of them, searching for a way to the ground. If one of the men in the chain happens to be barefoot, the route to the ground is found and men farther down the chain are spared the shock.

This is what's happening with a Leyden jar. The electricity is getting stuck inside with nowhere to go. The electrical "pressure" level in the jar is what we call the *voltage*, and it represents the electrical force difference between the inside of the jar and the outside of the jar. We'll be hearing a lot more about voltage as we proceed with our story. We'll get to that soon. But first, I had promised to tell the true story behind Franklin's kite.

It doesn't take a genius to realize that lightning looks like an enormous spark in the sky, so the notion that lightning storms are electrical wasn't particularly novel. The idea had been around long before Franklin arrived on the scene. It further seemed obvious to most people that sparks so big should represent a lot of electricity stored in storm clouds, most likely of exactly the same type produced by static electricity generators but on a much grander scale. Franklin articulated the case for lightning being an electrical phenomenon in 1749:

> Electrical fluid [electricity] agrees with lightning in these particulars: 1. Giving light. 2. Colour of the light. 3. Crooked direction. 4. Swift motion. 5. Being conducted by metals. 6. Crack or noise in exploding. 7. Subsisting in water or ice. 8. Rending [tearing apart] bodies as it

> passes through. 9. Destroying animals. 10. Melting metals. 11. Firing inflammable substances. 12. Sulphureous smells. . . . [We further know that] electrical fluid is attracted by points. We do not know whether this property is in lightning. But since they agree in all particulars wherein we can already compare them, is it not possible that they agree likewise in this? Let the experiment be made.[24]

In 1750, Franklin went on to publicly propose a specific experiment to explore the alleged electrical nature of lightning. In effect, it amounted to conducting the Flying Boy demonstration using lightning as the source of electricity, with two major differences in the experimental design: 1) the boy suspended from insulating silk ribbons was replaced with a man standing on an insulated platform [a stand with glass peg legs]; 2) the electrification of the body would be achieved not by touching the feet with an electrified glass rod, but rather by having the platform the man was standing on attached to a long *pointed* iron rod reaching 30 feet into the sky.[25] Franklin believed it essential that the rod be pointed, since it had been established that static electricity was attracted most strongly to pointed metal objects.

The idea was that the experiment would be performed at the top of a church steeple or other very high location during a thunderstorm, so that the pointed iron rod would draw electricity from the clouds and electrify the man, just as electricity from the rubbed glass rod had electrified the boy.

To be clear, Franklin didn't believe there would be an actual lightning strike to the iron rod. Rather, he believed the iron rod would simply bleed the cloud of its stored static electricity, as one might bleed the pressure from a shaken bottle of soda by slowly unscrewing the cap. But he realized some people might think it irresponsible to subject a person to such a lightning risk. So he suggested an alternative experimental design, where a grounded wire would be brought close enough to the iron rod to elicit a visual electrical spark as evidence of the rod's electrification by the passing cloud, thus eliminating the need for an electrified human to be part of the experiment.

Franklin was also concerned that it probably would be raining when weather conditions were ripe for conducting the experiment. So he specified that it should be conducted within a small wooden hut to protect everything from getting wet. Only the iron rod would protrude from the hut, into the sky. Because of the need for the hut, Franklin's proposal is often called the Sentry Box experiment.

Franklin published his Sentry Box idea and then waited for an opportune time to actually conduct the experiment. But the logistics were complicated. A church congregation needed to volunteer its steeple, a 30-foot pointed iron rod needed to be made, a sentry box needed to be constructed, etc. At some point in the process, Franklin decided that maybe a kite and a Leyden jar was all he really needed. Franklin reasoned that if his hypothesis about the electricity being in clouds was true, he should be able to capture some of the storm's electricity in a Leyden jar with just the use of a kite to get the electricity from the cloud. So that's what Franklin attempted to do with his kite; he sought to fill a Leyden jar with electricity captured from the sky.

His modified experimental design was to fly a kite, with a thin metal wire pointing up from its top, near the clouds during a thunderstorm. He hoped the kite's wire would allow him to bleed off some of the static electricity he believed to be trapped in the cloud. His expectation was that the cloud's electricity would flow down the wet kite string and into a Leyden jar. Conveniently, the wetness of the string would actually improve its electrical conduction. But how would he prevent himself from being electrocuted in the process? After all, wouldn't the electricity just travel from the string, through his body, and into the ground, perhaps shocking him to death in the process?

Franklin's idea was to insulate himself from the kite in way similar to the way the Flying Boy was electrically insulated from the ground: with a silk ribbon. He didn't directly hold the kite string in his hand. He held a silk ribbon that was tied to a door key. The kite string, in turn, was likewise tied to the same door key. In this way, the electricity in the kite would pass down the kite string only as far as the key, where it would accumulate because it couldn't pass through the silk ribbon into Franklin's hand on its way to the ground. The key could then be touched to the top wire of a Leyden jar, and electricity would flow from the key into the jar. As an added safety precaution, Franklin planned to stand within a rain shelter while he flew the kite to keep both himself and the silk ribbon dry, since water conducts electricity.[26] If he or the ribbon got wet, the electricity would have an alternative route, going through his body to the ground, thus jeopardizing his experiment and threatening his life. But if everything went according to plan, Franklin then would have captured some of the storm's electricity in the jar.

Franklin assembled his experimental materials and awaited a thunderstorm. The opportunity came in June 1752 when a storm descended upon Philadelphia. Franklin grabbed his kite and headed out to catch some electricity.

His modified kite went aloft just as planned, and Franklin started to realize the experiment was working when he saw the kite string's fibers beginning to stick straight out, just as static electricity can make the hairs on your head stand out. In his exuberance, he unwisely risked touching the key to his knuckle to confirm it would deliver an electrical shock. It did. Then he touched the key to the Leyden jar. He had imprisoned lightning in a jar!

There are a number of circumstantial reasons why some people came to doubt Franklin's account of the kite story. But most of the criticisms have been dispelled in recent years by historians who have fleshed out the details of the story and found them to be credible.[27] The general academic consensus now is that Franklin's account checks out; he performed the kite experiment as he claimed and when he claimed: June of 1752.

Ironically, at the time Franklin performed his kite experiment, a group of scientists in France had already been successful in performing his Sentry Box experiment. The French performed their experiment exactly as Franklin had previously specified. They had achieved their success one month before Franklin flew his kite, but Franklin was unaware of it. News traveled slowly at the time. Franklin didn't learn about the successful French experiment until a couple of months later. But it was of no concern to Franklin that he had been scooped. He just appreciated that his idea had been independently validated.[28]

One of the early criticisms of Franklin's claim had been that, if he actually had done the experiment, how did he live to tell about it?

Lightning kills! Everyone knows that. Russian scientist Georg Wilhelm Richmann was one such victim.[29] One year after Franklin's kite experiment, Richmann was in St. Petersburg using a metal rod to reproduce the Sentry Box experiment. He was trying to assess the electricity buildup in the rod, possibly by touching it with his hand, when the rod was abruptly and violently struck by lightning, instantly killing him. The electrical jolt was so powerful that it also knocked his companion down and split a nearby door frame. Up until then, Richmann actually had been a pioneer in the study of electricity. He was nearly as famous for his electrical studies as Franklin was at the time. But Richmann was as unlucky as Franklin had been lucky. We now remember Franklin as the first man to extract electricity from a storm, while Richmann's legacy is that he was the first person in history to die from an electricity experiment. Fate is cruel.

But maybe the difference between the fate of the two men wasn't just luck. Franklin had understood the problem of having your body "grounded" (i.e., connected to the ground) while conducting electrical experiments, and he took precautions, like the insulating silk ribbon, to prevent the electricity from passing through his body into the ground. Perhaps Richmann was more careless. We'll never know. But one thing is clear: *do not* try either man's lightning experiment at home. You likely won't enjoy Franklin's good luck.

Franklin's work with lightning didn't stop with his kite. He would do much more, as we shall soon see. But the kite experiment allowed him to establish that lightning behaves just like static electricity *because it is static electricity*. This, in turn, implied that the discovery of the fundamental physical laws that ruled static electricity's behavior would be applicable to lightning's behavior as well, and that realization had some very important implications.

Because static electricity moves from one place to another like a current of water and exhibits a type of pressure (voltage) when confined to Leyden jars, it isn't hard to understand why many people believed static electricity to be composed of some type of invisible fluid.[30] That seemed to be a logical conclusion. What people wondered about, however, was whether static electricity comprises just one type of fluid or two.

As mentioned already, the reason the two-fluid model came up was that electricity exhibited both attractive and repulsive properties. Could one fluid both attract and repel itself? That seemed incredible. It was easier to believe there were two different fluids, each repellent to itself but attracted by the other (or vice versa).[31] But Franklin had another idea. And his idea would hold up in substance, if not in detail, until the present day.

Franklin believed there was only one electrical fluid that sought to spread itself equally among all materials. In his own words: "We suppose . . . that electrical [fluid] is a common element, of which everyone . . . has his equal share."[32] In other words, the normal state of affairs is that the concentration of the electrical fluid among solid materials is the same; that is, it is constant. Rubbing two objects together means actually scraping the fluid from one of the objects onto the other. Franklin claimed the object that lost fluid was now fluid deficient, which he termed *negatively charged*. In contrast, the object that gained the electrical fluid was thus in fluid excess, so he called it *positively charged*. In Franklin's view, the attractive forces between electrified objects had to do with negatively charged objects trying to get their fluid back from positively charged objects. Thus, negative objects are attracted to positive objects because they are trying to redistribute the charge. In contrast, when two objects with the same charge are placed together, the effect is to exacerbate the charge distribution problem rather than relieve it, so the objects tend to repel each other.[33] It may be worth reading this paragraph again to fully absorb it, because this single-fluid theory of Franklin's reveals a fundamental truth of electricity. It is not correct in detail. There is no actual fluid, and the redistribution of matter actually moves in the direction opposite to the way he proposed (i.e., the charge usually moves from negative to positive rather than from positive to negative). Nevertheless, his perception of electricity was so insightful and provided such a useful model of how electricity behaves that we are still using Franklin's electrical terminology today.[34]

All these things are on my mind as I'm shown the beautiful collection of amber art and jewelry at the House of Amber. How can it be that one material teaches us so much about life? Its physical properties enable it to entrap and freeze life-forms in place for millions of years, and its clarity allows modern humans to actually see those entrapped inclusions and study their anatomy, telling us much about prehistoric animals and the evolution of species.[35] In addition, amber's physical properties revealed electricity to humankind and allowed us to perceive the invisible electrical world that is all around us, permeating our bodies and, as we shall see, enabling life itself to exist.

Yes, amber is an amazing gem that can do some very impressive stuff. In fact, I'm so impressed that I can't bring myself to leave the House of Amber without buying a souvenir. I have a wedding anniversary coming up. Could it be that amber is the traditional gemstone for our anniversary year? I check the internet. Nope . . . our anniversary year's gemstone is emerald. How boring! An emerald just won't do. So I find the most beautiful amber pendant I can afford and purchase it as an anniversary gift for my wife. Emeralds be damned! It would be hard to find an anniversary gemstone of more significance, or with a better backstory, than amber.

2

SHOCKING DEVELOPMENTS

ANIMAL ELECTRICITY

As the old fisherman remarked after explaining the various ways to attach a frog to a hook, it's all the same to the frog.

—PAUL SCHULLERY

Snakes, spiders, and bats are the stuff of many people's phobias. But for me, it's bears. And that causes a problem, because I spend a good deal of time in the great outdoors. Consequently, I've encountered my share of bears. Most of my sightings have been long range, but one was up close and personal. While I was hiking with my family in Glacier National Park in Montana, the epicenter for grizzly bears in the lower 48 states, a 400-pounder came trotting down the trail straight toward us.[1] We had the good sense to simply step off the trail, and the grizzly went barreling through paying us no mind, so no damage done. But oh, what could have been!

During daylight hours, bear risks are manageable. Travel in groups, make lots of noise, give them wide berth, carry bear spray, and you're likely to be fine. But at night, things get more precarious, particularly in the spring and summer when grizzlies tend to be nocturnal.[2] At night, you're vulnerable, because even a faint odor of camp food can attract grizzlies from miles away, and *because you are asleep*. And although your new L.L.Bean tent will likely keep you warm and dry while you slumber, it will provide no protection from a hungry grizzly.

That's why, for all my love of the outdoors, I like to spend my nights in a lodge, where I can plan the next day's hike over a hearty meal and a craft beer, and then retire to the comfort of a warm and safe bedroom. Yet, even lodges have their bear issues. Those smells of sautéing trout wafting from the lodge's kitchen attract more than just lodge guests to the dining hall. For this reason, lodges and even public campgrounds sometimes employ electric fences to keep hungry bears off their grounds. Towering as high as 10 feet, electric fences surround the perimeter, working through the night to keep bears out while people slumber safely within.

Although a 10-foot electric fence is an impressive structure, it doesn't obscure the view as a stockade fence would, because its thin wires are spaced well enough apart that the fence poses no optical obstruction. Yet, don't be deceived; those skimpy wires pack a powerful punch. Needless to say, lodges with electric fences have no bear problems.

Curious as to how these electric fences work and their safety to humans, I consult Kim Annis, a bear management specialist with Montana Fish, Wildlife, and Parks. She tells me electric fences are her chief bear management tool, and that she erects bear fences on a daily basis to separate bears from food attractants, typically livestock and fruit crops. Annis verifies that fences are very effective at keeping bears away and are very safe to bears, livestock, and humans—even small children. She says she's been working with bear fences in Montana for 11 years now and knows of no injuries to humans or wildlife from an electric fence. Annis has even intentionally shocked herself a few times by grabbing fence wires to test whether a fence was electrified, similar to Franklin's knuckle test. She exclaims: "It's not a pleasant feeling, but it's not dangerous either." And it leaves no mark or burn on the skin.

Bears hate the fences, but some have developed their own version of the knuckle test. Annis remarks, "Some of the smarter bears learn they can touch a fence very briefly with their noses, to test whether it is on or not." If it isn't on, they'll come barreling through. "For other bears, one shock in their lifetime is enough to keep them away permanently." It all depends upon the individual bear.

Originally designed as a cheap and easy way to keep cows in their pastures, electric fences can be as simple as a single electrified wire strung waist high around a field, or as complex as a multiwired structure 10 feet high. Electric fences work by delivering a powerful shock to any creature unfortunate enough to come in contact with both the wire and the ground at the same time. You can avoid the shock either by not touching the wire or not touching the ground, which is why birds perching on the wire don't get zapped. But until pigs fly, an electric fence will keep four-footed animals in their place.

The principle behind shocking bears with electric fences is the same as shocking monks with Leyden jars. Provide a strong source of electricity and a bodily route for that electricity to get to the ground—be it through cow, bear, or monk—and you will have a powerful deterrent to unwanted behavior. But don't expect too much. One commercial manufacturer warns its dairy farm customers that an electric fence designed to contain cows won't be adequate to keep a bull away from cows in heat. Bulls require higher fence voltages with more intense shocks to keep them from pestering the ladies.

Electric fences are safe because they are specifically designed to cause pain while *not killing*. It's the voltage that causes the pain. As you will recall, using our model of electricity acting like a fluid, voltage is the pressure behind the electric current; it is not the flow rate of the current. It takes a high flow rate to kill. Think of it this way. The water spigot in your yard and the fire hydrant on your street are both holding back the exact same pressure (voltage), because their water pressures originate from the same local water tower. But if you open their valves, the fire hydrant will deliver a lot more water flow (current) than the garden spigot because the hydrant's pipe diameter is much bigger. In fact, hosing people down with water from fire hydrants is sometimes used to disperse crowds by literally knocking people off their feet and pushing them out of the way, an impossible feat with a garden hose. You simply don't have enough water flow through a garden hose to threaten anyone.

While electrical pressure is called *voltage* (measured in the unit *volts*), electrical current flow is called *amperage* (measured in the unit *amperes*, which is usually shortened to *amps*).[3] Amperage gets its name from André-Marie Ampère, a French physicist who made major contributions to the scientific study of electricity. High amperage can be deadly. Homes in the United States typically run on electricity at 110 volts and between 10 and 20 amps, which are amperages quite high enough to kill an adult. Electric fences, in contrast, may run as high as 8,000 volts (ouch!), but usually have currents of just 0.12 amps. To provide additional safety, the fences are designed to deliver the low current in short pulses of less than one-hundredth of a second, spaced one second apart, thus further decreasing the current flow to the animal or person shocked. To kill someone this way would be as difficult as trying to drown a person by repeatedly squirting her with a water pistol. Thus, electric fences are pretty safe. Painful, always. Deadly, not so much.

The association of pain with electricity is apparent to anyone who ever has experienced a shock of any kind, which is just about everyone. As we've already seen, knuckle pain was the only way Franklin had to easily measure electricity, but he didn't yet know the pain he felt was specifically associated with the voltage. That understanding would come much later, when instruments were invented to distinguish voltage from amperage and measure them independently. In the interim, a better way was needed to measure electricity besides self-inflicted pain because pain was both unpleasant and hard to quantify. The solution to that problem came from another animal: the frog.

Frog legs aren't a fad food. They have been considered a great delicacy—true at least among the French and Chinese—for centuries. But before you try to cook up a dish of fresh frog legs, you need to prepare yourself to witness an unsettling phenomenon. The legs of freshly killed frogs can continue to move for quite some time after death. Even when the legs are cut away from the body, they continue to kick on their own for a while. Creepy!

Luigi Galvani, an eighteenth-century Italian scientist, thought he knew why.[4] He claimed nerve and muscle action was driven by something he called "animal electricity," a type of electricity allegedly created in the body, and that the twitching frog legs were a demonstration of internal animal electricity stimulating contractions in the frog's leg muscles.[5] He did a number of experiments running wires in various patterns to and from the nerves in severed frog legs to try to prove his point, but met with mixed results. Then, in one experiment, he thought he had hit gold. Galvani had taken a severed frog leg and impaled it on a brass hook that was mounted on an iron fence. Whenever the leg touched the fence it would contract, and move away from the fence. When the contracted leg would then relax and retouch the fence, sure enough, it would contract again. This would repeat itself over and over until the leg muscle was exhausted. Galvani believed this experiment convincingly demonstrated the existence of animal electricity, in that the iron fence was somehow bypassing the normal conduction of electricity in the leg and causing it to contract. But not all scientists agreed with his interpretation of the experimental results.[6]

Galvani's rival, Alessandro Giuseppe Volta, the electricity scientist who inspired the unit *volt,* thought the key to Galvani's results was actually the brass hook attached to the frog leg.[7] He suggested that while the brass hook was in contact with the iron fence, it resulted in a flow of electricity from one metal to the other. The frog leg, which was both impaled on the hook and also touching the fence, bridged both metals and provided a pathway whereby the electricity could flow from brass to iron through the leg, stimulating the muscle's contraction in the process. According to Volta, the leg wasn't producing any electricity; it was just responding to an electrical shock, as a finger recoils when shocked with static electricity. He predicted that, if the metal of the hook was changed from brass to iron, there would be no metal difference and thus no leg contraction. He was correct. Galvani's experimental results were dependent upon the hook being made of brass and the fence being made of iron.

Volta's discovery that electric current often flows between two dissimilar metals had a revolutionary impact on electrical science, because it opened the door for the invention of the *electric battery*. Yet, Volta didn't appreciate that what he had discovered was actually a chemical phenomenon. Most batteries work by bringing two dissimilar metals together, which thereby initiates a chemical reaction between the two, generating electricity in the process. If you don't have a frog, you can demonstrate the same electrical phenomenon with your own tongue. Your tongue is full of nerve endings, and saliva is a good conductor of electricity. You can actually feel a battery's electricity as a slight tingle just by touching your tongue to both terminals of a 9-volt battery—those little square batteries with the snap-on terminals frequently found in smoke detectors and other small household appliances. It's unlikely your tongue will be shocked or contract from the battery's weak electricity, but you will actually feel the current the battery produces.

We'll explore the nature of such *electrochemical* reactions later. For now, we just need to know that the frog's leg contraction resulted from its muscle being stimulated by an external electrical current from a battery, not by an internal one.

Although Galvani was wrong about why the muscle in the frog leg on the fence was contracting, he wasn't wrong about what stimulates muscles to contract: electricity. Even Volta had to acknowledge that electricity was stimulating the frog leg contractions. He just questioned the source of the electricity, and he showed it to be of external rather than internal origin. Nevertheless, electricity produces muscle contractions; that was not in doubt.

In an age before electricity measuring devices were invented, and pain to the knuckle was the only easy way to test for the presence of electricity, frogs became a viable solution. Michael Faraday, the scientist who was among the first to express skepticism about electricity being a fluid and who would later introduce his revolutionary concept of electric field theory, actually had a "froggery," a room where frogs are kept, constructed at the Royal Institution in London.[8] The froggery provided the frogs he required for his many experiments with electricity. To measure electricity, a decapitated frog was pinned to a board and electrodes were attached to each leg. If an electrical wire was "live," meaning that it is carrying electricity, it would cause the dead frog's legs to move.

In fact, the phenomenon that frog legs contract in response to electricity, and that the extent of contraction is proportional to the amount of electricity, allowed German scientist Hermann von Helmholtz to actually bump things up a bit.[9] He made an instrument for measuring electricity based on the contraction of frog legs. First, he constructed a chart recorder (a device that mechanically drags a piece of paper under an ink pen at a constant speed), and he attached the recording pen to a severed frog leg by means of strings. When he stimulated the frog leg with electricity, the pen was deflected and marked the timing, duration, and magnitude of the muscle contraction on the recording paper. If the leg was stimulated with more electricity, the pen was deflected more and it recorded a larger leg contraction event.

Helmholtz later used this simple device to show that the body's nerves were the anatomical equivalent of "wires" (conductors) that transported the electrical signals to the muscles, thus causing the muscles to contract. By electrically stimulating nerves at various points along their lengths, and measuring the time it took for a frog leg muscle to contract using the chart recorder, Helmholtz was able to deduce the speed at which electrical signals traveled through nerves: 27 meters per second. This value calculated by Helmholtz is similar to the signal speed of nerves measured today using modern electronic instruments.[10] The speed of electrical signals through nerves is much slower and weaker than the speed of electrical currents through metal wires, but they are both electrical signals nonetheless, whether they are contracting muscles or spinning the blades of an electric fan.

If you are tempted to scoff at Galvani's apparently naïve idea of animal electricity, don't. As it turns out, all animals generate some electricity, and some animals, like electric fishes, produce a lot of it. Even the ancient Romans knew about electric fishes.

Although there are hundreds of electric fish species worldwide, the one the Romans were most familiar with was the torpedo fish (*Torpedo torpedo*), a type of ray widespread throughout the Mediterranean. It has been the subject of experimentation since antiquity, and physicians had prescribed the shock from the fish to treat all types of diseases. Roman physician Scribonius Largus (first century AD) shocked his patients with torpedoes from head to toe; it was his preferred treatment for both headache and gout. Sometime later, another physician, Dioscorides, even used torpedoes to treat hemorrhoids. As gruesome as they sound, such electric fish treatments continued for many centuries.[11]

Two things about these fish intrigued the ancients. First, you don't have to actually touch the fish to get a shock because their electricity can act at a distance in water. It also can be carried through metallic materials, such as bronze and iron. So, if you spear a torpedo fish with a metal spear, you risk being zapped. Second, the shocking disappears when the fish dies. Since a dead torpedo won't shock you, it was thought its electricity must be some type of vital life force that was lost upon death, a supposition that would turn out to be not far off the mark.

It was difficult for the ancients to understand things that worked at a distance, so they put such things under the category of occult objects, having magical powers.[12] Both amber and electric fishes produced similarly feeling shocks that could work across a gap, suggesting they were both part of the occult world and were acting in the same way. But it wasn't clear whether amber shocks and fish shocks represented the same phenomenon, or different phenomena that just felt the same.

The main argument against their being the same thing was that, while amber produced visible sparks when shocking, electric fish shocked without sparks. No sparks?! How then could these fishes be *electricus*—"like amber"?

Over the centuries, people tried to explain what was going on with electric fishes, but little progress was made. In 1615, Jesuit priest Nicolas Godinho reported that although dead electric catfish could no longer produce a shock, when a live electric catfish was thrown on a pile of dead ones, "the [dead] fish thus brought in contact with it are seized with an inward and inexplicable trembling to such an extent that they actually appear to be alive."[13] This is similar to the response Galvani received when he touched severed frog legs to the iron fence; the dead leg kicked as if it were alive. External electricity evidently could stimulate muscle movement in both frogs and fishes.

In 1751, a French botanist, Michel Adanson, who had some experience with Leyden jars, encountered electric catfish while on an expedition to Senegal. He reported that when touching the catfish with an iron rod, one got a sensation indistinguishable from touching a Leyden jar, and he suggested the fish might represent a living Leyden jar that somehow collected and stored electricity from its environment and released it all at once when touched. At about the same time, Laurens Storm vans Gravesande, the Dutch governor of French Guiana and Surinam, reported home to Holland that there were local eels that, when touched, gave the same sensation as a Leyden jar, except there were no sparks. "But for everything else it is the same."[14]

These reports of living Leyden jars soon attracted the attention of Benjamin Franklin. Although he had no access to electric fishes himself, he encouraged his European colleagues to conduct experiments. Since the fishes were difficult to bring to the laboratory, some scientists went to the fishes. One electricity investigator, Jan Ingenhouz, actually took Leyden jars to a coastal area where fishermen routinely were shocked by torpedoes. He shocked the fishermen with the Leyden jars for comparison, and the fishermen attested the sensation from the Leyden jar was exactly the same as being shocked by a torpedo.[15] Still, the torpedoes produced no sparks, and sparks were thought to be fundamental to electrical discharge. So the jury was still out on whether torpedoes actually were producing electricity.

Finally, in 1775, English scientist John Walsh obtained a large specimen of *Gymnotus*, a genus including various electric knifefishes native to the Amazon region that can grow over 100 centimeters long (about 3 feet), and he was able to demonstrate a perceptible spark jumping through the air when the fish delivered a shock.[16] The elusive spark had been found. Electric fish do actually produce bona fide electricity.[17] Animal electricity is real; the fish are "like amber"!

Besides the similarities in the electric shocks they delivered, Leyden jars and electric fish had another thing in common. Shocking depleted them of their electrical reserves and it would take time to restore them. One famous story comes from the world-renowned naturalist of the eighteenth century, Alexander von Humboldt. Humboldt was a man of many interests, and is most celebrated for actually proposing the concept of nature as we now understand it—a complex and interconnected global force that has its own existence, of which humans are merely a part.[18] He was one of the first to argue that when humans perturb one part of the environment, the environmental balance is altered, which has ripple effects throughout the natural world.

In addition to his scientific skills, Humboldt was a well-traveled adventurer and explorer. And he was particularly curious about seeing the electric eels (*Electrophorus electricus*) that were said to inhabit the area near the village of Rastro de Abaso in present-day Venezuela. He visited the village and expressed an interest in obtaining live eel specimens. The locals were eager to oblige, and he witnessed them use an unusual but effective technique to safely catch large numbers of live electric eels, although the method cost the lives of a couple of horses.

The problem in catching these freshwater eels is that they spend a good deal of their time buried in the mud of ponds and rivers. So it isn't easy to catch them with a conventional net. The natives' solution was to drive a herd of about 30 horses into a shallow eel-laden pond where they

FIG 2.1. Electric eels wreak havoc on a herd of terrified horses. The famous eighteenth-century German naturalist and explorer Alexander von Humboldt described the technique the villagers of Rastro de Abaso, in present-day Venezuela, used to catch electric eels. According to Humboldt, the fishermen drove a herd of about 30 horses into a shallow muddy pond that was known to be inhabited by an abundance of electric eels. The horses' hooves drove the eels out of the muddy bottom in which they resided, and the eels retaliated by relentlessly and repeatedly shocking the horses, until the eels had exhausted all of their electrical reserves. The eels remained unable to shock any further for several hours, and this allowed the fishermen enough time to gather the eels without fear of being shocked themselves. (Chronicle/Alamy Stock Photo)

danced around and churned up the bottom, forcing the eels out of the mud. As Humboldt vividly describes: "The eels, stunned by the noise, defend themselves by the repeated discharge of their [electricity]. . . . Several horses [succumb to] the violence of the invisible strokes which they receive from all sides in the organs most essential to life; and stunned by the force and frequency of the shocks, they disappear under water."[19] But in shocking the horses, the eels had expended all of their electrical energy, so they then could be easily collected by hand, unable to shock their collectors. With the eels in their electrically discharged state, Humboldt was able to examine live specimens without being shocked. But over the next few hours, the eels recuperated and regained their shocking abilities, so it became too precarious to continue. Humboldt never was able to figure out the mechanism of the eels' shocking ability.

Electric eels don't really shock their prey to death; they just stun the prey and then attack. The eels can generate more than 800 volts (about seven times American household voltages), but they simply cannot produce enough amperage to kill. Each electric pulse from an eel lasts only a couple of milliseconds (one-thousandth of a second) and delivers less than one amp (one-twentieth of household amperage). This is similar to how electric fences work, delivering very short pulses of low-amperage electricity, thus shocking but not killing bears. It is also similar to a modern Taser electroshock weapon, which works by delivering a very high-voltage pulse (about 50,000 volts) carrying very low amperage (just a few milliamps).

But recently, scientists have learned eels can deliver more than just a stunning shock. They can actually incapacitate larger prey by biting and holding on, while using repetitive electrical shocks to exhaust the prey's muscles.[20] Just as a frog leg repetitively stimulated with electricity will stop contracting due to muscle fatigue, repeated muscle contractions caused by the eel's electrical pulses exhaust the prey's muscles.[21] With each shock, the muscles contract less and less until they are left exhausted, and the prey becomes too weak to escape or fight back. It may appear dead, but it is actually just paralyzed and completely vulnerable.

About 80% of an electric eel's body is devoted to producing electricity. The eel's main electrical organ can store electricity in a manner similar to a Leyden jar, but the electricity is collected in a different way. The eel relies on flattened cells called *electrocytes,* meaning "electricity cells."[22] These electrocytes produce voltage across their cell membranes by segregating negative and positive charges through chemical means. In doing this, an electrical *potential,* that is, an electrical charge difference between the inside and outside, is created across the membrane.[23] This is possible because an electrocyte's cell membrane is a very good electrical insulator. This chemically produced electrical potential is comparable to what happens when we mechanically make static electricity by rubbing amber with wool. The rubbing moves charge (in this case, negative charge) from the amber to the wool. The charge cannot move back to the amber through the intervening dry air because dry air, like the electrocyte's cell membrane, is a very good insulator. Thus, an electrical potential exists between the amber and wool, which can be released later in the form of a spark, capable of producing a shocking sensation just like the eel's, but obviously on a much smaller scale.[24]

Of course, a single electrocyte isn't going to produce enough electrical potential to shock anything—each one can produce only 0.15 volts. But the eel has thousands stacked next to each other, like a roll of coins, with each electrocyte in the stack contributing its 0.15 volts. It's easy to see how 0.15 volts can be turned into thousands of volts by summing them this way, and it's also clear why longer eels would have longer stacks and thus higher voltages.

So electric eels don't suck up static electricity from their environment and store it. Rather, they produce their own electrical voltage by chemically segregating electrical charge. They then release all that pent-up voltage at once to yield a highly shocking electrical pulse. As we shall see later, an electric eel's electrocytes do little more than what human nerve cells do, but they are able to do it on a much grander scale.

The details of how eels produce their electricity continues to be an area of active research to this day. A team of scientists from the United States and Switzerland is currently working on a new type of battery inspired by eels. They envision their soft and flexible battery might someday be useful for internally powering medical implants and soft robots. But the team admits they have a long way to go. "The electric organs in eels are incredibly sophisticated; they're far better at generating power than we are," laments Michael Mayer, a team member from the University of Fribourg, Switzerland.[25]

Kenneth Catania is another modern-day electric eel researcher. His laboratory at Vanderbilt University in Nashville, Tennessee, studies the behavior of electric eels. Specifically, he is interested in exactly how electric eels maximize their electric shocks to prey. Recently, he has found that eels either bite or touch their prey with their heads, while curling the end of their tails around the opposite side of the prey's body. Since the eel's head is positively charged and its tail negatively charged, the maneuver causes the electrical current to flow externally, in a straight line, from the eel's head toward its nearby tail, and thus directly through the prey's body. This head-to-tail maneuver maximizes the electric jolt the eel can deliver to the prey.

In effect, the eel is unwittingly making an electrical circuit. An *electrical circuit* is a direct pathway through which an electrical current can flow. If there is an interruption in the circuit, it is said to be "open." If there is no interruption to current flow, it is said to be "closed." In this case, the prey's body closes the electrical circuit by providing a direct route for the current to flow from the eel's head to its tail. As a result, the prey gets a major shock.

It is still a mystery how eels can shock their prey without shocking themselves. There are theories, but none of them seems completely satisfactory. Some researchers have found electric fishes share genes that could encode proteins that might insulate their organs from electricity.[26] Others say that, since the typical prey is smaller than the eel, an eel needn't emit a shock large enough to stun itself. Some say an eel actually does shock itself when out of the water, but underwater the water conducts the electricity away from its body. The truth: No one yet knows.

Dissatisfied with just observing the eels' behaviors, Catania wanted to make measurements of the electrical circuit between an eel and its prey, so he decided to wire up a small electric eel and use his own arm as the prey.[27] He made an open electrical circuit by attaching an electrode to the eel's tail and placing the opposing electrode in a jar half-filled with water. He then immersed his hand in the jar, exposing his bare arm above the jar's rim. The eel saw the hand but couldn't get to it because it was in the glass jar, so it protruded its head out of its tank water and above the jar's rim, touching the arm instead of the hand and delivering an electrical shock. Catania felt the pain and his arm involuntarily contracted from the shock, similar to a shocked frog leg.

Catania repeated the experiment multiple times, and he used the data to estimate the magnitude of the shocks that could be produced by eels of different sizes. He found very large eels—electric eels can grow to 8 feet long—could deliver a large enough jolt to temporarily paralyze a human, and potentially kill a person by drowning, if the shocked person fell in the water, unable to move.

Catania says he was surprised at how his arm involuntarily jerked away when shocked by the eel. He remarks, "The fact that there is an animal out there that can remotely control your nervous system, I think is a pretty amazing thing."[28]

Catania, like Franklin, isn't afraid of being shocked for the sake of advancing science. And he's likely to continue being shocked for some time to come because we aren't even close to fully understanding electric eels. In fact, nearly 300 years after Franklin's kite experiment, we still don't even have a complete understanding of lightning.

3

BOLT FROM THE BLUE

LIGHTNING

> If lightning is the anger of the gods, then the gods are concerned mostly about trees.
>
> —CHINESE PHILOSOPHER LAO TZU (SIXTH CENTURY BC)

Lightning is a mercurial beast. Just when you think you understand lightning, it does something you'd never expect. In fact, two somewhat recent lightning strikes—one in Oklahoma in 2007 and another in France in 2012—were so remarkable that they caused an investigating committee of the World Meteorological Organization (WMO) to recommend changing the very definition of lightning.

The Oklahoma strike was noteworthy in that it was the longest distance ever recorded for a single flash; it moved horizontally, spanning the sky for 200 miles (321 kilometers). No one knew lightning could travel that far before. But it was the lightning bolt in France that really was the game changer—it lasted for 7.74 seconds. Technically, it wasn't even lightning because the formal definition of a lightning discharge specifies that it involves a "series of electrical processes taking place over a time interval of less than one second." The long duration of the bolt in France caused the committee to recommend removing the time limit from the lightning discharge definition.[1]

Why hadn't this been seen before? Are long-distance and long-duration lightning bolts something new, yet another consequence of global warning? Not likely. It's probably because we've never been able to measure such things before. In the words of Randall Cerveny, the chief rapporteur of climate and weather extremes for the WMO: "This investigation highlights the fact that, because of continued improvements in meteorology and climatology technology and analysis, climate experts can now monitor and detect weather events such as specific lightning flashes in much greater detail than ever before. The end result reinforces critical safety information regarding lightning, specifically that lightning flashes can travel huge distances from their parent thunderstorms."[2] The implication is that the public should be even more careful about being struck by lightning because, even under clear blue skies, lightning from a thunderstorm that's miles away could come out of nowhere and strike you dead—literally, a bolt from the blue.[3] And exactly how should people protect themselves? Cerveny recommends this rhyming mnemonic: "When thunder roars, go indoors."

Yet, as an outdoor enthusiast, I know it's not always possible to go indoors. Once, while I was camping with a group of fellow backpackers deep in the Blue Ridge Mountains in North Carolina, we were awakened in the middle of the night by one of the worst thunderstorms I have ever experienced, with one brilliant lightning flash and deafening thunderclap followed by another. Our hiking guide made us all get out of our tents and stand on our backpacks in a crouched fetal position, legs held tightly together, with our heads down and our rain ponchos draped over ourselves to keep our bodies and the backpacks we were standing on as dry as possible. We stayed in this uncomfortable position for the duration of the storm, which to our chagrin lasted the better part of an hour, all the time trying to keep any part of our bodies from coming in direct contact with the ground.

The storm eventually passed without causing us any harm and we all crawled back into our tents, a little sleep deprived and wet, but otherwise no worse for wear. The following morning, we awoke to the sun brightly shining. We hung our wet things up to dry, settled ourselves around the breakfast campfire, and began debating the effectiveness of the prior night's backpack crouching position as a safety maneuver. Our leader explained the crouching down part was supposed to keep our bodies low and thus lessened the chance of a *direct strike*, while standing on the backpacks helped insulate us from the ground, thereby lessening the risk that a wave of charge produced by a *ground current* from nearby lightning would come up from the ground into our bodies. Lightning deaths can occur via either direct strikes or ground current, so the backpack crouching strategy was an attempt to simultaneously lessen the lethality risk from both types of strikes.

It has been many years since that stormy night, and I've sometimes wondered whether what we did back then was wise by today's lightning safety standards. Did the backpack crouching increase or decrease our risk of being killed? Are hikers now being instructed to do something different?

I decide to update my lightning safety knowledge by consulting the National Outdoor Leadership School (NOLS). NOLS is the world's foremost school for teaching best safety practices to leaders of outdoor expeditions. As such, NOLS has an outsize influence on the safety behaviors of people in the outdoors.

My contact is John Gookin, a longtime NOLS employee who had once served as NOLS' curriculum and research manager. Gookin is the world's top expert on backcountry lightning safety; in fact, he wrote a book on the subject.[4] According to Gookin, the backpack crouching position is still being taught but only as a last resort. He insists it is only a very small part of an effective risk management system for lightning in the backcountry. Since taking shelter in homes or vehicles—the very best option—is typically not a possibility in the backcountry, the next best alternative is to keep tabs on weather forecasts and avoid being on high-risk terrain when a storm hits.[5] This means staying away from tall trees and hilltops, of course, but also includes less well-known strategies, like setting up camp on the downwind side of a mountain when possible, since the upwind side gets the most lightning strikes. And avoiding caves, since cave entrances seem to be hot spots for lightning strikes for some reason.

Gookin says his early work on identifying backcountry lightning safety options was often stymied by the fact that lightning scientists refused to talk to him about safety alternatives—what they disapprovingly call "partial measures." They believed the public might receive a garbled message and think these alternative safety strategies could be substituted for seeking shelter. They considered shelter in buildings and vehicles to be the "full measure" of lightning protection, with everything else just "partial" and thus inadequate. They didn't want, in any way, to be involved in promoting partial measures to the public; it might result in increased deaths.

Frustrated, Gookin started attending lightning scientists' conferences to see what he could learn about lightning's behavior in the backcountry. He found the scientists had billions of lightning strike data points that allowed them to produce very high-resolution maps of lightning strike densities. Gookin tells me, "Their data had high enough map resolution to actually see that the lightning strikes were primarily occurring on the upwind sides of mountains." But the particular scientist who presented the mapping data hadn't recognized the strike pattern. When Gookin pointed the pattern out to the initially skeptical scientist, the scientist blurted, "Yeah, it is doing that!" Gookin pressed him, "Then, I think your data can tell us that as a storm is coming, people should run to the downwind side of a slope." The scientist said, "Yeah, that's what I would do!" This eureka moment amounted to scientific validation of the current lightning safety recommendation to set up camp on the downwind sides of mountains whenever possible. Prior to the map data that confirmed it, the recommendation had been based solely on theoretical and anecdotal considerations. Now it was backed by science.

Eventually, more and more lightning scientists began to see value in promoting partial measures in situations where seeking shelter was not an option, as long as the primary public health message of seeking shelter wasn't diminished. After Gookin had won over these scientists, they began to comb their lightning data for information that could be relevant to backcountry safety. With time, Gookin developed a close working relationship with many of them, and the result has been more scientifically based lightning safety recommendations that better protect people from lightning in backcountry situations. It is hoped that educating the public about the promotion of these "partial measures" in the backcountry hasn't resulted in a garbled lightning safety message, as the scientists feared, and instead actually has lowered rather than increased lightning deaths in places where shelter isn't available.

The truth is, notwithstanding 200-mile-long lightning bolts and partial measures, the absolute risk of being killed by lightning in the United States is extremely low and getting lower. In fact, 2017 broke the record for the fewest number of lightning deaths nationwide—just 16 people. And similar decreases in lightning deaths are occurring in other developed countries.[6] Some of this is due to better medical treatment of lightning victims—the death rate for lightning victims is currently only about 10%—but mostly it's because fewer people are being struck. The typical lightning victim lives in a rural area and spends a lot of time outdoors. As the United States becomes increasingly urbanized, a larger proportion of people are spending most of their time indoors, thus the potential for people being struck is decreasing. So the advice to go indoors, though wise, is largely irrelevant to most Americans; they are already indoors.

If even those 16 lightning deaths have you worried, move north; the most deaths occur in the high-lightning southern states from Texas through Florida. Or you could just stay inside all of July—it's the worst month for lightning deaths.

Staying inside buildings or motor vehicles is always the best decision for humans to protect themselves from lightning. But most wild animals never have the option of going indoors (or crouching on backpacks). So animals, particularly large animals, are at higher risk of lightning death.

A recent graphic example of how catastrophic lightning can be for animals comes from a reindeer herd. In the summer of 2016, an entire herd of 323 animals was struck dead by lightning.[7] The herd lived on a mountain plateau called the Hardangervidda in a remote area of a Norwegian national park that is home to more than 10,000 wild reindeer. Government wildlife officials discovered a large cluster of recently killed animals while patrolling the area for hunter activity. An investigation suggested lightning had killed them. Reindeer are known to huddle together for safety, and the thunderstorm may have caused them to huddle even closer than normal out of fear. Unfortunately, all that huddling put the whole herd at elevated risk. If the reindeer had spread themselves out, it wouldn't have been possible for a single bolt of lightning to kill them all.

But it isn't just wild animals that are threatened. You may have heard the expression "too dumb to come in from the rain." Although horses often will head for the barn when it starts to rain, cows seldom do. As a consequence, cows are the farm animals most likely to have an encounter with lightning. Cow casualties from lightning strikes in the United States aren't very high, but when such strikes do occur, they can be devastating for an individual farm. That's because cows, like reindeer, often huddle together. Losses of 10 to 30 cows at a time are regularly reported, which is a big enough threat to the survival of the business that some farmers carry insurance on their cows against lightning death.

There are a couple of other reasons, besides huddling, that put grazing animals at higher risk of lightning death. First, they are typically standing in a treeless field (or huddled under a lone tree), making them the highest objects around. Their height above ground means they are more susceptible to a direct strike by lightning, which often hits the highest object in the vicinity.

Given the risk associated with height, you might suspect that among grazing animals, giraffes would be at particularly high risk. Although no one is methodically tracking giraffe lightning deaths, anecdotal evidence suggests giraffes are more susceptible to lightning. All three of the giraffes on the Rhino and Lion Reserve near Krugersdorp, South Africa, were struck by lightning at one time or another between 1996 and 1999, resulting in the deaths of two of the three.[8] In 2003, a quickly moving thunderstorm descended upon Walt Disney World's Animal Kingdom in Florida, leaving little time for the keepers to move the animals indoors. The only animal struck by lightning was Betsy the giraffe who, unfortunately, didn't survive the experience.[9] And in June of 2019, two giraffes were killed by a thunderstorm passing over Lion Country Safari

Park in Loxahatchee, Florida. A park official called it a "billion-to-one" freak accident, and the park's wildlife director Brian Dowling remarked, "This was just an act of Mother Nature that we couldn't stop; we couldn't alter how that came about."[10] But a "billion-to-one" seems like a gross underestimate of the risk. Experience tells us giraffes are likely targets for lightning, the state of Florida tops the nation for the frequency of thunderstorms, and summer is the riskiest season for lightning deaths. The inability to intervene is also a questionable claim; we now have very effective ways to reduce the risk from lightning strikes. The park needs to call in a lightning protection consultant to help prevent another "billion-to-one" accident from killing any more giraffes.

The second reason grazing animals are at risk is that they are standing on four well-separated legs. The further the legs are apart, the greater the difference in ground voltage between one leg and another. A difference in voltage is what drives amps of current through the circuit. When lightning kills a large group of animals, such as those reindeer in Norway, it is typically ground current, rather than a direct strike, that's the culprit. In effect, the animals' legs can act like electrodes to complete an electrical circuit from the ground, through their bodies, and back to the ground. A portion of a pulse of electricity moving across the surface of the ground first encounters one foot and may then decide to take a little side trip up one leg and through the animal's torso before exiting back down to the ground via another leg. The greater the distance between any two legs, the greater the chance of death or injury.

A standing human luckily has but two legs, and consequently can make just a single circuit with the ground. And humans can reduce their risk of death from ground current even further by bringing both legs tightly together, thus forming just a single electrode with no circuit back to the ground.[11] Humans can pull off this merged leg trick with little difficulty.[12] But as you know if you've ever seen a calf-roping event at a rodeo, when a cow has its four legs brought together, it falls over.

The ancient Greeks believed lightning bolts were the weapons of Zeus, the god of the sky. If Zeus was unhappy with you, he would send a lightning bolt your way to wreak some havoc. But even among the Greeks, there were skeptics who thought thunderstorms and their lightning bolts were best explained as natural weather phenomena rather than retribution from the gods.

John Jensenius, a lightning safety scientist with the National Oceanographic and Atmospheric Administration (NOAA), has a more modern explanation for lightning strikes.[13] I ask him to explain the mechanism of lightning strikes. He tells me lightning is basically a very long spark produced when static electricity trapped in a cloud reaches a voltage (pressure) so great that even the air, which is a very good electrical insulator, cannot contain it. As a consequence, the cloud's excess charge jumps directly to the ground.

In order for a lightning discharge to occur, either the cloud must be close to the ground or the voltage must be extremely high. It turns out that the approximate threshold for a lightning strike is about 2 million volts per meter.[14] That is to say, the cloud needs to contain about 2 million volts for every meter of distance between it and the ground in order for a strike to occur. Understanding this, it's easy to see why a cloud with very high voltage, moving over the terrain at a constant elevation, would be spontaneously triggered to strike by the first high object it encounters. The sudden decrease in the distance between the cloud and the high object shrinks the gap between cloud and sky and pushes the cloud over the 2 million volts per meter threshold, triggering a lightning bolt. This characterization of how lightning is triggered is generally true, but Jensenius warns that all such depictions of how lightning works should be taken with a grain of salt. He admonishes, "Over time I've realized lightning does whatever it wants to, and that I should never be surprised by anything!"

Despite Jensenius's caveat, given the mechanism for inducing the more typical lightning strikes, one might expect anything that bleeds the charge out of the cloud will reduce the probability of strikes. That's because bleeding the charge should reduce the cloud's voltage. Franklin's 30-foot (9-meter) pointed rod from the Sentry Box experiment, which we discussed earlier, allegedly was bleeding the cloud of its voltage. Suppose a similar rod was electrically reconnected such that the electric charge it stole from the cloud was redirected to the ground rather than the insulated platform. The ability of the insulated platform to absorb electricity is limited, but the ability of the ground to absorb electricity is unlimited. Presumably, a ground connection for the rod would be able to drain the overhanging cloud of its charge and lower the cloud's voltage to the point where it would no longer have the potential to deliver a lightning strike. In other words, the voltage would drop below the threshold of 2 million volts per meter required to trigger a strike.

Franklin, of course, knew nothing of voltages or thresholds for lightning strikes. To his mind, the cloud had simply, by some unknown means, become oversaturated with electric fluid from the earth, and the surplus charge was just trying to find its way back to the ground. The long, grounded iron rod provided an ideal route for the electric fluid to get from the cloud to the earth. Franklin made the bold assertion that attaching long, pointed iron rods to the tops of buildings, and connecting the rods to the ground using metal wire, would drain clouds of their electric fluid, reducing the number of lightning strikes and thus avoiding the consequent damage they inflicted on people and property.

While modern Americans aren't at significant risk of death from lightning, this was not so for the American colonists of Franklin's day. The colonies were largely agrarian, meaning lots of people were working outdoors in the summer—the months of greatest lightning risk. Also, their barns and farmhouses were usually placed, for water drainage reasons, on the highest ground of the farm—the places at highest risk for lightning strikes. Even for city dwellers, high buildings, particularly church steeples, frequently were struck by lightning. Indeed, churches were prime targets for lightning strikes, and many church fires were the consequence of lightning. As a result, most people in the eighteenth century saw lightning as a major threat to both their property and their lives.

In the summer months, colonial newspapers were peppered with anecdotal stories about catastrophic lightning encounters, destroying property and killing people. We don't have statistics to know exactly how much greater the lightning death risk was in colonial times or the economic burden of lightning damage back then. But we do know from the writings of the time that the colonists were concerned about lightning, and many were very afraid of it.

Although the colonists distinguished between "lightning" (the light) and "thunder" (the sound), they often used the two terms interchangeably and, knowing nothing about the lightning's electricity, often attributed the casualties to the thunder rather than the lightning, as can be seen in this newspaper story from 1732:

> On Wednesday last about 12 o'Clock, a sudden clap of thunder struck upon the House of Ebenezer Prout near this place [Trenton, New Jersey]; he was sitting at the Front Door and one William Pearson at his right hand; his only son a Boy about 9 years old, who stood 3 Foot of them with his back toward the Door, was struck down dead, the hair of his head burned off closely, his Jacket, Shirt and Breeches were torn all to Pieces, but no part of his body touched. The Posts of the House were split, the Rafters shattered as small as you can imagine. The Woman of the House being in the new Room (where the Thunder did the greatest Execution) was so much hurt that we despair of her Life, her youngest Daughter in the same condition; William Pearson is much hurt but likely to recover. The Man of the House is not hurt. The Boy was buried yesterday.[15]

These unfortunate colonists were struck despite being inside their home. This is because it is only modern buildings—structures with indoor plumbing and electrical wiring systems—that afford lightning protection. The modern utility systems provide routes for conduction of lightning's electricity to the ground that didn't exist in colonial homes. Without the utilities found in modern buildings, there is scant protection afforded by being indoors. That's why, in contrast to the

United States' current annual lightning death toll of about two dozen, Africa has a lightning casualty rate of about 10,000 deaths per year because millions of Africans still live in primitive huts and other structures having no modern utilities. They are as vulnerable to lightning now as the American colonists were over two centuries ago.

Understandably, the colonists saw Franklin's discovery of a remedy for the lightning problem as a godsend, and his rods were quickly installed on many public and private buildings throughout the colonies as well as in Europe. In fact, some of those original rods are still in use today. The Maryland State House is currently protected by one of Franklin's original rods, which was installed in 1788.[16] Protruding 28 feet (7.5 meters) into the sky above the building's high dome, it is the tallest Franklin-designed lightning rod ever created during Franklin's lifetime. On Friday, July 2, 2016, the building was saved from lightning damage by its 228-year-old lightning rod.[17] When the rod was hit, it triggered a sprinkler system in the building's dome, but there was no smoke or fire, and not a single person was injured. Governor Larry Hogan, who was nearby when the lightning struck, remarked that "it somehow feels fitting for the 4th of July [American Independence Day] weekend that the oldest State House in the nation was hit by lightning but saved by [an] . . . original Ben Franklin lightning rod."

The lightning strike to the Maryland State House illustrates another point. Lightning rods protect buildings very well, but not for the reason Franklin had thought. The rods can bleed static electricity from clouds, just as he proposed, but in most cases not nearly enough to prevent a strike. Franklin did recognize an alternative mechanism for protection, but he didn't think it very likely. He said in the rare case a building couldn't be prevented from being struck by a lightning bolt, the lightning rod would "*conduct* it [to the ground], so that the building should suffer no damage."[18] We now know the conduction of lightning bolts to the ground is actually the major mechanism through which lightning rods protect buildings. They do not lessen the likelihood of a strike, as Franklin had erroneously believed.

At this point, you must be asking yourself: What's the big deal with the ground, and why is it that electricity always seems to be trying to find a path to it? Good question.

To answer the question, try this little exercise. Look at all the various things currently in your surroundings. All of these objects have the capacity to hold, or absorb, electrical charge to some extent. Some objects absorb better than others, but no matter what their capacity to absorb electrical charge, they all have a limit to their absorption. Their limit largely is determined both by how well they conduct electricity and by their total mass. A metal plate will absorb more charge than a wooden board, and a big metal plate will absorb more than a small one.

Now consider the ground—the surface of the Earth. The ground, particularly wet ground, is a very good conductor of electricity. And the ground is part of something very big: planet Earth. In comparison with just about anything else in your surroundings, the Earth is infinitely larger, meaning it has the capacity to absorb as much electrical charge as you throw at it.

So whatever charge you're trying to get rid of, whether from lightning or anything else, it will find a good home in the ground. Yes indeed, the Earth can absorb a massive amount of charge and (this is the important point) can do so *without influencing its own overall charge in any significant way*. A little more electrical charge is just a drop in the bucket for the Earth. That makes the ground, for all intents and purposes, the charge constant of our daily environment. For this reason, we use the charge of the Earth as a reference point for all other charges. We define the overall charge of the whole Earth as zero and compare everything else to the charge of the Earth.[19] Since the ground is "zero charge" relative to any other object, the ground looks like a great option for a cloud to dissipate its charge and relieve itself of its elevated voltage. What this means to you is that you should never put yourself in a position between high voltage and the ground; the excess charge may very well decide going through your body is the best way to get home to planet Earth.

Although the cure for the lightning problem was fairly well understood—simply relieve the cloud of its static electricity by providing its charge a direct route to the ground—no one in Franklin's day understood exactly how all that static electricity got into the clouds in the first place. In a letter from 1755, three years after his kite experiment, Franklin lamented to a friend: "I wish I could give you any satisfaction on clouds. I am still at a loss about the manner in which they become charged with electricity."[20] As it turns out, just like when static electricity is produced from amber, some rubbing must happen for clouds to become charged.

The physics of thundercloud activity is very complex, and the details are still a bit controversial. But there are a few conditions that must be present in order to set the stage for the accumulation of static electricity in a cloud. First of all, the cloud must have interior temperatures well below freezing—cold enough for ice particles to crystalize from the cloud's water vapor. In addition, there must be upward-moving interior air currents. The upward air currents within the cloud push the ice particles up with them, but not all particles move at the same rate. Since the particles differ in size, they move at different speeds and thus tend to bump into each other as they move up.[21] Think of how smoothly traffic flows on a highway when all the cars are traveling at the same speed. If some of the cars are traveling more slowly, they provide obstacles to the faster cars that must pass them by, and some fender bumping likely will occur. The net effect of many ice particles bumping into each other is that a lot of static electricity is produced. Applying the single-fluid model, we could say the smaller particles are scraping the electric fluid off of the large particles, resulting in the larger particles becoming negatively charged (electric fluid deficiency) and the smaller particles positively charged (electric fluid excess).

Also, the force of gravity has the effect of separating the charged particles within the cloud. Because the larger particles (negatively charged)

FIG. 3.1. Lightning in Arizona. As a storm cloud passes over the landscape, negative charge near the bottom of the cloud induces the accumulation of an area of positive charge on the surface of the terrain below. "Leaders" of negative electrical charge start to emerge from the cloud's underside and make their way in jagged ("stepped") fashion in the direction of the ground, attracted to the positive charge concentrated there. These stepped leaders appear faintly lit (as can be seen on the left and right sides in this photograph), and most never actually make contact with the ground. But those that do reach their destination thereby close an electrical circuit, with the consequence being a massive flow of current accompanied by a brilliant flash of light, called a *return stroke,* along the entire length of the leader (as seen near the center of the image). (Photo © David Blanchard)

are heavier, they tend to accumulate near the lower part of the cloud, while the smaller particles (positively charged) are less affected by gravity and thus tend to accumulate at the top of the cloud. What this means for the cloud as a whole is that it becomes electrically segregated, with the positive charge at its top and the negative charge at its bottom. So looking up from the ground below, the passing thundercloud appears to be a massive ball of negative charge, giving the cloud extremely high voltage relative to the ground with its zero charge. All that negative charge in the cloud is seeking a place to go, and the ground is looking pretty appealing. The only barrier to making the move is all the intervening air—a poor conductor of electricity—between the cloud and the ground. But as we discussed before, if the ground suddenly became closer (decreased distance) or the amount of charge in the cloud grew larger (increased voltage), the obstacle presented by the air could be overcome and the negative charge could simply jump straight to the ground despite the air.[22] When that happens, we have a lightning strike.[23]

You may have heard some people claim lightning jumps from the ground to the sky, not from the sky to the ground. It was actually Franklin who first proposed this contrarian idea based on his misunderstanding of the direction that electrical current flows. In a 1753 letter to his friend Peter Collinson, he remarked, "For the most part, in thunder-strokes [lightning], *tis the earth that strikes into the clouds, and not the clouds that strike into the earth.*"[24]

Franklin came to this odd conclusion when he discovered that the static electricity he drew from thunderclouds was negatively charged, rather than positively charged as he had expected. (He didn't know that only the lower parts of the clouds are negatively charged; their tops are positively charged.) Franklin was perplexed by this revelation. According to his fluid model, a negative charge for the clouds meant they were *deficient* in their electrical fluid. Since Franklin's fluid model specified that electrical current always flows from positive charge to negative charge, this meant the visible lightning flash must represent electrical fluid jumping up from the ground to the negatively charged cloud. The opposite current flow didn't fit his model. Was everyone's perception of the direction of lightning wrong, or was Franklin's model wrong about the direction that electrical current flows?

For the most part, Franklin's original assignment of positive or negative to the two charge states was random and inconsequential; he just as easily could have called them charge A and charge B, and claimed that current moved from A to B or from B to A. But he preferred to call them positive and negative because he believed them to represent an excess or deficiency of a single electrical fluid. This was the very foundation of his single-fluid model of electricity. The implication of the model and his nomenclature was that a current of electrical fluid would move from an area of excess (positive) to an area of deficiency (negative), and not the other way around, because to do so would make no sense in terms of relieving the electrical pressure (voltage).

Although Franklin never explained his rationale for determining which charge would be called positive and which negative, he must have had his reasons, and he never expressed any doubts about the correctness of his positive and negative assignments. That is why his realization that the thunderclouds evidently were negative was so startling to him.[25] It suggested the clouds were sucking up excess electrical fluid from the ground, and the lightning bolts thus were emanating from the ground rather than the sky, a very counterintuitive idea. Franklin was, nevertheless, willing to accept this upside-down lightning bolt explanation rather than reverse his assignment of positive and negative charges.[26]

We now know Franklin was wrong about the movement of charge. In most cases, charge doesn't flow from positive to negative; it flows from negative to positive. This is simply because the negative charge is able to move. (The positive charge usually remains stationary for reasons we'll get to later.) And that is why electricity jumps from the negatively charged lower part of the cloud down toward the ground. Under most circumstances, however, the only thing essential for a functioning electrical current is that some type of charge be on the move; the exact direction of the flowing charge is usually less important to know. (It makes no difference if you put the batteries into an ordinary flashlight backwards; it will still light up.) That's why, in deference to Franklin, when dealing with electric devices we still define electrical "current," with a wink and a nod, as moving from positive to negative, while all the time knowing that the charge, specifically the negative charge, is really moving in the other direction.

But with regard to Franklin's claim that lightning went from the ground to the sky, you could say he was partly correct. Although the negative electric charge moves from the sky down to the ground, it does so as a relatively faint *step leader*, a downward-traveling spark. When the step leader reaches the ground, the negative charge discharges to the ground, and an extremely bright light is produced, illuminating the air channel through which the step leader has just traveled.[27] This brilliant light, called the *return stroke*, originates at ground level and the illumination migrates up the channel all the way to the cloud. So the electrical charge comes down from the sky, while the brilliant light comes up from the ground.[28] But don't try to see the light progress upward. The whole process happens so fast that the human eye cannot discriminate the light's direction of travel; special high-speed cameras are required.[29]

You might have thought Franklin's newly invented lightning rod would have spread like wildfire around the world, and every owner of a tall building would have installed one. And, in fact, the use of lightning rods did spread very quickly, with one exception: church steeples. Surprisingly, adoption of lightning rods for churches was slow despite the lightning rod's near 100% effectiveness in protecting buildings. The reason was that church officials believed they had something better than lightning rods for protection; they had bells.

To this day, you can see inscribed on the tower bells of some medieval churches the Latin inscription *Fulgura frango*, meaning "I break the lightning." This came from the widespread belief that a ringing church bell would ward off lightning. Some thought the ringing of the bell beckoned God to protect the church from oncoming lightning. Even nonreligious people shared the belief that the ringing of bells was somehow protective. Because the practice had been around for so many years and was so widespread, most everyone surmised there must be something to it. Perhaps the loud sound had the effect of breaking up the clouds.

One thing is clear; ringing bells didn't work very well. During one 33-year period in France (1753–1786), 386 church bell towers were hit by lightning, and 103 bell ringers were killed by electrocution through the bell ropes.[30] And during one remarkable night in 1718, a thunderstorm passed over Brittany, France, its lightning striking 24 church steeples between Landerneau and Saint-Pol-de-Léon.[31] The fate of the churches' bell ringers is unknown, but presumably there were significant casualties.

Although we have no testimonials from the lightning-struck bell ringers themselves, we do have a highly credible witness to one such event. As a young schoolboy, the future Nobel laureate Santiago Ramón y Cajal saw the mortal consequences of a lightning strike when his local priest decided to ring the bell of his town's church during a thunderstorm rather than run for cover. Cajal cited the incident as one of the most dramatic, intense, and influential events of his life. He described the immediate aftermath of the lightning strike in his autobiography:

> A voice coming from among the crowd called our attention to the strange, blackish figure hanging on the railing of the bell tower. In fact, there, beneath the bell, enveloped in dense smoke, his head hanging over the wall lifeless, lay the poor priest who had thought that he would be able to ward off the threatening danger by the imprudent tolling of the bell. Several men climbed up to help him and found him with his clothes on fire and with a terrible wound on his neck from which he died a few days later. The bolt had passed through him, mutilating him horribly. . . . Little by little we took in what had happened. The bolt or flash of lightning had struck the tower, partly melting the bell and electrocuting the priest; afterwards, continuing its capricious path, it had entered the school through a window, and pierced the ceiling of the lower floor where we children were, shattering a great part of the ceiling, had passed behind the mistress, whom it deprived of sensibility, and, after destroying a picture of the Savior hanging upon the wall, had disappeared through the floor, by a gap, a sort of mouse hole close to the wall.[32]

For Cajal, the lightning incident changed his view of both the natural world and life in general. Afterward, he saw Mother Nature as fickle and sometimes extremely cruel. He also began to appreciate the capriciousness of who dies and who lives, and it scared him. Despite the trauma of the lightning event, he believed it to be a transformative experience in his life, and it likely contributed to his asserted belief in the value of "strengthening the mind by continuous observation of the spectacle of nature."[33] As we shall soon see, Cajal did indeed have a very strong and observant mind. It was observant enough for him to discover the *neuron,* the cell that is the fundamental signaling unit of the nervous system, and thereby secure his position as the father of neuroscience. It's tempting to speculate that, for both Cajal as well as Franklin, a large part of their scientific greatness may have originated from their transformative encounters with lightning.

If the lightning bolt that struck the Spanish priest was typical, it would have delivered somewhere between 40,000 and 120,000 volts. That seems like a massive amount of voltage, considering the voltage of a modern household electrical outlet runs around 110 volts (or 220 volts) and even shocks from household electric outlets are often lethal. Yet, the static electricity shock you get by touching a conductor after rubbing your feet on a carpet can be as high as 25,000 volts, which is nearly as high as the lower end of lightning's voltage range, and no one has ever died from a carpet shock.

The reason for the difference is this: compared with the huge amount of electrostatic charge built up in the thundercloud that provides the lightning's voltage, the electrostatic charge you build up in your body by rubbing your feet on the carpet is virtually nothing. You'll never be able to get enough charge in your body to shoot lightning bolts from your fingertips. And since current really is just the flow rate of charge, inadequate overall charge means a very limited and brief current. As we've already seen, high voltage accompanied by low and brief amperage may be sufficient for producing pain even in large animals like bears, but you're never going to shock one to death that way.

So exactly how much amperage is needed to kill a human? Not much. Even 0.01 amps of sustained current can interfere with muscle movement, including the diaphragm, and thus affect breathing. Amperage 10 times higher (0.1 amps) starts to interfere with the beating of the heart and can thereby result in cardiac death. At increasingly higher amperages, all of the organ systems eventually become involved, but the nervous system and the muscles are particularly vulnerable.[34]

You might ask, given that even low amperage can kill, how is it anyone survives a direct hit by a lightning bolt? A good question. The primary reason seems to be a phenomenon known as *flashover*—a situation where the electrical charge moves over the surface of the body without significant penetration into the internal organs.[35] When the lightning encounters the body, its current initially starts to move internally, but internal resistance thwarts its movement and it finds a less resistant route over the skin, particularly if the skin is sweaty or otherwise moist, as might be the case for someone caught out in a thunderstorm. Consequently, a flashover often occurs, with electricity passing over the body but not actually entering it. A flashover diminishes the electrical energy deposited within the body and often results in survival.

The second reason is that a lightning strike is extremely brief, typically lasting only milliseconds. This short duration of exposure also reduces lightning's lethality and distinguishes a lightning strike from a shock from a high-power electrical wire, where the current flow into the body typically lasts much longer. The flashover phenomenon is the major reason why lightning seems so capricious, killing some victims while sparing others.

We've seen that voltage is responsible for pain, while amperage causes death. Nevertheless, the pain versus death dichotomy isn't quite as simple as it first seems. That's because you need a certain minimal voltage just to push the amperage. In other words, if we go back to our water analogy, there needs to be enough water pressure (voltage) to drive the flow of water (amperage) through the hose, or all you'll have is a hose full of water dribbling insufficient flow to wash your car.

Speaking of cars, automobile batteries are a good example of the trade-off between volts and amps. They typically put out between 1 and 8 amps of current, which should be sufficient to kill people, if amps alone could determine death. But most car batteries are only 12 volts, or about the same voltage as eight flashlight batteries connected end to end (1.5 volts each). That voltage usually doesn't produce enough electrical pressure to threaten anyone's life, so car batteries aren't considered particularly hazardous to handle. But if 12 volts won't do it, what is the threshold for voltage to become lethal to humans for any given amperage? That information, unfortunately, isn't exactly known. And the question is more complicated than you might think, because the voltage value will be influenced not just by the amperage but also by the body's innate ability to conduct the current through critical tissues. Nevertheless, there have been documented cases of electrocution deaths[36] at voltages as low as 47 volts.[37] And theoretical considerations suggest that, under just the right conditions (or better said, wrong conditions), as few as 25 volts could do it.[38]

You'll notice that the theoretically lethal voltage (25 volts) and the allegedly safe automobile battery voltage (12 volts) are different by only twofold. If we were talking about some type of drug therapy, a twofold *therapeutic ratio* (sometimes called a therapeutic index) would be considered an extremely dangerous drug. It would mean taking one pill would relieve your headache, but two pills would put you in a coffin. That's a pretty narrow safety margin.

So the best practical advice is to treat all voltage with respect because, if you happen to stumble into the wrong conditions, yesterday's safe shock might be today's lethal jolt. Ironically, even though modern life, with its indoor existence and its buildings protected by lightning rods, might make you fear lightning less, it may be wise to fear your toaster more. The bottom line: Don't take any chances with electricity, regardless of voltage. But I digress. Let's now get back to the electrocuted priest.

The shame is that the incident Cajal describes happened in Spain in 1860, over 100 years after Franklin had invented the lightning rod. Had a lightning rod been installed on the church, the priest's life would have been spared even if he was ringing the bell during the thunderstorm. It needn't have been one or the other. The priest still could have rung the bell while under the protection of a lightning rod. God wouldn't have minded his having a backup plan.

Contrast this unprotected Spanish church's tragic experience to an Italian church that installed a lightning rod in 1776. The records of St. Mark's Basilica in Venice show lightning had damaged its 100-meter bell tower in 1388, 1417, 1489, 1548, 1565, 1653, and 1745.[39] Once a lightning rod was installed in 1776, the tower remained undamaged by lightning until it finally just died of old age. The centuries-old creaking tower spontaneously collapsed on July 14, 1902. Fortunately, just prior to its collapse, the sudden appearance of a large crack on the tower's exterior as well as the sound of stones falling inside alerted people to clear the area. The only fatality when the tower fell was the caretaker's cat, which evidently had been enjoying its ninth life.[40]

Because of the undeniable success of his lightning rod, Franklin became world famous, not just within the scientific community but also among the general public. One unforeseen outcome of his worldwide lightning fame was that many people saw him not only as an electricity expert but also as an overall genius (which he certainly was). So they consulted him on all sorts of matters, including their illnesses. Asking Franklin for a medical consultation wasn't very farfetched. Franklin actually had a strong interest in medicine, and he wrote often and knowledgably on a variety of medical issues.[41] He had been a big proponent of inoculation against smallpox, and he even invented a flexible catheter to help his ailing brother, John, get relief when he was having difficulty passing urine due to a bladder stone.

Franklin frequently dispensed medical advice to friends and relatives while at the same time including the disclaimer that he wasn't a physician: "I hope you consider my advice, when I give any, only as a mark of my good will, and put no more of it in practice than happens to agree with what your Dr. [physician] directs."[42]

Despite his lack of medical credentials, Franklin wasn't averse to personally administering experimental treatments to the sick. Ever since the Leyden jar had been discovered, people knew an electrical shock from a fully charged jar resulted in powerful muscle contractions. So a reasonable question was whether electricity could be an effective treatment for people with paralyzed muscles. Given Franklin's famed electrical expertise, it wasn't long before paralytics sought him out for possible treatment with electricity. And treat them he did.

Franklin described these treatments in a 1757 letter to his physician friend John Pringle:

> [Many] paralytics were brought to me from different parts of Pennsylvania and the neighboring provinces, to be electris'd, which I did for them, at their request. My method was to place the patient first in a chair on an electric stool. . . . Then I fully charg'd two 6 gallon [Leyden] jars, each of which had about 3 square feet of surface coated, and I sent the united shock of these thro' the affected limb or limbs, repeating the stroke commonly three times each day. The first thing observed was an immediate greater sensible warmth in the lame limbs that receiv'd the stroke than in the others. . . . The limbs too were found more capable of voluntary motion and seem'd to receive strength. . . . These appearances gave great spirits to the patients, and made them hope a perfect cure; but I do not remember that I ever saw any amendment after the fifth day: Which the patients perceiving, and finding the shocks pretty severe, they became discourag'd, went home and in a short time relapsed; so that I never knew any advantage from electricity in palsies that was permanent.[43]

With that, Franklin went back to his lightning rod and continued to tinker with it for several years, constantly trying to improve on its design. He closed the door on the potential of electricity to effect medical cures and pursued electrotherapy no further. But others would.

4

FOR ALL THAT AILS YOU

ELECTROTHERAPY MACHINES

> It is better to have recourse to a quack, if he can cure our disorder although he cannot explain it, than to a physician, if he can explain our disease but cannot cure it.
>
> —CHARLES CALEB COLTON

I step with trepidation onto the foot-high platform that is electrically insulated from the ground by its four peg legs made of solid glass. As instructed by my "electrotherapist," I put my left hand on a brass rod that will serve as one electrode. The electrotherapist then lowers the other electrode—an inverted brass "crown" suspended from the end of a metal rod—down from above until it hovers about 16 inches (40 cm) over my bald head. I am now ready for my electrotherapy to begin.

I am undergoing a treatment simulation for an imaginary head ailment with a Toepler Influence Machine.[1] Electricity machines manufactured by Toepler were considered state-of-the-art medical therapy machines in 1900. Toepler produced machines of various sizes, but the one that I'm hooked up to was its biggest and most expensive. Imagine a large, ornately decorated marble-topped desk, drawers and all. In the rear of the desktop is a huge glass "fish tank," spanning the entire width of the desk and rising to a height of about 3 feet (1 m) above the desktop.

The tank could easily accommodate 150 gallons (568 L) of water, but there is no water and there are no fish to be seen. Instead, what you see in the tank are multiple round plates of translucent brown-colored glass, resembling large amber disks, with diameters of about 3 feet, mounted side by side on a single axle. A hand crank that looks like it may have been commandeered from the starter of a Model T Ford is mounted at the right-hand corner of the desktop. Turning it sets the glass wheels in motion, generating electricity.

My electrotherapist begins by turning the crank to get the glass disks spinning. He assesses voltage output of the machine by moving two brass balls on sticks, resembling big round lollipops, across the desktop, toward each other. The balls are connected to the positive and negative electrodes of the machine. The idea is to move the balls closer and closer to each other until visible sparks just start to fly between the two. If the sparks fly when the balls are still far apart (>12 inches; 30 cm), it means the voltage is very high. If the balls must be almost touching to produce sparks, it means the voltage is very low. Thus, the distance between the balls is an indirect gauge of the magnitude of the voltage. When the gap required to produce the sparks is just about 6 inches (15 cm), the voltage level is deemed suitable for my treatment. The electrotherapist flips a switch, and the electrical circuit is rerouted away from the metal balls . . . toward my head!

The crown suspended above my head looks just like something King Arthur of English folklore might wear to a roundtable meeting with his knights, with one exception. Instead of having a half-dozen high peaks protruding above its rim, this crown has 100 small points facing downward, resembling the dentition of a wide-mouthed shark about to bite my scalp.

The whole experience seems surreal, and I would be lying if I said that I'm not a little afraid for my head. After all, seeing the leaping sparks and having the electrode crown suspended over my skull seems like a scene from a Frankenstein movie. I brace myself to be shocked. But I feel no shock. Instead, I feel a cool breeze coming down from above, the skin of my scalp and face begins to tingle, and my shirt clings to my chest. In a word, it feels pleasant.

This is what I feel, but what I cannot see is the shimmering halo of blue light that envelopes my dome, like the northern lights that crown the Earth's Arctic regions. So my therapist dims the room lights and takes my place on the platform. Then I crank the machine for him, so that I can see the blue halo surrounding his head. Fantastic!

This re-created medical treatment of the nineteenth century was called "static breeze" therapy, and I can understand why.[2] The flow of static electricity coming down from the suspended crown to my head does feel like a cool summer breeze, just like that from a slowly turning ceiling fan. Static breeze treatment wasn't restricted to the head; it could be used on other body parts, simply by substituting the inverted crown with other electrode attachments. But when static breeze was used for head ailments, it was often called an "electric head bath." The bath was thought to cleanse the patient of everything from headaches to bad thoughts. If the head problem persisted or returned later, the patient was given more baths.

My electrotherapist is Jeff Behary, a collector and restorer of antique electrotherapy machines of all sorts. Behary, in his mid-thirties, is boyishly animated and enthusiastic as he shows me his multiple electrotherapy devices of various vintages, spanning three centuries. His zeal for his hobby is infectious, and I find myself drawn into his magical world of medical sparks and shocks. Behary has a lot of amazing stuff, but the Toepler machine is the star of his collection. There probably are less than half a dozen such Toeplers still in existence, and his is the only one in the world that actually works. It works because Behary restored it to its original state, with the help of a variety of different craftsmen, including glaziers who refurbished the glass disks.

There are a number of other collectors of antique electrotherapy machines in the world, and Behary knows most of them. He laments that the growing popularity of collecting such machines has driven up the market to the point that he wouldn't be able to afford to reassemble his current collection at today's prices. But, despite the competition for rare pieces, Behary's collection is unique because most of his machines actually work. That is, they produce electricity. Whether they cure any diseases, as their manufacturers had claimed, is a whole other matter.

As we've already seen, the idea of using electricity to treat disease can be traced all the way back to amber. So when static electricity machines became widely available in the latter part of the eighteenth century, it is not surprising that people would want to see if they could cure diseases.[3] As we've seen, even Franklin gave it a go, but finding no lasting benefit for palsy patients, he soon abandoned electrotherapy altogether. However, Reverend John Wesley, the famous founder of the Methodist Church and a huge fan of Franklin's electricity work, was not as easily deterred. Wesley had a strong interest in health and disease issues. He had read everything Franklin had published about electricity, and he was convinced electricity had curative properties. Over just a few years, he attempted electrotherapy treatment on many different people with all sorts of diseases. In 1760, he published his own electricity book, called *Desideratum: Or, Electricity Plain and Useful.*[4] In the first part of the book, he faithfully recounted Franklin's experimental findings, in an effort to make electricity understandable to the public; that is, "made plain." But in the second part of the book, he endeavored to explain how electricity might be made "useful," and its true usefulness, as Wesley saw it, was in curing diseases.

Wesley was particularly optimistic about diseases of the nervous system, claiming electricity to be a "rarely failing remedy, in the nervous cases of every kind (Palsies excepted)."[5] Perhaps his exclusion of palsies shouldn't be seen so much as a confirmation of Franklin's negative findings with palsies but more of an effort not to contradict Franklin. (Wesley remains highly deferential to Franklin's work throughout his book.)

In any event, Wesley proposes a neurophysiological mechanism to explain why the nervous diseases respond so well to electrotherapy:

> Perhaps if the nerves are really perforated (as is now generally supposed), the electric[ity] is . . . the only fluid in the universe, which is fine enough to move through them. And what if the *nervous juice* itself be a fluid of this kind?[6]

In essence, Wesley is speculating in 1760—well before Galvani ever shocked his first frog leg (1786)—that electricity is the "fluid" that drives the nervous system. In effect, he is equating nerve function to electrical activity—an idea that we now understand to be fundamentally correct.[7] He further speculates, "[Electricity] is subtle and active enough not only to [be the] cause of motion, but to produce and sustain life throughout all of nature, as well as animals and vegetables."[8] In other words, Wesley is proposing electricity is responsible not just for nerve function and muscle movement in animals but also as the vital force of all life on earth, which was another idea way ahead of its time.[9]

Although Wesley is sometimes dismissed as an electrical amateur and electrotherapy quack, he deserves credit for being among the first to link electrical science with neuroscience. He was modest about his own capabilities and his ideas. He considered his book just an overview of the subject of electricity, and he expressed the desire that "someone with more . . . ability than me, would consider it more deeply, and write a full practical treatise on electricity, which might be a blessing to many generations." That wish ultimately would be fulfilled over a century later when the writings of the brilliant British scientist Michael Faraday, covering his entire life's work with electricity, were collected and published together as a three-volume set entitled *Experimental Researches in Electricity*.

At the beginning of the nineteenth century, electrotherapy was moving ahead at full speed, not because of any insightful breakthroughs into the mechanisms of neurological diseases, but rather in spite of the lack of such knowledge. The fact is, nineteenth-century physicians had a very poor understanding of human physiology. If they had fully understood that the neuromuscular system works largely on electrical principles, they might have focused their electrical treatments on diseases of nerve and muscle tissues, as Wesley had suggested. But, instead, many physicians who used electrotherapy subscribed to the notion that diseased tissues and organs suffered from some kind of localized nutritional deficiency, and they believed electricity enhanced blood flow to the diseased tissues, thereby improving their nutritional state.[10] This wasn't a totally ridiculous idea since treated areas often appeared to redden due to vasodilation, so the reddening actually did mean greater blood flow. Given this notion that increased blood flow reduced localized nutritional deficits, it is no wonder they thought all tissues and organs could benefit from electrical treatments. The electrotherapy practitioners believed it to be self-evident that all organs required adequate nutrition to function properly, otherwise they would become diseased.

In his 1901 textbook, *Electricity in Medicine and Surgery*, William Harvey King, a prominent physician and devotee of electrotherapy, describes the development of electrotherapy and the latest state-of-the-art treatments. The book displays King's impressive command of fundamental electrical principles but an extreme level of misunderstanding and naïveté about even basic physiological principles and the causes of disease. King seemed mostly concerned about precisely mapping "motor points," the places on the body surface where an electrode should be placed in order to elicit a contraction of the underlying skeletal muscle. These motor points likely represented places where nerves enter muscle just beneath the skin. He saw the ability of the physician to find and electrically shock these motor points as a finely honed clinical skill.

An electrical shock delivered at a specific motor point produces a contraction in a specific muscle group. If disease is suspected in a particular muscle on one side of the body, King believed the key to diagnosis of that disease was a comparison of the strength of its motor point contraction with the corresponding motor point contraction of the healthy muscle on the other side. Greater or lesser contraction of the putative diseased muscle relative to the normal muscle was seen as evidence of pathology. Thus, the motor points were used for diagnosis, but they were subsequently targeted to deliver jolting therapeutic shocks as well.

King was a big fan of Toepler machines, and he describes their inner workings in great detail in his book.[11] But his treatments weren't limited to the electric breeze, nor to low voltages. He recommended the use of large Leyden jars in combination with Toepler machines "to store up a quantity of electricity and let it off in large volumes."[12] In fact, these treatments with Leyden jars delivered very strong electrical shocks to the patients.

King also seems to appreciate the value of controlling currents as well as voltages when treating patients, and credits Adolphe Gaiffe for inventing the "first useful and reliable [amp measuring] instrument" for clinical medicine.[13] He notes, however, that physicians were slow to realize the importance of controlling currents in their treatments until Georges Apostoli, an influential Parisian gynecologist and practicing electrotherapist, "insisted that his treatments could not be intelligently given without [the use of an amp meter]."[14] Still, most electrotherapists ignored current (electrical flow rate) and focused exclusively on the voltage (electrical pressure) needed for treatment.

Surprisingly, despite his near obsession with mapping motor points, King devotes a large section of his book to treatments of gynecological and urological conditions. It seems like few gynecological disorders couldn't benefit from a shock to the vagina. He delivered the shock by positioning one electrode in the vagina and the other in the rectum. Men with urogenital complaints might be treated with two alternative electrode placements. Both involved sticking a long narrow electrode up the urethra of the penis, but the second electrode might be placed either up the rectum or alternatively attached to the testicles. King laments, "It is not an easy matter to administer electricity to the testicles." But he goes on to suggest a practical solution: "Take an ordinary oblong gravy dish—one of common earthenware is best, as the plainness of the outlines makes it more convenient than the fancy shape of the expensive ones." One should then lay a copper plate in the bottom of the gravy boat and cover the plate with a sheet of asbestos "to prevent the testicles from coming in direct contact [with the plate]." The copper plate is connected with electrodes and the gravy boat is partially filled with salt solution. The naked patient is then instructed to straddle the gravy boat and immerse his testicles in the salt solution, at which point the electrical current is turned on. Ouch!

Yes, the pain of treatments was a problem. King counsels his fellow electrotherapists:

> It is your duty at all times to try to cause but little pain. This is sometimes impossible, but it should be a rule not to hurt or distress the patient in any way until he has become acquainted with the method and acquired confidence in you, when you will have no difficulty in influencing him to bear the necessary burning or other disagreeable sensations.[15]

Such treatments sound more like tortures, especially when you consider they were administered two to three times a week. The treatments were continued either until the patient was "cured" or until he failed to return for more treatments. Ludicrously, missing patients were also frequently counted as cures. After all, if they weren't cured, why didn't they come back for more treatments?

Before we jump to condemn the physicians who ignorantly practiced such electrotherapeutics, let's not forget that very few of the medical treatments at the time had any benefit and many were equally cruel. The best one could expect to gain from most medical treatments was a *placebo effect*—a positive health response that cannot be attributed to the treatment itself but rather to the patient's faith in the treatment.[16] Many medicines were highly toxic and made people even sicker. Surgery was restricted mostly to amputations that, until the latter part of the nineteenth century, were typically done without anesthetics. If patients survived the surgery, they might later die of infection, because antiseptic surgical techniques, first introduced in 1867 by Joseph Lister,[17] weren't widely in use and antibiotics were yet to be discovered.[18]

Practitioners of electrotherapy, like all physicians of the day, suffered from ignorance about certain aspects of human physiology and the underlying mechanisms of disease. However, the physicians at the time did not allow their ignorance about the underlying causes of an illness to prevent them from aggressively treating an illness, even if they lacked evidence for a benefit from their treatment. And how could they not in the face of so much disease and human suffering? If nothing else, they provided patients with hope. Few physicians were trying to con their patients. They were not selling snake oil. Most sincerely believed in the treatments they were delivering even though their evidence was entirely anecdotal, based merely on testimonies from their patients and other physicians. Eventually, the pitfalls of accepting anecdotal evidence in medicine would become all too apparent, and a more scientific approach to evaluating therapies would emerge. At the time, however, withholding these unproven treatments would have been considered unethical.

Although some physicians claimed profound insight into how electricity allegedly cured disease and based the details of their treatment regimens on the purported first principles of their unproven theories, others weren't so confident. George Pitzer, a physician and educator practicing electrotherapy, candidly acknowledged in 1883: "Just how the favorable change is wrought we do not pretend to explain, but we do know that with [electricity] we can frequently relieve, and many times speedily and permanently cure people who had lingered for months under different plans of treatment, medical and hygienic."[19] If you really believed you were seeing improvement and cures, even in just a small group of your patients, while no other types of treatment were helping, wouldn't you keep on treating with electricity?

But these physicians lacked a deeper understanding of the pathologies of the various diseases they were treating. And a few, not appreciating how each disease is unique, employed their electrotherapy universally and indiscriminately. Such physicians believed in unproven comprehensive theories about the mechanisms of disease and that there should be a universal cure, such as electrotherapy. And, consequently, there were few ailments they didn't treat with electrotherapy. For example, in King's textbook on the use of electrotherapy, he describes treatment procedures for everything from acne to cancer. Gout is the only disease that he reluctantly acknowledges to be refractory to electrotherapy: "After many trials, and many methods, we have never seen any material reduction of gouty deposits from its use."[20]

To truly understand the rise and fall of electrotherapy, we must look at it in its larger context. Electrotherapy was but one of many questionable medical practices that existed at the time. As medical historian David Wootton explains, "In eighteenth-century America there was something close to a free market in medical training, different therapies competed against each other without hindrance, and regulation only became the norm late in the nineteenth century."[21] It was hard to distinguish one therapy from another by its effectiveness because all were ineffective. Most diseases either killed the patient outright or resolved on their own, but every physician took credit for curing those patients who didn't die.

It wasn't until *germ theory*, the idea that many diseases were caused by microscopic creatures infecting humans, began to take hold in the late 1800s that dramatically different outcomes were seen for different types of treatments. In particular, surgeons who rejected germ theory had many postsurgical deaths due to wound infection, while the patients of surgeons who practiced antiseptic techniques mostly lived. The different outcomes were so obvious there was no denying it. No formal clinical trials were necessary; you could just compare death rates at hospitals that used surgical antisepsis to those that didn't. As Wootton succinctly puts it, all you needed to know to determine which was best was to take "the ratio of those patients who were still above ground to those who were now below ground."[22]

It was the triumph of germ theory that caused people to look at diseases differently. And when they looked at diseases differently, they thought of treatments differently. Little by little, each type of therapy would come under scrutiny by both the medical community and the public as to its effectiveness. By 1900, the notion that electrotherapy could cure all ailments was living on borrowed time.

As early as 1831, however, there was a particular electrotherapist who rejected the idea that electrotherapy was a panacea, believing instead that electrotherapy should be targeted at specific diseases. For him, the problem could be reduced to identifying the specific diseases where electrotherapy would succeed. Furthermore, he thought you should be able, by identifying the diseases that were responsive to electrotherapy, to define a class of organs and tissues where electricity played a central role in normal physiology. In effect, he believed that electrotherapy could be used not only to cure disease but also to find the underlying cause of disease . . . as long as physicians chose their cases wisely. This visionary physician seeking specific disease targets for electrotherapy was Guillaume-Benjamin-Amand Duchenne.[23] Targeted therapy—the concept that each disease and each disease stage warrants its own special treatment—is now the mantra of medical practice. At the time, however, this clinical concept was a contrarian view.

Dr. Duchenne was a French physician who was the first to apply electricity in a scientific way to the study of neurological diseases. The scientific and medical communities at the time generally ignored his work. In fact, King makes no mention of Duchenne or his work in his electrotherapy textbook of 1901. Nevertheless, Duchenne is now recognized as one of the pivotal early researchers in the field of *neurology*.[24]

Duchenne was the scientific heir of Galvani, who, as we've already seen, studied the physiological mechanisms underlying frog leg contraction by electrical shocking. Duchenne brought those studies to the level of the human body.[25] He was most interested in understanding the *pathology*, the characteristic features of a disease, for each of the many human neuromuscular illnesses. He thought electricity might be used as a powerful probe to identify the specific muscle tissues involved for each particular disease.

In 1835, Duchenne started to employ an "electropuncture" technique in his research, whereby very fine and sharp electrodes were inserted underneath human skin and used to probe specific muscles and shock them with electricity to assess their contractile function. In doing this, he was able to compare muscle function with muscle biopsy tissue that he had acquired, using a medical device of his own invention, the so-called Duchenne trocar—a puncturing device within a tube that is inserted through the skin in order to take fluid or tissue samples from internal tissues.

Using these and other novel approaches, Duchenne was able to identify and describe a number of neuromuscular diseases, some of which still bear his name. One is *Duchenne muscular dystrophy* (DMD), a severe inheritable type of muscle disease of young children, particularly boys. DMD is one of nine different types of muscular dystrophy. All nine involve progressive weakening or degeneration of muscle tissue in one way or another.

Duchenne also had an interest in hearing. He experimented by sticking electrodes in his own ears and experienced various sounds induced by electricity. He also tried the same experiments on deaf and hearing-impaired patients to compare what each would experience. Most of them reported hearing similar sounds. For some completely deaf patients, it may have been their very first experience having any type of hearing sensation. Some of the hearing-impaired patients claimed their hearing had actually improved after the experiments. Although they were likely just experiencing a placebo effect, it still caused quite a sensation among the deaf community in France. Just as paralytics, nearly a century earlier, had sought out Franklin for electrical treatments because they had heard Leyden jars could induce muscles to contract, Duchenne was similarly deluged with deaf people seeking an electrical remedy for their problem. Duchenne later lamented, "These [alleged cures] were quickly known, and then deaf and deaf-mutes were sent to me in numbers, and whether I would or no, I was obliged to continue these empirical experiments."[26] He considered them mere experiments rather than treatments, because he wasn't convinced they did the patients any good.

Duchenne published the findings of his clinical experiments on hearing in 1851. His major conclusion was that there were two forms of deafness, one where some hearing experience could be simulated using electricity and one where hearing was unresponsive to electricity. The type responsive to electricity he named "nervous deafness" because he believed it was caused by some unknown defect of the nervous system. He was optimistic that nervous deafness might one day be curable because it was amenable to electrical stimulation. For other types of deafness, he offered no hope. "I am," he said, "so little encouraged by the success obtained in those deaf patients who have not, under electrical exploration, given evidence of the characteristic sign [i.e., an electrically induced hearing response] that I now absolutely refuse to treat those cases."[27]

All of his work with muscle diseases and hearing loss should have locked in Duchenne's reputation as a great medical scientist. Unfortunately, Duchenne's scientific work in neurology became somewhat tainted by one of his other medical interests. Regrettably, Duchenne was also a subscriber to the concept of *physiognomy*, and that interest caused his research to take an odd turn. You may be familiar with the defunct practice of *phrenology*, where a person's mental health is allegedly diagnosed from the bumps on his skull. Well, physiognomy is something similar except that, instead of skull bumps, it's facial expressions that are supposedly telling. Physiognomists claimed facial expressions were windows to the character of a person. Although facial expressions can obviously be indicative of a person's thoughts or reactions to events (e.g., normal people don't smile in reaction to pain), physiognomy claimed much more. Facial expressions were supposedly rooted in a person's soul. So understanding the physiological mechanisms underlying different facial expressions was thought to provide insight into the fundamental psyche of a human being, thus bridging physiology and psychology.

Because of his interest in physiognomy, Duchenne started shocking people's faces to involuntarily induce different facial expressions and

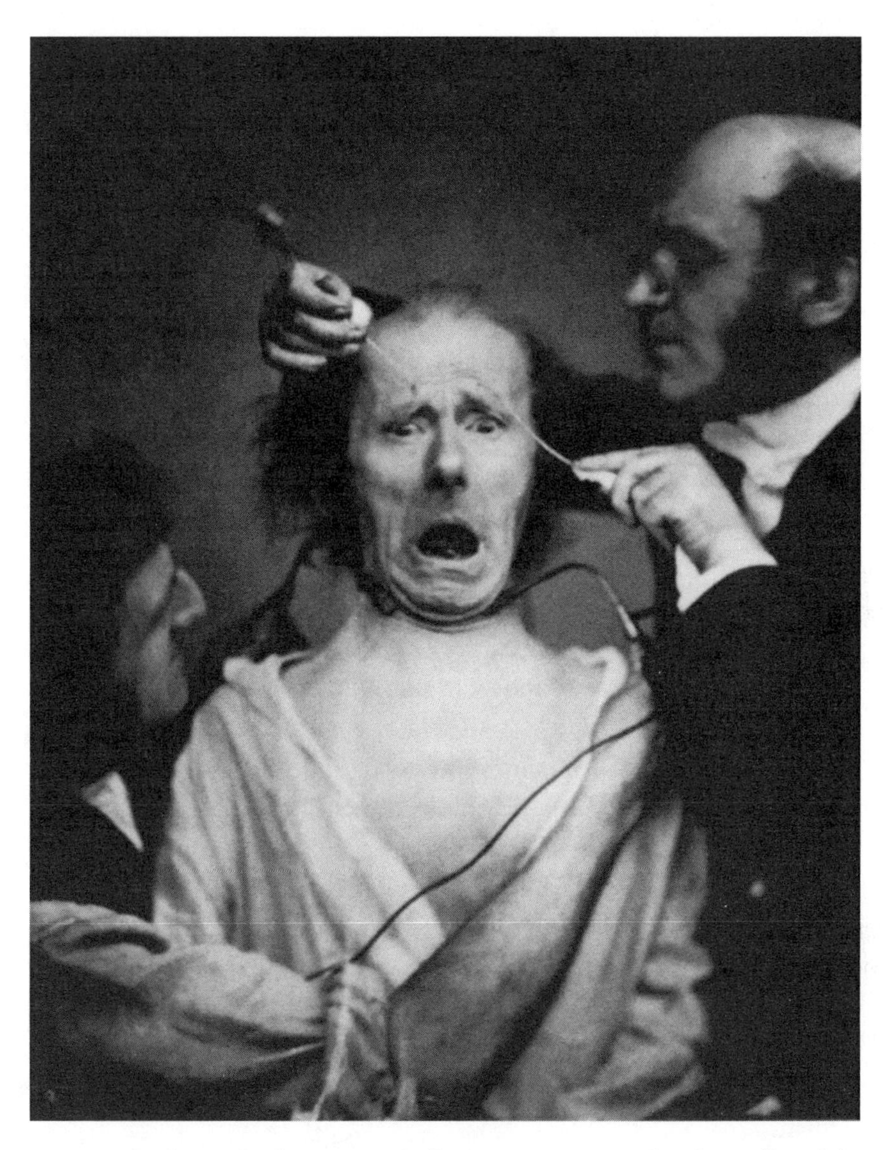

FIG. 4.1. Duchenne shocks a patient. Guillaume-Benjamin-Amand Duchenne (far right) was a nineteenth-century French physician who was among the first to use electricity as a diagnostic probe to investigate the mechanisms of various neuromuscular diseases. He described a number of distinct diseases, including Duchenne muscular dystrophy, a severe inheritable type of muscle disease that mainly affects young boys. Duchenne also shocked the faces of normal people to involuntarily induce different facial expressions and determine the exact nerve and muscle groups that were involved. He primarily used one subject, shown here, for his experiments because the man had some preexisting facial numbness and was, therefore, only mildly susceptible to the pain from the electrical shocks.

determine the exact nerve and muscle groups that were involved in smiling, frowning, grinning, etc. He primarily used one particular older man as a subject, because the man had some facial numbness to begin with, so he wasn't as sensitive to the pain from the facial shocks. And Duchenne documented his research with photographs.

Duchenne published a book in 1862 entitled *The Mechanism of Human Physiognomy*, in which he identifies thirteen fundamental emotions and describes the muscle groups involved in each emotion. Although physiognomy of course is nonsense, the book attracts attention even to this day, primarily for two reasons. First, his photographic subject, the elderly man, is a comical looking character, and the photographs of him with widely varied and exaggerated expressions is reminiscent of a subject straight out of a Norman Rockwell painting. Second, Duchenne was a pioneer in the use of photography as a means to collect and document scientific data, so his photographs of facial expressions are considered milestones in both the history of the scientific method and the history of photography. Original editions of the book are now collectors' items.

Although Duchenne's achievements in neurology were overshadowed, and somewhat tainted, by his side trip into physiognomy, he was a medical visionary nonetheless. His ideas about targeting therapies to specific diseases were revolutionary. And he truly understood that failures in medicine were more related to ignorance in pathology than to lack of an appropriate treatment. In the words of his biographer, G. V. Poore, "Pathology takes higher rank than treatment, and one is more ready to blame the insufficiency of the treatment than find fault with the notions of the pathologist, and it is the former which is ever in need of reform."[28]

Although the physicians who used electricity to shock people back to health may have been naïve and misguided, most weren't intentionally swindling the public. Other people, however, were.

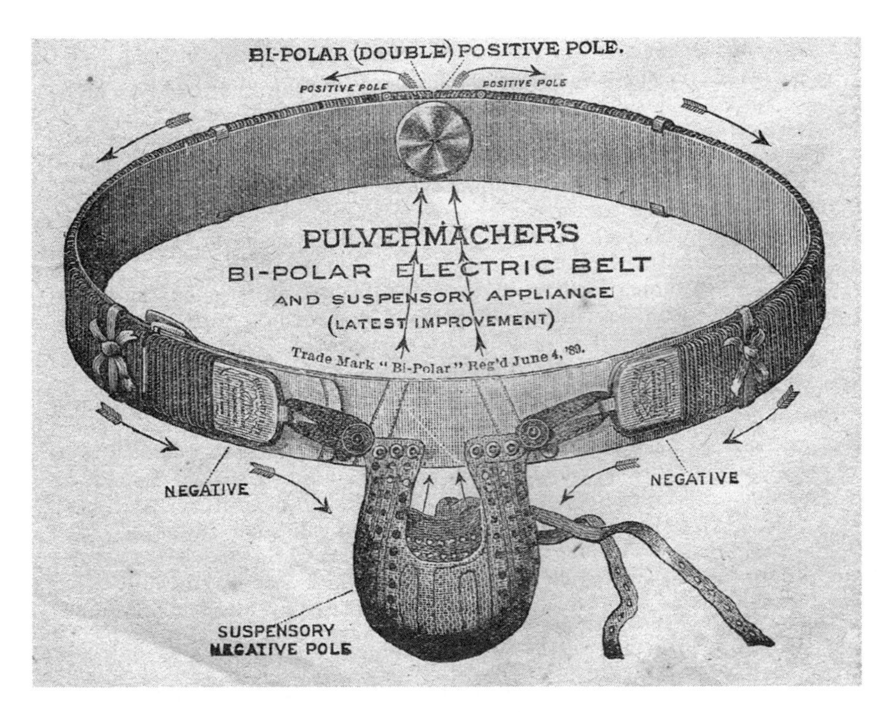

FIG. 4.2. Pulvermacher's belt. One of the most successful products of the Pulvermacher Galvanic Company of San Francisco was its electric belt, introduced in 1889. Although marketed to both men and woman as a panacea for a wide range of health ailments, its main customers were men seeking relief from erectile dysfunction and other sexual complaints. Treating such ailments required an attachment that enclosed the scrotum in an electrified pouch. A major appeal of the belt was that confidentiality could easily be maintained. It could be worn discretely under clothing during the day (or while sleeping at night), was sold through the mail, and did not require a medical prescription.

At the turn of the twentieth century, an odd set of circumstances led to the meteoric rise of a variety of mail-order electrotherapy devices that were marketed directly to consumers by means of newspaper advertising replete with fraudulent health claims. Most prominent among them was Pulvermacher's Electric Belt. Patients wore the belt around the waist, in direct contact with the skin. The belt was made up of batteries that released a steady electric current that the wearer would feel as a rather pleasant skin tingling under the belt.

Shortly after its introduction to the market, an attachment was sold with the belt. It was a pouch that connected to the front of the belt to hold the testicles, like a modern-day athletic supporter. The pouch was likewise electrically wired so the electrical skin tingling produced by the belt was extended to include the scrotum. The claim was that the scrotal attachment improved "sexual vitality," which was a nineteenth-century euphemism for sexual performance. With the release of the scrotal attachment, the belt manufacturer started to focus all its advertising on men. The fact that the belt could be purchased individually and privately through the mail, and didn't require any prescription by a physician, made it appealing to a huge population of American men who were suffering from a sexual problem peculiar to their time.

The late 1800s was a challenging time for male sexuality. For various social and economic reasons, the average age for an American man's first marriage had crept up to nearly 30. In addition, premarital sex was socially unacceptable, meaning respectable men were expected to practice complete sexual abstinence between puberty and marriage, a period lasting about 15 years. Even more frustrating, masturbation, which had long been vilified on religious grounds, was now being strongly condemned by physicians because it supposedly sapped the energy from a man and threatened normal sexual function in later life. To make matters worse, those who did manage to abstain from both premarital sex and masturbation were highly prone to nocturnal ejaculations, commonly referred to as "wet dreams," which we now know is the body's normal way of ridding itself of old sperm. But, at the time, any ejaculations outside of sexual intercourse were considered abnormal by the medical profession, so wet dreams were considered to be a symptom of the disease known as *spermatorrhea*. Sufferers of this imaginary disease were thought to require immediate medical treatment to their genitals, similar to the horrific "gravy boat" treatment we've already heard about. These were the sexual pressures on young American men at the turn of the century.[29] But older men faced their own sexual challenges.

Impotence (now called erectile dysfunction) is a common male sexual problem that increases in prevalence with age. At the time, impotence was thought to be a long-term health consequence of masturbation.

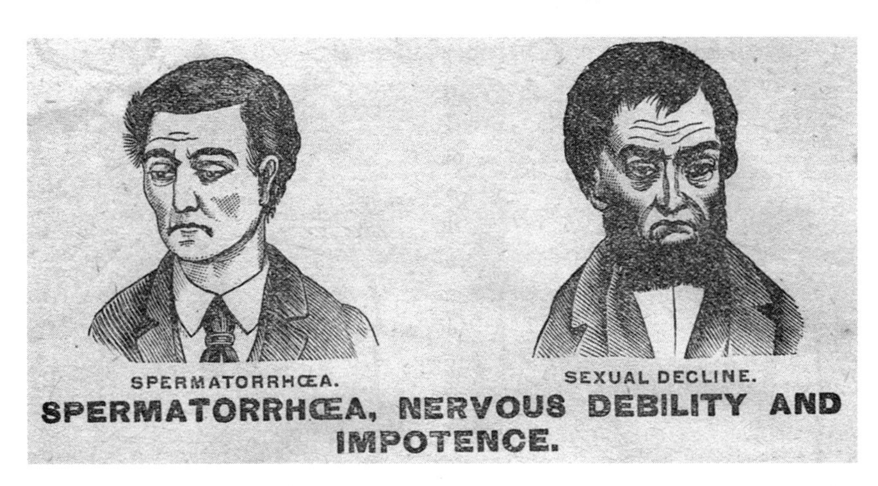

FIG. 4.3. Faces tell all. One of the scare tactics used by the Pulvermacher Company to sell more belts was the claim alleging that sufferers of sexual problems could easily be identified from their facial characteristics: "The eyes and countenance of most men are their own accusers." According to Pulvermacher's marketing material, the young man pictured on the left has the characteristic countenance of a sufferer of spermatorrhea (spontaneous ejaculations during sleep, now recognized as normal for most young men), and the older man supposedly has the countenance indicative of sexual decline (a euphemism for age-associated erectile dysfunction). The prevalences of these two "conditions" among younger and older men, respectively, were very high, so the potential market for the belt was enormous. Fearing that their faces were betraying their purported sexual inadequacies to the whole world, many men were motivated to purchase the belt.

Masturbation, commonly referred to then as "self-abuse," was taboo. Self-abuse was additionally supposed to result in bags under the eyes and pallor on the skin, two facial characteristics also associated with aging. This meant many men suffering from impotence typically displayed all three of the characteristic symptoms of self-abuse. So seeking treatment for impotence from a physician was essentially the equivalent of admitting to be a self-abuser, and your guilt was written all over your face. Not good.

If you add up all the young men dealing with masturbation or wet dream worries, and the older men thinking they were dealing with masturbation-induced impotence, you can see that a huge percentage of American men erroneously believed they were sufferers of an unmentionable sexual dysfunction. Some men just suffered with it. Some sought the attention of physicians despite the social stigma. And others sought remedies they could employ in secret, in the comfort and privacy of their own homes. It was these men who were the main market for Pulvermacher's Electric Belt. As promotional literature attested, "[It is] important, indeed indispensible, that self-applicable treatment should be contrived, so that the vast and valuable properties of electricity could be accessible to every patient without the intervention of a medical electrician [electrotherapy physician]."[30]

The belt offered a discrete remedy for all of these male sexual problems. No one needed to know. The belts could be worn under clothing or while sleeping, and you could administer the treatment in the privacy of your own home. And it was certain to be effective in solving the problems. After all, wasn't electrotherapy being used by physicians to treat the very same types of problems? The do-it-yourself belt was just a means to take the physician out of the loop when getting electrotherapy. All this made the patients happy, but the physicians not so much. Dr. George Pitzer, a prominent electrotherapist, cautioned, "The common use of electricity, all over the country, mostly by means of cheap and inefficient instruments [Pulvermacher's Electric Belt] can only result in disaster to nearly every patient who suffers such quackery to be imposed upon them."[31]

Some physicians even attacked the battery manufacturers that sold batteries directly to the public, because ready access to batteries enabled the patients to treat themselves.[32] Of course, attempts by physicians to make batteries a "controlled substance" were doomed to failure. In fact, electrical batteries were destined to soon proliferate beyond anyone's control because of the many advantages they conveyed beyond electrotherapy. The development of the battery is an interesting story in itself. But that story will need to wait for chapter 5.

The truth was that, despite the associated stigma, physicians also were enjoying the increase in revenue from patients seeking treatment for sexual disorders. By 1901, King's electrotherapy textbook was dominated by treatments for gynecological and urological problems, devoting twice as much of the book to these disorders as to diseases of the nervous system.[33] However, home remedies for sex troubles were vastly more popular, and for patients taking that route, Pulvermacher's Electric Belt, with its low-level current delivery, was really the best option.

As the manufacturer claimed, "[The] easy yet effectual use of mild, continuous currents, by means of Pulvermacher's Electric Belts and Bands, is indispensible for self-treatment; since the quantity of electricity generated and conducted through the body in a few hours is comparatively more than can be brought to bear upon the system in ten or fifteen minutes by the ordinary batteries [typically the high-powered batteries used by physicians]. The latter treatment, by powerful shocks, is not only difficult to manage and regulate, but is even dangerous."

Although the belt was marketed as a treatment for a wide host of diseases, half of the testimonials published in the company's promotional literature had to do with cures of sexual ailments. It is worth noting that the published testimonials of cures from former sufferers of nonsexual problems, like rheumatisms, were signed with the patient's full name and town of residence; however, the testimonials from people allegedly cured of their sexual problems were always anonymous. It's hard to know whether this difference in testimonial attribution speaks to the stigma associated with sexual therapy or the fraudulence of the sexual testimonials.

Given that Pulvermacher's Electric Belt removed the intermediaries—the physicians—from the electrotherapy business, linking the electricity machine manufacturers directly with the patients, it is easy to see why electrotherapy physicians hated the use of the belts. But it wasn't easy for the physicians to convincingly condemn in-home electricity treatments as bogus and dangerous while at the same time claiming their own in-office electric treatments were beneficial and safe. That distinction rang hollow with patients. Besides, with a few exceptions like electric breeze therapy, the electrotherapy that physicians practiced was painful, while electric belt therapy was not. If you were a man with a sexual function problem and you faced either the electric belt or the gravy boat treatment, which would you choose? I, for one, think gravy boats are best used to hold gravy.

As mentioned earlier, Dr. William Harvey King was a committed electrotherapist, judiciously mapping dozens of motor points all over the human body and sharing their locations with other physicians in his 1901 book.[34] What King fails to mention in his book, however, is that the patient need not necessarily be alive for the electrical muscle stimulation to work. As we've seen with frogs and fishes, an animal doesn't need to be alive, or even intact, in order to have its muscles stimulated into contractions by electricity. The same is true for humans.

Most people are familiar with Mary Shelley's 1818 book *Frankenstein*, although few have actually read it. The multiple movie versions of the book have taken some liberties with the storyline, leaving some viewers with the erroneous impression that the monster in the story is named Frankenstein. The fictional monster actually has no name; rather, the book is named after the monster's fictional creator, Victor Frankenstein. In the book, the sudden and untimely death of Frankenstein's mother, and the witnessing of the dramatic destruction of a tree hit by lightning, has young Frankenstein brooding about the vulnerability and brevity of life. It sets him to wondering whether death is actually a reversible process, and he devotes himself to the goal of "reanimating" corpses; that is, bringing dead bodies back to life. Although Shelley never reveals exactly how Frankenstein manages to do this—presumably, it might enable readers to replicate the dastardly deed—she hints electricity is somehow involved. So, henceforth, depictions of the monster typically have electrodes protruding from his neck, and there is always a lightning storm to provide the electricity that puts the spark of life into him. But how did Shelley come up with such a bizarre story idea?

The Italian scientist Galvani, who did the original work with shocking frog legs impaled to a fence, had a nephew, Giovanni Aldini, who decided to scale up. Aldini, a highly productive scientist in his own right, was trying to prove his uncle's theory of animal electricity.[35] But Aldini wasn't content just to shock dismembered frog legs and watch them twitch, as his uncle had done. Rather, he went around jolting dead human bodies—mostly executed prisoners—stimulating contractions in their muscles and getting their bodies to move. His favorite public demonstration was to shock decapitated bodies in such a way as to get them to promptly sit upright for the audience, which freaked out most everyone who saw it.

Aldini's most famous demonstration occurred at London's Newgate Prison in 1803. A prisoner named George Foster had just been executed by hanging. Aldini applied electrodes to Foster's body and connected them to a high-voltage battery. A witness described what happened next:

> On the first application of the process to the face, the jaws of the deceased criminal began to quiver, and the adjoining muscles were horribly contorted, and one eye was actually opened. In the subsequent part of the process, the right hand was raised and clenched, and the legs and thighs were set in motion.[36]

Some of the less informed in the audience thought Aldini was actually attempting to bring Foster back to life.[37] He was not, but he sure put on a good show—better than all the Flying Boy and shocked-monk demonstrations put together.

Shelley was only three years old when Aldini "resurrected" Foster. But she likely became aware of Aldini through her father, who counted among his friends some of the leading electricity researchers of the day. They were well aware of Aldini's "science" demonstrations and had described them to Shelley's father.[38] Presumably, her father relayed them to her. So when Shelley, at age 19, entered into a friendly fiction-writing competition as a distraction from a disappointing summer vacation plagued by inclement weather, it is perhaps not surprising that she produced a story about creating a monster by assembling dead body parts and animating them with electricity. The story won the competition, and she later developed it into a full-length novel. Perhaps the real-life Aldini was her model for the fictional Victor Frankenstein, and the unfortunate George Foster was his monstrous creation, with electricity playing a central role. Art imitating life . . . and death.

In 1908, Abraham Flexner embarked on a whirlwind tour of all 155 medical schools in the United States. Within 18 months he had visited them all, and then he sat down to write a report on his findings. Published in

1910, his report would prove to be one of the most important medical investigations of the twentieth century.[39] It ultimately would result in the closure of 80% of American medical schools and change the face of American medicine forever. It also would spell the demise of electrotherapy.

Flexner was an education reformer.[40] He already had gained fame (some would say notoriety) for his book *The American College: A Criticism*, which was a brutal critique of the state of American college education at the time. His work caught the eye of Henry S. Pritchett, the president of the Carnegie Foundation for the Advancement of Teaching. Pritchett liked what he saw.

Pritchett had been approached by the American Medical Association (AMA), an organization of physicians concerned about bringing professionalism into the field of medical practice.[41] The AMA's Council of Medical Education had an ongoing concern about the complete lack of standards among American medical schools, but had decided that any investigation on the AMA's part would be seen as biased and politically motivated. They wanted an outside investigation, and they wanted it done by an independent and respected organization. The AMA chose the Carnegie Foundation, and Pritchett chose Flexner. Flexner wasn't a physician—he had never even been inside a medical school—but he did know education. As it turned out, Flexner's lack of medical expertise became his biggest asset: he was able take a fresh look at the medical school situation without any preconceived notions. Flexner did take a look and did not like what he saw.

The chief agitators for reform among American doctors were those who had received their own medical educations in Europe, particularly in Germany. In Germany, medical schools had an entirely science-based curriculum. But in the United States, medical schools with an exclusively science-based curriculum were in the minority; rather, it was common for only the initial years of coursework to be science based. When it came time for clinical training, science often took a backseat. In its place, untested treatments based on unsubstantiated disease theories were taught with a near religious fervor. And from school to school, the clinical teaching varied depending upon the specific disease theory to which the school subscribed. Flexner was flummoxed by this ridiculous dichotomy of rigorous science followed by unsubstantiated dogma. He vigorously protested: "But the scientific method cannot be limited to the first half of medical education. The same method, the same attitude of mind, *must permeate the entire process.*"[42]

A major problem, as Flexner put it, was that clinical training in American medical schools often was governed by different medical "sects" that ran their own schools and indoctrinated their students with preposterous disease and treatment theories.[43] These sects—most notably homeopathy and allopathy but also many others[44]—were based on ill-conceived and unverified theories of disease and the workings of the human body. Their clinical practices thus were driven more by the delusions of the sect and the desire to promote ideas of the sect's founder than on any scientific evidence. Flexner said all types of sectarianism in medical education needed to go, to be replaced with only science-based training. He was particularly antagonistic toward sects that rejected germ theory, saying they contributed to the spread of disease throughout the population and were thus a menace to public health. Although he never specifically singled out electrotherapy in his report, electrotherapy as it was practiced at the time looked an awful lot like what Flexner was calling a sect. And that meant things weren't going to end well for electrotherapy.

It is difficult to overstate the impact of Flexner's investigation into American medical education. Every major news outlet in the United States covered the release of his report, particularly the newspapers of cities that had medicals schools, which was every major American city. The *New York Times* announced the report's release with the headline "Factories for the Making of Ignorant Doctors." In addition to the bad press for doctors, the report provided all the ammunition university presidents needed to crack down on their own medical schools, because the report called out the problem schools by name.[45]

It wasn't just the Flexner report that was putting the squeeze on electrotherapy. There was also the Pure Food and Drug Act of 1906, which was instituted partly to protect the public from quackery, that brought federal government oversight and regulation into the picture. The Act's ban against "mislabeling" was broadly interpreted to include unsubstantiated claims for drugs and later extended to include medical devices. There was also a 1909 amendment to US postal regulations that made it a federal offense to use the mails to perpetrate a "scheme or artifice" on the public. Shipping fraudulent medical products through the US mail system thus became a federal crime. All this scrutiny was not good news for the physicians practicing electrotherapy, who typically referred to themselves as "electricians," or to electrotherapy device manufacturers, like the Pulvermacher Galvanic Company, which made mail-order electric belts.

Unable to support its health claims with scientific evidence, electrotherapy's golden age began to wane. Ironically, electrotherapists had been their own worst enemies. By claiming they could cure everything, electrotherapists' failures were used by their critics as evidence that electrotherapy cured nothing. Electrotherapy no longer was taught in medical schools, and the public started taking their ailments to doctors using other treatments, such as surgery, X-rays, and drugs.[46] Soon electrotherapy machines went the way of buggy whips, and the few that avoided the dump found their way into museums of medical history and private electrotherapy machine collections, like Behary's. Electrotherapy died, or at least went into hibernation. If electricity was used at all for treatment, it was used to cauterize surgical wounds by virtue of the heat it produced or to remove body hair by killing hair follicles (i.e., electrolysis). No scientific studies were needed to support those uses. Yes indeed, the wounds were cauterized and the hair was gone.

As Behary shows me each of the various machines in his electrotherapy collection, he explains that the irony of the electrotherapy story is that, although electrotherapy itself died because of a lack of underlying science to support its health claims, the field spawned most of the early advances in electrical generator technology. This is because the constant demand for electricity-producing machines by physicians funded electricity research and development. Physicians wanted state-of-the-art electricity equipment. And only physicians could afford to pay for the machines because only physicians had a means of generating revenue from them: billing patients. In an age before ubiquitous electrical appliances satisfied all our needs, the machines didn't have many practical uses except for treating patients. The physicians were, in fact, funding electrical engineering research by providing a ready market for all the electricity machines that the manufacturers could produce. And, of course, the patients were funding the physicians through the billings for the electrical treatments of their diseases. In a sense, human suffering and disease drove electrical progress and allowed new discoveries about electricity to be made. With time, the prices of the machines would come down, and electricity then would find other profitable uses.[47]

I ask Behary about his background in electricity. How was he able to restore the machines? Does he have electrical engineering training? "Hardly," he exclaims. "I went to culinary school! I learned everything I know about electricity on the fly." I follow up, "What then started you on collecting these old electrical therapy machines?" He tells me that while he was in culinary school, he and his father had a plan to open a café together after he graduated. They decided the décor of the café would be based on antiques, so they started collecting them for their future business. One day in an antique shop, Behary found an electrotherapy machine that was actually still operational. When he started it up, the screen on an old television in another part of the room started flickering. He decided he'd better unplug the television before it got damaged. But when he went to unplug the television, he found *it wasn't plugged in*! The electricity therapy machine somehow was affecting the screen of the unplugged television. At that moment, he thought, "Wow, this electrotherapy stuff is really cool!" From then on, he combed antique shops specifically seeking old electrotherapy machines to stock his growing collection. And the rest is history.

Out of curiosity, I close our conversation by asking Behary if his father had also gotten the bug for collecting electrotherapy gizmos. He shakes his head. "No," he tells me, "by the time I graduated culinary school, my father had lost interest in both antiques and the café business. His interests drifted toward musical instruments and he opened a violin shop instead."

5

A CIRCUITOUS ROUTE

THE BATTERY

> The electric organs of fishes offer another case of special difficulty; for it is impossible to conceive by what steps these wondrous organs have been produced.
>
> —CHARLES DARWIN

If I were to tell you electric batteries are useful things and enumerate all of the various ways that batteries are valuable to modern society, you would think me trite. So, I'll skip that. But I will tell you there once was a time when I was completely obsessed with my flashlight batteries.

In my youth, my friends and I did a lot of scuba diving. Sometimes we dived at night. Why at night? Because night diving provides a completely different experience than diving during the day, primarily for two reasons. The first has to do with the sea life you see at night. Just as for land animals, the animals of the ocean tend to be either diurnal or nocturnal (active during the day or at night, respectively). So diving at night allows you to see a different community of animals than you would during the day.

The second reason for diving at night has to do with colors. As white light coming from the sky moves down through the water column, it begins to get filtered out. The longer wavelengths, representing red light, get filtered out more quickly with depth than the shorter wavelengths, such as blue light. This means things appear bluer the deeper one dives. Worse yet, fish and coral that are actually colored brilliantly red appear to be an ugly dark brown in the deep due to the lack of red light. The fact is, diving deep during the day makes everything appear chromatically washed out, devoid of vivid coloration.

Night diving restores the colors to their full majesty because night divers use flashlights to see. In effect, they are bringing their light source along with them, down into the deep. Since the light from the flashlight is traveling only a short distance through the water, the full color spectrum is retained. So night diving is a color extravaganza.

There is a problem, however. Flashlights and seawater are not a good combination. If even a small amount of the very corrosive and highly conductive seawater should leak into the battery compartment, it will short out the stack of batteries inside and instantly plunge you into total, and I mean *total,* darkness. I'm not one who is particularly afraid of the dark, but I can tell you the black holes of outer space cannot be any darker than scuba diving in the ocean, on a moonless night, without a working flashlight. It really isn't dangerous, since you can simply swim up to the surface, but it is a very eerie experience indeed.

Consequently, I pampered my diving flashlight and the 10 flashlight batteries that powered it. I always used fresh batteries before every dive, carefully inspecting each one as I put it in the battery compartment. Then I lubricated and carefully examined the entire stretch of the rubber O-ring that sealed the battery compartment. I inspected for even the tiniest grain of sand that might produce a fissure in the O-ring when the flashlight came under the very high water pressure at my typical 60-foot diving depth (about 3 atmospheres, or 300 kPa). All of this effort paid off. I never had a flashlight failure while diving at night, although some of my less fastidious diving friends can't say the same.

It's impossible to scuba dive at night without electrical batteries. It is a whole realm of human experience that would be shut off from people, if we didn't have batteries. There are many other human experiences that would be impossible if we didn't have batteries . . . but I promised I wouldn't say any more about that.

Today, we think of batteries chiefly as a means of storing electricity that was generated in another way. We plug our batteries into power outlets, and they soak up electricity that was generated elsewhere by hydroelectric, coal, nuclear, solar, or other means of power production. Then we use the batteries as portable sources of stored electricity to run our electronic appliances at another time and location. But this wasn't always the case. Originally, the value of batteries was that they actually generated their own electricity through *electrochemical reactions,* a more convenient alternative than the bulky static electricity generators that ran largely on human muscle power. As we've already seen, the portable battery-powered Pulvermacher belt was much preferred over treatment with large hand-cranked static electricity machines, which were only available in physicians' offices.

By simply combining the right chemicals in the right way, within a container called an *electrical cell,* it is possible to generate—at least temporarily—an electrical current without relying on any other outside source of energy. By linking the cells, it is possible to produce a lot of electricity. This energy is liberated by chemical reactions within the cells—no need for hand-cranked static electricity generators. An electrochemical battery is truly a remarkable thing. And what's more remarkable: We owe it all to biology.

The person we credit with the discovery of the battery is Alessandro Volta.[1] You'll recall he was the eighteenth-century scientist who was so highly critical of Luigi Galvani's animal electricity theory.[2] Volta believed the twitching of the frog leg hooked to Galvani's fence could be best explained by the fact that the hook was made of brass and the fence made of iron; he argued it had nothing to do with the frog leg producing any electricity on its own. According to Volta, when two dissimilar metals are brought together, they release very small amounts of electricity, but enough to stimulate a frog leg into twitching. Volta had previously discovered this electrical phenomenon of dissimilar metals by accident while doing his own frog leg experiments, but he didn't understand how it worked. While trying to figure it out, he encountered many difficulties, particularly with measuring small quantities of electricity.

Gauging pain to the knuckles or measuring the lengths of sparks weren't approaches that would work for the very tiny amounts of electricity Volta was interested in. So he set out to invent a device to measure very low levels of electricity, and he eventually succeeded. It was a simple device called a *straw electrometer*. It measured the separation between two negatively charged pieces of straw, caused by repulsive force of the like charges, in a way similar to how the weak electric charge of rubbed amber can repulse small pieces of straw. But the amounts of electricity sufficient to make a frog leg twitch were still too low to be measured by his straw electrometer. It simply didn't have the electrical sensitivity of a frog leg.

You might wonder why Volta didn't just forget his straw electrometer and simply use the frog leg as an electricity measuring tool. We've already seen that the neurophysiologist Helmholtz was less concerned about having a physical means to measure electricity. He chose to sidestep the problem by using the frog leg contractions themselves as a surrogate for direct measurement of electricity. Employing this biological approach, he was able to measure not only the amount of electricity but also the speed of nerve impulses, without the need for any sophisticated electricity meters. Volta, however, was more of a physicist than a neurophysiologist. He preferred a simple, accurate, and precise physical instrument that could be calibrated to some universal physical standard.

Volta was frustrated because no matter what kinds of physical meters he came up with, none was as sensitive to electricity as the leg of a frog. In fact, he was even able to get a frog leg to twitch by using a supposedly empty Leyden jar. When he charged up a Leyden jar and then completely discharged it, the "empty" Leyden jar could still stimulate a twitch in a frog leg, suggesting there was still a slight but biologically relevant amount of static electricity in the jar. He pressed on, trying to make an electricity meter as sensitive as a frog leg, using the frog leg as the "gold standard" for sensitivity.

But early on in his experiments with frog legs he encountered a problem: frog legs were *too sensitive*. Frog legs were so sensitive to static charges that he had to take special precautions to insulate his experiments from the ambient static electricity in his surroundings. Comb your hair and you become electrified, pet a cat and you're electrified, wear a wool sweater and you're electrified, etc. He even found some of his metallic experimental probes themselves gave off enough electricity to stimulate a frog leg. Investigating further, he found that *bimetallic instruments*—those instruments constructed with two different types of metals—stimulated a frog leg. In contrast, instruments made of just one type of metal would not. Apparently, bimetallic instruments were generating their own electricity without the need to rub the two metals together. Quite extraordinary.

It was at about this time that Volta heard of Galvani's claim of animal electricity coming from frog legs. Although he initially congratulated Galvani on his discovery, Volta soon came to suspect Galvani wasn't taking the same precautions to minimize the ambient static electricity while conducting his frog leg experiments. So, when Volta learned Galvani had used brass hooks attached to iron fences—two different metals—to impale his frog legs, he knew Galvani's results were erroneous. They were just a *scientific artifact*—the name scientists give to data caused by an external interfering agent. Galvani had gotten artifactual results because he used a flawed experimental design. Volta concluded Galvani's experimental findings were merely due to electricity being generated by the connection of two dissimilar metals, not because frog legs produce electricity on their own. But Volta couldn't design an experiment to definitively disprove Galvani without an extremely sensitive electricity meter.

To be clear, it's not that Volta didn't believe animals could produce electricity; he knew all about electric fishes. It's just that he believed that, in order to produce electricity, the animal had to have an electricity-generating organ, just as electric fishes do. Frogs have no equivalent organ, so he didn't think it possible they could produce electricity.

In fact, over time Volta eventually came to believe the secret to understanding electricity would come precisely from research into the electricity-generating organs of electric fishes. So, despite his misgivings about the role of electricity in biology, he became an electric fish researcher. He thought that, by studying the fish, he might be able to deduce the underlying electrical physics that was involved, and perhaps even make an artificial electrical organ similar to the fish's natural electrical organ.

Galvani, in contrast, stuck with his frogs. However, once while vacationing at the sea, he had the opportunity to conduct an impromptu experiment with torpedo fish. After obtaining a couple of live torpedo specimens, he did an experimental surgery to test his animal electricity hypothesis. Somehow he managed, presumably without shocking himself, to quickly sever the brain stem of one of the fish, thus separating the brain from the rest of the nervous system. For the other fish, he quickly severed its heart. This of course sent both fish into rapid death spirals. But he noticed the fish with its brain severed from the body immediately lost its ability to shock, even before it died. But the fish with the missing heart retained its shocking ability until it was completely dead. Galvani believed these results supported his idea that the brain is the body's reservoir for storing animal electricity, making it logical that severing the brain from the body should immediately stop the body's electricity while severing the heart (the experimental control) would not.[3] So the involvement of the brain seemed to be key. Galvani contended, in opposition to Volta, that any animal with a brain was capable of producing at least some electricity, whether it had a specialized electricity-producing organ or not.[4]

Volta's quest for electricity meters with increasingly higher sensitivity was partially satisfied when William Nicholson, an English chemist and inventor of some renown, came up with an instrument that bested Volta's straw electrometer.[5] Nicholson's instrument worked by concentrating small amounts of charge until the level of concentrated charge was high enough to be measurable. Since the charge level increased in multiples of the number of times the instrument's handle was cranked, it was called a *multiplier*. Using Nicholson's multiplier, Volta was able to investigate the electric output of bimetallic materials without the need for any frogs.

Armed with Nicholson's highly sensitive electrometer, Volta readdressed his bimetallic electricity hypothesis. To do this work, he made coin-size disks, each of a different type of metal, and paired them up in different combinations to compare their electrical outputs. Some pairings produced more electricity than others, but the outputs for all the pairings were disappointingly weak. Of them all, copper paired with zinc was one of the best.

But the Nicholson instrument proved to be of no use for putting Galvani's animal electricity hypothesis to the direct test because, as Volta soon discovered, the friction produced by cranking the instrument's handle made enough static electricity within the instrument to stimulate a frog leg.[6] If you tried to use it to measure the alleged animal electricity coming from the frog leg, the instrument itself would make the leg twitch! Again, the most sensitive means of measuring electricity proved to be the frog leg.

Volta soon learned his tongue was nearly as sensitive to the electricity produced by dissimilar metals as a frog leg. Although his tongue didn't twitch when exposed to two different metals, if two different pieces of metal were put on the tip of his tongue, they produced a distinct taste and tingling sensation. He correctly concluded bimetallic electricity was affecting his tongue. Likewise, if two metals were placed on the conjunctiva (the white area) of his eye, he could perceive a light sensation. His described what he saw as arrays of light that moved around his visual field as he changed the positions of the electrodes on the surface of his eyes. So his tongue and his eye appeared to be as sensitive as a frog leg to bimetallic electricity. He then performed a particularly ambitious experiment. He connected his tongue, his eye, and a frog leg all together in one single open circuit. When he closed the circuit by adding the two metals, he simultaneously induced tingling of his tongue, light in his eye, and twitching of the frog leg.[7] This suggested electricity was mediating not only muscle contractions but also the senses of taste and sight. Curious about hearing, he stuck wires deep into each of his own ears (just as the French neurologist Duchenne would later do), and connected them to the bimetallic electricity. When the electricity was running, he heard "a kind of crackling, jerking, or bubbling as if some dough or thick material was boiling." Perhaps electricity was mediating all five bodily senses: taste, sight, hearing, smell, and touch. An intriguing idea.

But human body senses were no better than frog leg twitching as a tool for disproving Galvani's animal electricity theory; they were all imperfect biological measures of electricity output. The lack of a nonbiological alternative to frog legs as an electricity detector was beginning to seriously hurt Volta's argument that frog legs don't produce their own electricity. In addition, Galvani's animal electricity idea was getting indirect support from other investigators. In 1771, Nicholas-Philippe Ledru (sometimes called Comus), a physicist and member of the French National Academy of Medicine, built a static electricity machine where the normally rubbed glass disks were replaced with disks made from animal nervous tissue that had been baked solid in an oven. The nerve disks likewise produced some static electricity when rubbed, which Ledru interpreted as evidence that nerves can produce electric fluid.[8] Volta claimed, in rebuttal, that Ledru's static electricity device proved nothing of the sort and was an irrelevant finding; nevertheless, Ledru's report was starting to cloud the main issues regarding the alleged existence of animal electricity.

The fundamental problem was Galvani's experimental design, in which the frog leg served as both the supposed source of the animal electricity as well as the detector of the electricity. If one needed a frog leg to detect electricity emanating from bimetallic sources, it wasn't possible to rule out the frog leg as the actual source of the electricity. Volta needed a physical electricity meter as sensitive to electricity as a frog's leg, so that he could remove the frog leg entirely from the experiment and directly demonstrate that a brass hook hung on an iron fence produces electricity. Until he had done that, thus settling the debate forever, his feud with Galvani over animal electricity would continue to rage. Frustrated, Volta decided to redouble his efforts on the metal disks and try to figure out what exactly was going on with bimetallic electricity generation, and to revisit frog legs later.

As some point in 1799, Volta became aware of a journal article by Nicholson—the same Nicholson who had invented the multiplier electricity meter—that proposed a model of how an electric fish's electrical organ might work.[9] The paper argued that the stacked coin-like electrocyte structures in the fish's electrical organs were highly conductive subunits, filled with a substance that was "electrified oppositely," whatever that means. These electrified conductors allegedly were separated from one another by insulating tissues, in a repeating sequence of conductor, insulator, conductor, insulator, etc. Each subunit held very little charge, but when taken together—Nicholson estimated there were 500–1,000 subunits in the fish's electrical organ—they could produce a substantial electrical discharge. It was analogous to linking Leyden jars together to boost electrical discharge.

Volta realized the metal disks he used for his bimetallic testing were similar to the coin-like electrocytes of the electric fish, and they too produced only small amounts of electricity. Perhaps stacking them like the electrocytes of the fish might increase their output as well. So he started playing with the disks and stacking them in different orders to explore whether something like bimetallic electricity generation was happening in the fish organ. To simulate the intervening insulating tissue that Nicholson had claimed was essential to the mechanism, Volta used disks of dry paper, including them in his trial-and-error stacking sequences. Eventually, he happened upon the repeated sequence of copper-zinc-paper-copper-zinc-paper-copper-zinc-paper. . . . But this arrangement too produced no significant electricity.

Disappointed with his negative results, Volta decided to reexamine the anatomy of the fish's electrical organ. He soon found there actually were no electrically insulating materials at all within the fish's electrical organ. Every tissue of the organ conducted electricity. He realized at this point that his stack of bimetallic disks separated by insulating paper didn't emulate the fish's electrical organ at all. There were no insulators between the electrocytes of fish.

Volta, therefore, decided to try the opposite approach. He substituted the insulating dry paper disks with conducting wet paper disks.

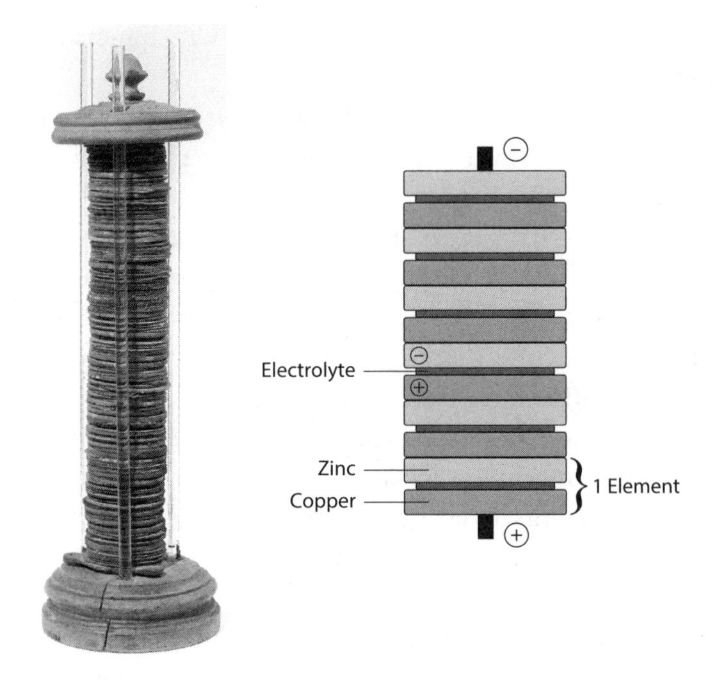

FIG. 5.1. The first battery. The predecessor of the modern electrochemical battery was the Voltaic pile. Invented by Italian scientist Alessandro Volta, it amounted to a pile of alternating copper (or silver), zinc, and paper disks. The paper disks were wet with an electrolyte (salt) solution. Volta was trying to produce an artificial version of the electrical organ of an electric eel. He stacked the coin-like disks because the electricity-producing organ of an eel also is composed of a series of coin-like disks known as electrocytes. Volta accidentally happened upon a combination of disks that produced an electric current and called his invention an "artificial electrical organ." But what Volta found had nothing to do with eels (an eel's electrical organ works by an entirely different mechanism), and the device soon became widely known as a *voltaic pile*. The voltaic pile shown here is purported to have been made by Volta himself. It was publicly exhibited in 1899 in Como, Italy, to commemorate the one hundredth anniversary of Volta's remarkable discovery. (Wellcome/WikiMedia Commons)

He did this by simply disassembling his stack, dipping the paper disks in a solution of salt water, which is highly conductive, and reassembling the stack. He then tested the electrical output of this new stack containing the wet paper disks with his reliable, though insensitive, straw electrometer. Wow! The electricity output had jumped off the charts. He concluded that he had discovered the secret of how electric fish produce electricity.[10] He actually had not, but what he had accidentally discovered was the electrochemical battery. This discovery would far eclipse all of Volta's previous work, and the issue of animal electricity no longer seemed important to him. Let Galvani do what he pleased with frogs. Whatever electricity they may or may not have produced was insubstantial compared to his new discovery. The future of electricity belonged to the battery.

Because Volta called his stacked disks "piles" when he published his results, the world came to know his discovery as the *voltaic pile*. Later, the name would change to *voltaic battery*—something that produces electricity via one or more linked electrochemical cells. The term *battery* was used because the individual cells in a multicell battery were likened to a battery of artillery cannons firing in unison.[11]

Volta had serendipitously discovered the battery by swapping a conductor for an insulator within a stack of alternating metal disks. Volta thought the wet paper disks functioned by decreasing the electrical *resistance* between the metallic disks, and he focused on lowered electrical resistance between bimetallic materials as the central explanation of the mechanism of his battery. Volta was among the first to explore the electric parameter known as resistance, but he had very crude ideas about how it worked and contributed little to its underlying science.[12] Resistance turns out to be an extremely important electrical parameter. It might be useful to pause here and contemplate what, specifically, is meant by resistance.

If there is an electrical property nearly as intuitive as current, it's resistance. Electrical resistance is just what you'd expect it to be: something that obstructs, or resists, the passage of current. It seems simple enough. But as it turns out, resistance is a little more complicated, and more interesting, than you might think.

Electrical resistance is often compared to friction because it does have a lot of similar attributes. If you've ever played the tabletop game called air hockey, you know the puck is riding on a thin film of air, which reduces the friction between the puck and the playing surface tremendously. This means there is little resistance to the forward motion of the puck when you try to slam it into your opponent's goal, and that enables the puck to pick up tremendous speed. Turn on the game board's air pump and you have low resistance to the movement of the puck; turn off the air pump and you have high resistance. So the analogy between electrical resistance and friction works. But we are using a flowing water model for electricity now, so we'll leave the friction analogy to others, forgoing the air hockey game and alternatively relying on our much less exciting garden hose to describe resistance.

How might we alter the resistance in a garden hose? Well, we could pinch the hose with our hand to reduce the flow and call that "resistance," but the pinching force would be hard to measure and replicate. Rather, suppose we put at the end of the hose a mechanical nozzle where we could vary the restriction to water flow in a measured way. When near fully constricted, the nozzle almost stops water flow. While fully open, the water flows as freely through the hose as though it had no nozzle.[13]

Since our water flow model of electricity calls for voltage to be the equivalent of the water pressure at the tap and amperage to be the flow rate of water through the hose, resistance amounts to a constriction in the diameter at some point in the hose, such that it reduces the water flow. We have placed our nozzle at the end of the hose, but alternatively it could be placed in the middle to the same effect.

If we dial the nozzle down to the point that it restricts the flow rate by half, we have doubled the resistance and accordingly halved the amperage. As you can see, there is an inverse relationship between resistance and amperage. But suppose we don't like this loss of amperage because it means it will take twice as long to fill our water bucket from the hose. We could either reduce the resistance by fully opening the nozzle or increase the pressure (the voltage) at the tap.[14] How much must we increase the voltage to restore the amperage? Since the added resistance of the nozzle halved the flow, we would need to double the voltage to compensate. This is because amperage is proportional to voltage and inversely proportional to resistance. If you would forgive my introducing an equation into our discussion, we can describe this relationship as

$$Amperage = Voltage/Resistance$$

Electrical resistance is an intrinsic property of the conductor of the current. You can disconnect the hose from the faucet while keeping its nozzle attached and connect it to another water source; and, provided you don't change the nozzle setting, it will have the same resistance. Resistance is not a property of the water flow; *it is a property of the hose.* And the hose equates to the material conducting the electricity.

As we know, volts and amps are the units for the electrical parameters of voltage and amperage, respectively. But it's not that simple for the parameter we call resistance. The units of resistance are not, as we might hope, called "resists." No, that would be too easy. The unit of resistance is the *ohm,* after Georg Simon Ohm, the nineteenth-century German scientist who provided us with our modern understanding of resistance. But, unlike for Volta and Ampère who were awarded both the parameter name and the unit name, Ohm was denied the parameter name and only given the unit name. That's why we don't call the resistance parameter "ohmage."

Luckily, despite the somewhat inconsistent nomenclature, at least the meaning of the unit itself is very simple. Recognizing how resistance, voltage, and amperage are all interconnected, the resistance parameter is defined as the quotient (i.e., the value obtained by dividing one number by another) of the other two parameters. To mathematically describe the ohm, the standard unit for resistance, the equation above is simply rearranged as

$$Resistance = Voltage/Amperage$$

And the ohm unit is just defined as 1 volt divided by 1 amp:

$$1\ ohm\ of\ Resistance = 1\ volt/1\ amp$$

So, if you can measure the voltage and amperage of an electrical circuit, then you can easily calculate the amount of resistance in the circuit in ohms.

What Ohm had provided the world was a unified understanding of the relationship between volts and amps: they are related to each other by the level of resistance within the conductor through which the current flows. We call this *Ohm's law*.

Now that we better understand the meaning of resistance, we should update our understanding of conductors. Earlier, we had defined a conductor as a material through which electric current flows (e.g., a paper disk soaked with salt water) and an insulator as a material though which current doesn't flow (e.g., a dry paper disk). It would be more accurate now to say that conductors are materials with low resistance and insulators are materials with high resistance. Better still, we could drop the insulator term altogether and just say all things are conductors of electrical current to some extent, depending on their level of resistance. Some things conduct electricity very well and some very badly; the ohm level tells us how badly.

But it wasn't just the decreased resistance of the wet disks that allowed for the voltaic pile to produce its electricity. It was the fact that the salt water also contained free-flowing *ions*. Ions are molecules with an electrical charge. When you dissolve common table salt (sodium chloride), which is electrically neutral and a very poor conductor, in water, it separates into its two chemical parts—positively charged sodium and negatively charged chlorine—producing a solution we call an *electrolyte*. Other types of salts likewise dissociate into pairs of oppositely charged ions. When electricity flows through such salt solutions, the ions themselves become part of the electrical current, facilitating the flow of the current and also participating in the chemical reactions that produce it.

If you are having trouble grasping this concept now, don't despair. As you are quite aware, you don't need to know the chemistry behind an electrochemical battery in order to make use of its electrical output. In fact, at the time, Volta had no idea he was witnessing an electrochemical reaction. It wasn't until later, when gas bubbles were observed forming at the battery's electrodes while it was producing current (and only while it was producing current), that a chemical reaction was suspected.

You now know at least as much as Volta did at the time he discovered the battery. I say "discovered" rather than "invented" because electrochemical reactions are happening around us all the time; we just weren't aware of them before Volta discovered the battery. Prior to that, there was no understanding that things as mundane as rusty metal were the result of electrochemical reactions. And why should there have been any understanding? Electrical scientists' standard theory of electricity—the single-fluid model—did not involve any chemistry.

At this point in our story, the single-fluid model of electricity was starting to fail scientists; it could not account for all the electrical phenomena they were beginning to see. But lack of understanding didn't stymie the further development of the battery. This is because Volta's newly discovered battery was considered to be a scientific instrument, and the scientific instrument makers—who were more artisans than scientists—rushed into the field. These instrument makers were tinkerers; they improved the batteries incrementally by changing this and that and then seeing how it affected the electrical output. They had no fundamental understanding of how batteries worked, nor did they care. According to technology journalist Henry Schlesinger, who wrote a book about the technological revolution that was started by the discovery of the battery, "It's [remarkable] just how little desire there was to explore the way in which the battery produced its charge. Even the most serious experimenters focused almost solely on what that charge could be made to do in the laboratory. The fact that it worked was enough."[15] Understanding the way it worked wouldn't come until much later.

Volta's discovery of the battery gave him worldwide fame and catapulted his prestige beyond that of his rival, Galvani. And for that, the highly competitive Volta was pleased. But there are also two ironic consequences to his story. The first is that the scientific community at the time had begun to distinguish between strong electrical outputs—those that could actually produce powerful shocks, as from Leyden jars and electric fishes—and the more insidious and weak electrical outputs reported by Galvani for frog legs and by Volta for his bimetallic instruments. The thinking was that they might represent two fundamentally different electrical phenomena. And because Galvani's reports of weak currents preceded Volta's reports, they gave credit for the discovery of weak currents to Galvani. Hence, they called all weak currents, produced by any means, *galvanic currents*. So the "Volta" battery was said to produce "galvanic" current, a name that paid homage to Volta's archrival! This terminology is sometimes still used today to describe the electrical output of batteries.

The second irony is that the original supposition that the fish's electrical organ worked by storing electricity in its many subunits, like multiple little Leyden jars, was fairly close to the truth. There actually is an insulating layer separating electrical charges within the fish's electrical organs. Volta couldn't find it because he was looking on the wrong size scale. The insulator isn't *between* the electrocytes; it is *within* the electrocytes. In other words, Volta was looking for the insulator at the size level of the tissues, while the insulator is found at the size level of the individual cells, which are hundreds of times smaller. The missing insulator is the *cell membrane* of the nerve cell, and it insulates the cell's internal electrical environment from its external electrical environment, analogous to the way glass separates internal and external charge for a Leyden jar. The important role of the cell membrane in biologically generated electricity wouldn't become apparent until much later when nerve cell research caught up with electrical research.

So, Volta's bimetallic pile worked, but not for any of the reasons he thought. His was a completely serendipitous discovery of which he had little understanding. But it's all right that Volta had more luck than insight. Many of the world's best scientific discoveries are a product of good luck favoring a prepared mind, and Volta definitely had a prepared mind. In fact, it would have been impossible for Volta to reason his way to the battery because the science of electricity hadn't yet developed to the point that novel discoveries could be deduced based on the first principles of electrical science theory. The electricity model of the time—still Franklin's single-fluid model—was designed to explain static electricity phenomena. It was not meant to account for the yet unknown electrochemical reactions. A model that included both wouldn't come until much later. Volta's discovery of the battery truly had leapfrogged, if you'll pardon the expression, ahead of the science. But electrical science would soon catch up.

The fundamental design of the battery—two conductive terminals made from different materials, at least one of which is a metal, with an intervening electrolyte between them—hasn't changed in over 200 years. Though major improvements to battery design have certainly ensued, all modern batteries still include these basic components.[16] But electrochemical science has provided us with a better understanding of the basic principles that underlie battery output, so we are now less dependent on a trial-and-error approach for the advancement of battery technology.

One might think of the evolution of battery designs over time as a type of Darwinian "survival of the fittest." The fittest batteries are those that produce the highest electrical output in the smallest possible size—electrical engineers would say they have a high *energy density*—while simultaneously being safe to use and cheap to manufacture. And for many applications, it is also important that the battery be lightweight. If a particular battery design is lacking in any one of these selection criteria, it is likely to become extinct, replaced by a fitter design. Currently, the more fit batteries are based on the lightweight element lithium. Among the lithium-based batteries, the *lithium-ion* batteries dominate because they have all the desirable features mentioned, plus they are rechargeable.[17] And among the lithium-ion batteries, the "fittest" is the 18650.

A visit to a scuba-diving shop illustrates the value of the 18650. My old diving light required 10 size D batteries in an aluminum alloy housing the size of a 1-pound coffee can, and it weighed nearly 8 pounds. The modern diving flashlights that the shop offers are lightweight, compact, and run on one 18650 battery, which is just slightly larger than a classic

AA-size alkali battery.[18] In addition, the 18650 is fully rechargeable. No more disassembling and reassembling the flashlight to change batteries. You can simply recharge the flashlight through an external USB port. Recharging is possible because the same electrochemical reactions that release the battery's electricity can run in reverse when an outside external electricity source is connected. Unfortunately, the modern diving light still has the same Achilles' heel as my old light: the battery compartment requires a watertight O-ring. One manufacturer even includes three spare replacement O-rings with its light. (Isn't that telling?) Some things never change.

Although it's similar in size to Volta's pile and has the same basic components—two dissimilar electrodes separated by an electrolyte—the 18650 has a very different internal anatomy.[19] Instead of being a series of stacked disks resembling a fish's electricity organ, the 18650 has an internal anatomy more like a jelly roll-up. A layer of electrolyte material is sandwiched between two flexible sheets of electrode material—one made of some type of lithium compound (negative electrode) and the other made of carbon (positive electrode). This thin sandwich is rolled up and then slipped into a small cylinder.

The 18650 is such a robust battery that it is even used in electrically powered automobiles. While a modern diving flashlight requires just one 18650 battery, the Tesla Model S electric car is powered by 7,104 of them.

Additionally, when the 18650's cylindrical shape is an obstacle to use, as it is for smartphones, electronic tablets, and laptop computers, a flattened version of the lithium battery is available. Instead of rolling up the lithium-electrolyte-carbon sandwich and slipping it into a metal cylinder, it can be left unrolled and simply slipped into a flexible plastic envelope, thus providing a totally flat version. Yes, the lithium battery is truly the workhorse battery of our day, and it all started with an electric fish.

Before we close this story of the battery, we should mention one last irony and one last posthumous slight to its discoverer, Volta. Remember the experiment that started the feud between Galvani and Volta, the one where Galvani impaled a frog leg on a brass hook connected to an iron fence? With modern electrochemical hindsight, we can now give that experimental finding a more modern interpretation.

Rather than the twitching of the frog leg being due to the electrical output from the bimetallic combination of brass and iron alone, as Volta contended, the true explanation is more nuanced. In retrospect, we can now see that the wet frog leg was not simply responding to very weak electrical output from brass and iron electrodes to which it was connected. The frog leg was actually essential to producing the required electrical current in the circuit that resulted in the leg's twitch. That's because the wet frog leg, with its ample tissue electrolytes, equates experimentally with the salt-solution-soaked paper of Volta's pile; *the leg actually provides the electrolyte that bridges the different metals.* As all Voltaic battery cells require two different metals and an electrolyte to bridge them, there would be virtually no electrical output from the brass hook on the iron fence had it not been for the circuitry that included the electrolytic frog leg. In other words, by connecting the frog leg to the fence in the way he did, Galvani had unwittingly created the first single-celled voltaic battery. So while history ironically credits *Volta's* battery as producing *galvanic* current, you could also make the equally ironic argument that it was actually *Galvani's* frog leg, hung as it was from a brass hook on an iron fence, that created the very first *voltaic* battery.

Despite Volta's frustration with all the difficulties of trying to conduct research on very weak electrical currents, there was at least one upside: he was at no risk of being accidentally shocked to death. But as technical advances allowed for ever-increasing voltages and currents, accidental high-intensity shocking became an increasing threat to anyone working with electricity. It would have been very unwise for Volta to use his tongue and his eye to detect electrical current had the amperages been any higher. The wise scientist needs to treat all electricity with respect. To do otherwise could result in sudden death by electrocution.

6

JOLTED BACK TO REALITY

ELECTROCUTION

> My first acquaintance with killing by man-made lightning came at Dannemora.
>
> —ROBERT G. ELLIOTT

On the evening of January 2, 2008, US Army Staff Sergeant Ryan D. Maseth, age 24, disrobed and stepped into what would be his last shower. A short time later, a barracks mate found Maseth unresponsive on the bathroom floor, his body half in and half out of the shower stall with the water still running, and called for help.

One of the soldiers responding to the incident got a strong shock when his clothed arm accidentally brushed a water pipe connected to the shower, indicating the shower's water system was electrically charged. As one soldier went to cut off all power to the building, others attempted to resuscitate Maseth. Unfortunately, their efforts to revive him were in vain.

An Army investigation confirmed what everyone had already suspected. Maseth died by electrocution when he touched the shower faucet. The wayward electric current was traced to a water pump on the roof of the building. The investigative report concluded the following:

"Failure to ground the pump and improper grounding of the building's electrical system allowed the metal pump housing and the water distribution pipes in the building to energize."[1] Maseth's naked and wet body had served as a convenient ground for the electrically charged pump. This lethal accident was due to faulty workmanship by contractor personnel who had "inadequate training and expertise." Maseth paid the ultimate price for their incompetence. But he was not alone.

Maseth was stationed in Baghdad as part of Operation Iraqi Freedom, a United Nations–authorized multinational military invasion of Iraq that began in 2003 and ended in 2011. His quarters were constructed on the grounds of the Radwaniyah Palace complex, one of several captured presidential palaces that had once served as resorts for former Iraqi president Saddam Hussein. As it turned out, Maseth's electrocution was just 1 of 18 electrical deaths that had occurred among American soldiers during Operation Iraqi Freedom. The investigation of the Maseth electrocution soon was expanded to include the other 17 deaths as well.[2]

Of the 17 deaths, 9 were clearly attributable to situations where the victim had touched a live electrical power line, for one reason or another, but the remaining 8 were not as clear-cut. They apparently involved some type of equipment malfunction that resulted in an electrical leak, and one of these prior incidents seemed eerily similar to Maseth's death.

On the morning of September 11, 2004, a marine lance corporal entered a shower facility to find Hospital Corpsman Third Class David A. Cedergren, US Navy, unconscious in a shower stall. The corporal suspected Cedergren may have been electrocuted because he had previously heard several marines complaining that they had experienced shocks while showering there. The corporal, with the assistance of other marines, found a plastic poncho and used it as an electrical insulator to protect themselves while they pulled Cedergren out of the shower stall. He was taken to a first aid station where he was pronounced dead.

Remarkably, Cedergren's autopsy report said his death was due to natural causes, specifically, lymphocytic myocarditis, a type of heart disease. Although a weakened heart may have contributed to his electrical death, his heart condition most certainly wasn't the primary cause. His death had been caused by electrocution. How did the autopsy miss that?

Unless an electrocution is severe enough to cause actual tissue burns, electrical deaths can be very difficult to ascertain at autopsy. Burns are even less likely to appear if the victim was shocked while in water, since the water itself absorbs a lot of the electricity's heat. So, in the absence of visible burns, coroners typically use a combination of both a postmortem examination and an investigation of the death scene to deduce whether the cause of death was electrocution.[3] Apparently, the pathologist overseeing Cedergren's autopsy wasn't aware of the circumstances of his death and hadn't visited the accident scene. Cedergren's underlying heart ailment evidently was severe enough that it plausibly could have caused his heart to suddenly stop. In the absence of any other apparent morbidity or trauma that could produce sudden death, Cedergren's death had erroneously been attributed to his heart problem.

An investigation into the circumstances contributing to Cedergren's death found multiple breaches of electrical safety protocols in the electrical wiring system of the shower facility, including wiring that was not grounded. It is likely one of these wires had shorted to the water supply system and Cedergren was killed when he touched or grabbed a faucet or pipe, just the same as Maseth.

The worst-case scenario for Maseth and Cedergren would have been that they had grabbed the faucet with their hand rather than merely touching or brushing against it. In grabbing the faucet, their hand muscles would have strongly contracted, causing them to clench down hard on the electrified faucet, unable to let go, thereby subjecting themselves to prolonged and continuous electrocution.[4] This "no-let-go" phenomenon makes the handling of live electrical wires particularly hazardous.[5] This is especially true since people don't start to feel an electrical current until it exceeds about 0.0002 amps. When it gets as high as about 0.0070 amps, they feel it but cannot let go.[6] Thus, the range between perception and let-go amperages is extremely narrow, which means, if you can feel the electricity, it's probably too late to do anything about it.

So Cedergren's case was another example of a showering soldier who was dead due to someone's negligence. In fact, it is likely all of the eight unexplained electrical deaths of US soldiers in Iraq were related to leakage from electrical appliances and equipment. In this day and age, it's hard to classify such deaths purely as "accidents." They are completely preventable.

As we've seen with lightning, electrical charge seeks a way to the ground. You can protect a building by affixing a lightning rod to its roof so the electricity can reach the ground without going through, and thereby damaging, the building. Likewise, electrical fixtures and appliances need their own little "lightning rods"—wires that connect their metal housing to the ground—in case the housing becomes charged by an electricity leak. If someone touches the ungrounded housing, or anything connected to it, he becomes a human "grounding wire," and electric current runs through his body on its way to the ground. The consequences are not good. But properly grounded electrical appliances are entirely safe to their users.

In 2004, the same year as Cedergren's death, scientists at the University of Aarhus, Denmark, published a report on all electrocution deaths in Jutland, the peninsula that comprises the mainland of Denmark, from the period 1916 to 2003.[7] Jutland is a small and relatively sparsely populated region, so the numbers weren't large. But the death trends and autopsy findings from such a comprehensive study spanning electrocutions over nearly a century were revealing.

From 1916 to 1975, there was a steady increase in electrocution deaths, which isn't surprising because this was a time when electricity and electrical appliances were coming into greater use. But after 1975, electrocutions fell precipitously, and from 1996 to the time the study ended in 2003, there were no electrocution deaths in Jutland.

The drop in electrocutions was attributed to the introduction of residual current devices (RCDs) in the mid-1970s. These relatively simple devices come in different forms, but all work on the same ingenious principle: the amount of current coming out of a normally working appliance should be equal to the amount going in. This should be true for any intact electrical circuit . . . *unless* the current is somehow leaking from the appliance to the ground. In that case, the amount of current that returns would be less than what's going in. So RCDs are designed to compare the current going in to the current coming out. When they detect a difference, they break the circuit, stopping the current flow altogether. If your body happens to be the current's route to the ground, the RCD stops the current from going into your body.

In the United States, RCDs are most often seen incorporated directly into a home's electrical wall outlets. Such outlets are called ground-fault circuit interrupter (GFCI) outlets and can be recognized by the small reset buttons plainly visible on the face of the outlet. (There is also often a tiny light on the outlet that glows green when it's working.) They cost about $10 USD (roughly 9 euros), which is much more expensive than a $2 non-GFCI outlet, but those extra 8 bucks very well could save your life.[8]

RCDs are very sensitive to current leaks. They can detect leakages as low as 30 milliamps (0.03 amps) and shut off the current in response. The shutoffs are extremely fast: they stop electric current flow in less than 40 milliseconds (0.04 seconds). It's their quickness that saves lives. That's because, as we've seen for electric bear fences, the longer-duration shocks are the ones that kill. A bear fence has an output current of about 120 milliamps (0.12 amps) delivered in 10-millisecond (0.01-second) bursts, which is much too short to do any damage to a bear. Likewise, these RCDs don't allow long shocks, so they are very effective in saving lives—nearly 100% effective, if properly installed. In 1985, the Danish government made RCD use mandatory, and this regulation drove the electrocution incidence rate in Jutland down to virtually zero. The researchers estimated that 69% of Jutland's electrocution deaths over the last century could have been prevented had RCDs been in use.

But the Jutland study also gleaned valuable information from the people who weren't saved. The scientists reviewed autopsy results and accident scene reports to try to determine exactly how the electrocuted victims had come to die. Very few electrocution deaths (11%) involved high voltage (which they defined as greater than 1,000 volts); most all were at lower voltages. This is because only lower voltages are used in homes and businesses, so people are seldom exposed to higher voltages. The researchers also found most of the electrocutions involved indirect contact (e.g., shower faucets) rather than direct electrical contact (e.g., live electrical wires). And some type of wet surface was involved in about a third of all cases. In these two respects—low voltage and wet surface—the circumstances surrounding the electrocutions of Maseth and Cedergren were quite typical of the electrocutions in Jutland.

Regarding the route of the lethal current, 69% of the time it had gone through the victim's heart, 6% of the time it went through the brain, and 9% of the time the route had been through both the brain and the heart. So a shock to the heart seems to have been involved in 78% of all death cases. Thus, the heart appears to be a major target organ for accidental low-voltage electrocution death. This finding is consistent with the notion that electricity directly interferes with the function of cardiac cells, nodal tissues, and nerve conduction tracts of the heart, all of which are important to maintaining the heart's normal beating.[9]

Interestingly, 81% of the victims had no apparent pathological changes to their internal organs. This is also in agreement with the idea that low-voltage electrocution primarily kills just by stopping hearts, not by destroying tissue. The lack of any apparent internal organ damage likely contributed to the erroneous conclusion that an ordinary heart attack had killed Cedergren.

All of this accidental electrocution stuff hits a little too close to home for me. During my foolish youth in the 1960s, before any RCDs appeared on the scene, I used to spend hot summer days poolside with my neighborhood friends. We had an old phonograph—old even for that time—which we used for playing our vinyl music records, enjoying our favorite pop tunes while we swam in the pool. The phonograph sat on a wooden picnic table that stood on the concrete deck surrounding the in-ground pool.

There must have been an electrical short in the phonograph because every time we tried to change a record, we got a shock from the tone arm. Being very "smart" kids, we realized the problem was that we were standing with our wet bare feet on an electrically grounded concrete surface. But we also knew rubber was a good electrical insulator, so we "fixed" the problem by wearing our rubber-soled flip-flop sandals whenever we touched the tone arm. It worked. No more shocks. As a "safety precaution," we taped a warning note above the phonograph: "Wear flip-flops or get shocked!" It scares me now to realize how close we had come to becoming just another electrocution statistic. My only consolation for my foolhardy behavior is that I am in good company. Benjamin Franklin, the "genius" of electricity, once did something just as stupid, and he too lived to tell about it.

One of the earliest firsthand testimonials about a self-inflicted electrocution comes from Franklin himself. As you probably know, Franklin was a big fan of turkeys. He even suggested the wild turkey (*Meleagris gallopavo*) rather than the bald eagle (*Haliaeetus leucocephalus*)—both large birds native to North America—be officially named the national bird of the United States. His love of the turkey primarily stemmed from his gastronomic interests. He was very fond of food, and turkey was one of his favorite dishes.

For some reason, Franklin believed a turkey killed with electricity made better eating than one dispatched by conventional means: decapitation. He claimed, "the birds kill'd in this manner [i.e., by electric shock] eat uncommonly tender."[10] So he set out to develop a standard procedure for electrically preparing turkeys for the table using static electricity collected in Leyden jars. One day, while performing a demonstration of the proper way to electrocute a turkey, he mistakenly touched the electrified wire intended for the turkey while his other hand was grounded, thereby diverting the full brunt of the turkey-killing charge into his own body. I'll let him explain for himself what happened next:

> I have lately made an experiment in electricity that I desire never to repeat. Two nights ago, being about to kill a turkey by the shock from two large glass jars, containing as much electrical fire as forty common phials, I inadvertently took the whole through my own arms and body, by receiving the fire from the united top wires with one hand, while the other held a chain connected with the outsides of both jars.[11] The company present . . . say the flash was very great, and the crack as loud as a pistol; yet, my senses being instantly gone, I neither saw the one nor heard the other; nor did I feel the stroke on my hand, though afterward I found that it raised a round swelling where the fire entered as big as half a pistol bullet, by which you may judge of the quickness of the electrical fire, which by this instance seems to be greater than the sound, light or animal sensation.
>
> What I can remember of the matter is that I was about to try whether the bottles were fully charged by the strength and length of the stream issuing to my hand, as I commonly used to do, and which I might safely enough had done if I had not held the chain in the other hand. I then felt what I know not how to describe—a universal blow through my whole body from my head to my foot, which seemed within as well as without; after which the first thing I took notice of was a violent, quick shaking of my body, which, gradually remitting, my sense as gradually returned, and I then thought the bottles must be discharged, but could not conceive how, till at last I perceived the chain in my hand, and recollected what I had been about to do. That part of my hand and fingers which held the chain was left white, as though the blood had been driven out, and remained so eight to ten minutes after, feeling like dead flesh; and I had a numbness in my arms and the back of my neck, which continued till the next morning, but wore off. Nothing remains now of this shock but a soreness in my breast bone, which feels as if I had been bruised. I did not fall but suppose I should have been knocked down if I had received the stroke in my head. The whole was over in less than a minute.[12]

Franklin appears to have been just as embarrassed about his foolish behavior with the turkey as I am about fooling with that energized phonograph. He ended his letter that told the above tale by saying, "You may communicate this to Mr. Bowdoin [Franklin's Boston friend who also was experimenting with electricity] as a caution to him, but do not make it more public, for I am ashamed to have been guilty of so notorious a blunder." I think it is probably safe to say that all the turkey-eating enthusiasts who witnessed Franklin's accident that day likely decided decapitation was still the best way to prepare turkeys for the table. But some people may disagree.

When electrocution is associated with lower-voltage accidents, like Maseth's and Cedergren's, it usually doesn't leave any telltale traces; hence the problem determining Cedergren's cause of death. Franklin must have also received a relatively low-voltage shock because he too had suffered no burns, other than perhaps that welt on his hand. But as voltages increase, electrocutions are more likely to involve burned tissue along the path of the current. This is because high voltages typically drive a lot of current flow and high current (i.e., amperage), in turn, can produce a lot of heat.

The amount of heat produced by an electric current is proportional to the current level times itself; in other words, the amperage squared. So 3 amps of current would produce nine times ($3 \times 3 = 9$) as much heat as 1 amp ($1 \times 1 = 1$). But the electrical resistance of the tissue also proportionally increases the heat produced. So 5 ohms of resistance would produce five times as much heat as 1 ohm.[13]

The bad part for the body is that the heating effects of current and resistance aren't independent. They interact with each other to drive heat production even higher. That is to say, they have a multiplicative relationship, rather than just additive. So in the example above, 3 amps and 5 ohms would produce 45 times as much heat as 1 amp and 1 ohm:

$$\textit{Relative heat increase} = \textit{3 amps} \times \textit{3 amps} \times \textit{5 ohms} = \textit{45-fold}$$

Such high amperage combined with high resistance is a very bad thing for the body. It introduces a whole other mode of injury: burning. And in some cases, particularly when the heart is spared by not being in the direct route of the current flow, it is the severe burning of tissue that can cause death. The bottom line is that high voltages, by virtue of the high current flows they often drive, can quite literally "cook your goose" as well as your turkey.

Franklin had big plans for electricity use in the "modern" kitchen. He described his vision of the future of cooking like this: "A turkey is to be killed for our dinner by the electrical shock, and roasted by the electrical jack, before a fire kindled by the electrified bottle [i.e., Leyden jar]."[14] Dinner on the table in minutes, centuries before the advent of the microwave oven. And he very well might have achieved this instant turkey dinner had he persisted. Fortunately, he seems to have reconsidered the wisdom of cooking with electricity after he nearly killed himself trying.

But is it humane to kill a turkey by electrocution? Colonists weren't much concerned about animal welfare and probably wouldn't even have understood the question. But our contemporary society does worry about how animals die.[15]

There has been great concern among animal welfare groups that decapitation, the standard method of slaughtering turkeys, chicken, and other poultry, is painful and cruel. To remedy this concern, *stunning,* the process of rendering animals immobile and unconscious immediately prior to slaughtering them for food, is now required throughout the European Union. Typically, this is done by an electrical shock to the head. There are no similar regulations in the United States; nevertheless, virtually all of the nine billion chickens slaughtered annually in the United States are stunned by immersion in an electrified water bath prior to slaughter.[16] Although gases such as carbon dioxide are sometimes used for stunning poultry, an electrical shock to the head is an inexpensive, efficient, safe (i.e., for the workers), and effective method of stunning, preferred by many poultry-slaughtering operations.[17]

Cost, efficiency, and safety to workers are rather easily assessed in an industrial situation. But how do we know the method is effective in making the birds immediately unconscious and pain-free? It's the bird's brain activity or, more accurately, the immediate loss of brain activity, that tells us. In order to feel pain, a human or animal must have a consciously functioning brain, and a conscious brain gives off electrical brain waves—a topic we address in chapter 15. Birds stunned by an electrical shock to the head either have no subsequent electrical brain activity, or activity so profoundly suppressed that it is incompatible with consciousness.

But an electrically stunned bird isn't necessarily a dead bird. It could recover consciousness. So immediate decapitation is still required to finish the job. And even decapitation doesn't always work. Botched decapitations of chickens have sometimes occurred. One such famous bird was Miracle Mike, the headless chicken. In 1945, a farmer's failed attempt to slaughter a chicken had left the bird headless but alive.[18] And it remained alive and walking around for another 18 months, sustained by a milk-and-water mixture fed into its esophagus through an eyedropper. Evidently, the axe had severed the bird's neck just above its brain stem, which is located at the base of the brain. A functioning brain stem was all Miracle Mike needed to keep on going. And he likely lived the rest of his life pain-free, since he had lost his *somatosensory cortex*, the pain-sensing part of the brain.[19]

Most people don't think there are any nutritional effects from the electrical stunning of poultry, but one research group has claimed the meat from electrically stunned birds can be somewhat tougher.[20] So Franklin's contention that an electrocuted turkey is tastier doesn't seem to be supported by modern data.

Even before the animal data existed to support it, the complete loss of consciousness and the quickness of death caused by accidental high-voltage electrocutions suggested to many people that it was one of the least painful ways to die. This was largely based on chance observations of accidental human electrocutions and the testimony of eyewitnesses.

In 1881, a drunken man in Buffalo, New York, stumbled near an electrical generator and grabbed its housing to try to break his fall. He received a strong shock. Alfred P. Southwick, a local dentist, was a witness to the incident. He immediately ran up to the man but found him already dead. Southwick was impressed by the quickness of the man's death and deduced that he likely died without any suffering. Southwick later came up with the idea that electrocution might be the most humane way of executing condemned prisoners.

To test his idea, Southwick began electrocuting Buffalo's stray animals and found that animals, like the electrocuted man, seemed to die instantly, supporting his idea that electrocution was a quick and painless death. Being a dentist, Southwick soon got the idea that some type of electrified chair, similar to a dental chair, might be the best way to deliver the lethal current to people. Southwick started lobbying politicians to get the State of New York to adopt electrocution as the state's mode of capital punishment, as an alternative to the sometimes slow and agonizing death by hanging that was typically employed to do the job.

The governor became interested in the electrocution option, and he appointed a three-member commission, which included Southwick, to study the matter and issue a report. The commission enjoyed some notable social prestige because its chairman, Eldridge T. Gerry, was the grandson of a signer of the Declaration of Independence. And the third commissioner, Matthew Hale, was a New York politician and grandson of Nathan Hale, the American revolutionary patriot who was hanged as a spy by the British in 1776. Nathan Hale had become famous for his last words on the gallows: "I only regret that I have but one life to lose for my country." So the Hale family had been personally affected by execution by hanging, and we can surmise that Matthew Hale brought that unique perspective to his work on the commission.

The commission was nothing if not thorough. They investigated electrocution, as well as numerous other methods humans had used to execute people throughout history, and weighed their pros and cons. In case you're curious, they identified 34 different methods.[21] The commissioners discarded most of the methods immediately because they were specifically designed to cause slow and painful deaths as part of the punishment (e.g., flogging to death and burning at the stake). The commission's task was to find a humane way to execute, not an inhumane way, so anything designed to cause suffering was out. By process of elimination, they ended up choosing electrocution as the best method, although decapitation by guillotine was a close contender. But the commission ultimately passed on the guillotine because they found it to be shockingly bloody, and it was thought to be socially untenable because Americans associated the guillotine too closely with the horrific mass political executions during the French Revolution. Also, it was impossible to know how long a decapitated head retained consciousness. Based on expert testimony from electrical scientists, the commission reasoned that, since electricity was known to travel much faster than nerve impulses, the victim's brain and nervous system would be electrically destroyed before any pain signals could reach the brain. Thus, electrocution should be painless.[22]

The commission did have some serious misgivings, however, about the reliability and predictability of electrocution. The commissioners acknowledged too little was known about the mechanism by which electricity killed, and they noted that people struck by lightning did not always die. Some remarkably survived for unknown reasons, and they postulated that some people might have bodies with an innate ability to repel electrical charge. They asked, given that natural lightning didn't always kill, whether man-made lightning could be expected to be any more reliable a killer than the natural version.[23]

Notwithstanding these concerns, electrocution ultimately was deemed the most humane way to execute prisoners. The commission issued its report on January 17, 1888.[24] It recommended the use of electrocution, specifically done with an electrified chair, for carrying out capital punishment.[25]

The New York Medico-Legal Society was charged with taking Southwick's concept of an electric chair and making it into an actual device that could be practically used in prisons for executions. In short order, the society completed its task. An electric chair was built, installed, and used for the first time at New York's Auburn Prison on August 6, 1890, to execute William Kemmler, a produce merchant convicted of murdering his girlfriend by hacking her to death with a hatchet. He had confessed to the crime and there were witnesses, so he was convicted and sentenced straightaway. But Kemmler's electrocution did not go well.

The *New York Herald* published a firsthand report from a journalist who was an eyewitness to the electrocution:

> The scene of Kemmler's execution was too horrible to picture. Men accustomed to every form of suffering grew faint as the awful spectacle was unfolded before their eyes. Those who stood in the sight were filled with awe as they saw the effects of this most potent of fluids [i.e., electricity], which is only partially understood by those who have studied it most faithfully, as it slowly, too slowly, disintegrated the fiber and tissues of the body through which it passed. The heaving of the chest which, it had been promised, would be stilled in an instant of peace as soon as the circuit was completed, the foaming of the mouth, the bloody sweat, the writhing of shoulders and all other signs of life. Horrible as these all were, they were made infinitely more horrible by the premature removal of the electrodes and the subsequent replacing of them for not seconds but minutes, until the room was filled with thc odor of burning flesh and strong men fainted and fell like logs upon the floor.[26]

No, it certainly had not gone well. Kemmler was first shocked with 1,000 volts for 10 seconds, which was expected by the physicians on hand to be more than sufficient to do him in. His body jerked violently and he lost consciousness. But to everyone's surprise, when they took his body out of the chair, he seemed to gasp for breath. So, he was put back in the chair and shocked a second time with 2,000 volts for "several" minutes. It was during this second shocking that Kemmler's body began to burn.

As we've previously noted, amperage and resistance are important drivers of electrical heat production, but the length of the electrical

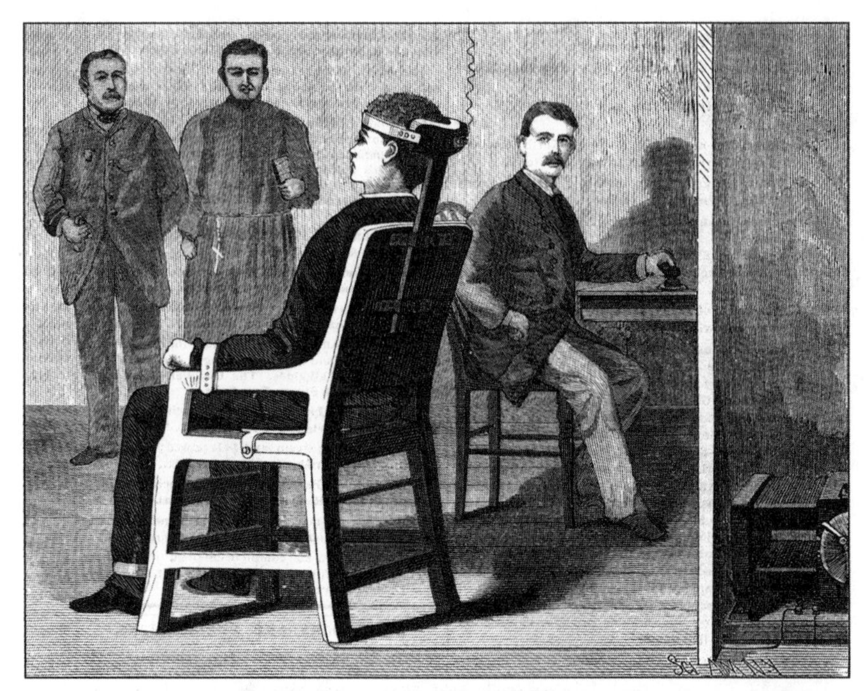

EXECUTION BY ELECTRICITY. SHORTLY TO BE INTRODUCED IN N. Y. STATE.

FIG. 6.1. The first electric chair. The electric chair was first used in New York on August 6, 1890, to execute William Kemmler, who had been convicted of murdering his girlfriend with a hatchet. Prior to Kemmler's execution, the magazine *Scientific American* ran this illustration of the chair to be used and described how it worked to its readers: "The criminal is seated bound to a chair having a metal seat connected with one pole of the current. At the back of the chair there is an adjustable head rest having a metal plate on its face and a metal band which passes around the forehead of [the] criminal. The wires may be connected with [a] dynamo . . . or the current may be supplied from [an] electric light plant. . . . Sponges or dampened cloths should be at the points of contact with the convict to render connection more perfect. At the proper moment [a] switch is turned by the officer and instant death ensues. The current passes along the spinal column and the brain and nerve centers. The current may be [a] few moments to bring about complete exhaustion" (*Scientific American*, June 30, 1888). (WikiMedia Commons)

shock is also a major determinant of tissue burning. The amount of heat damage is directly proportional to the time of exposure to the current. Increasing Kemmler's shock time from 10 seconds to 3 minutes (or 180 seconds) would have increased heat output 18-fold. No wonder his body began to burn.

If Kemmler was, in fact, alive after the first jolt of 1,000 volts for 10 seconds, it likely had to do with the chair's design. The electrodes were connected to the top of the head and the lower back near the base of the spine. The most direct route for current to flow between the two electrodes might have skirted the heart. Upping both the voltage and length of the second shock raised the heating output enough to also cause the body to burn.

Of course, it was the high amperage, not the voltage, that was producing the burning effect. But prison officials at most electric chair executions weren't initially focused on the amperages, only the voltages. The close association between electricity's amperage and its lethality wasn't yet fully appreciated. After gaining more experience with the electric chair, however, the role of amperage in driving lethality became better understood; so amperage as well as voltage started to be monitored, with the goal of finding the ideal combination of volts and amps to hasten death while minimizing tissue burning. But burning was never completely eliminated.

The three doctors on the scene at Kemmler's electrocution tried to downplay the apparent horror of it. One of the doctors pronounced, "The man was killed instantly, I think. Those were only muscle contractions, and the fellow never suffered any pain. That's one sure thing about it." The doctor might have been correct. As we already have seen with Giovanni Aldini's demonstrations with shocked human corpses, a body doesn't have to be alive to be jolted into movement by electricity. Kemmler was probably stunned insensible, and likely rendered brain-dead, by the first shock to his head.[27] But we will never know for sure.

Nevertheless, the doctors present at Kemmler's electrocution were heavily criticized for bungling the execution. They took the lion's share of the blame for the fiasco. They had prescribed the length of the shock without any scientific understanding of how electricity kills, and their decision to put Kemmler back in the chair and shock him again seemed to prove their ignorance, particularly since they later claimed he had been killed by the first shock. The *Buffalo Evening News* condemned the doctors for overseeing an electrical execution of a human being even though they had virtually no understanding of electricity's effects on the human body.[28] The *Auburn Daily Advertiser* similarly defended the prison staff while condemning the doctors, saying that, due to the doctors' insufficient knowledge of the biological effects of electricity, they could only have guessed at how long the current should have remained on.[29]

Thomas Edison also attacked the doctors for their ignorance of even a basic understanding of electrical current.[30] He said that although attaching an electrode to the skull to target the brain sounds like a good idea, it has major practical limitations. In particular, he claimed the skull bone impeded the flow of electricity because of its high electrical resistance. He further contended the blood-filled hands and arms, in contrast, were good conductors of electricity because of the low electrical resistance of blood. He pointed out that accidental electrical deaths most frequently involved someone unwittingly touching a live electrical wire with both hands and being killed instantly, thus proving that current entering the body through the hands is lethal—likely because the current passes through the heart as it flows from one arm to the other. He suggested, therefore, that it was only common sense that the best way to electrocute a prisoner was through his hands, and he recommended that a better approach would be simply to connect an electrode to each hand and forget about the skull. The doctors countered that Edison didn't understand that the purpose of the head electrode was to instantaneously render the brain insensible to pain, something that couldn't be achieved by electrocution through the hands because little current would pass through the brain.

The doctors who oversaw Kemmler's electrocution not only had botched the delivery of the electrical shock; they also dropped the ball on his autopsy. First, they decided to allow the body to sit for hours at room temperature to cool down, at which point *rigor mortis* had set in, with the corpse assuming a sitting position. So the doctors had to struggle to straighten out the body before they could begin to autopsy it. Internal organs appeared for the most part normal, including the heart.

They dissected Kemmler's brain and spinal cord but saw no visible damage to either. They then divided the tissues and put them in jars for later microscopic inspection in the laboratory. But since there had been no attempt to collect or quantify the parameters of the electrical current that the body tissues may have received, no useful information was obtained about the effect of the current on body tissues. Thus, all autopsy findings were inconclusive.[31]

Despite the debacle, prison officials declared the Kemmler electrocution a success. They claimed that a few glitches were to be expected on the first run, but that future electrocutions, with a few modifications and the development of standardized procedures, should go flawlessly.

Still, the issue of the best electrode placement persisted, and strong criticism by a man like Edison, widely regarded as the most knowledgeable electrical expert in the country, couldn't be easily dismissed. The pressure was on to test Edison's recommendation of electrocution through the hands.

Dr. Carlos F. MacDonald, the chief medical counsel for Kemmler's electrocution, had suffered the bulk of the criticism for its mishaps. When he was put in charge of another electrocution, he took the opportunity to test Edison's idea. A special electric chair was constructed at Sing Sing Prison in New York. This chair differed from typical electric chairs in that the prisoner's upper arms were strapped to sloping armrests, with the hands dangling over the edge. This allowed each hand to be immersed in a glass jar of salt solution, with each jar containing an electrode. The salt solution served as an electrolyte to maximize the electrical contact with the hands. But the chair also was equipped with the standard skullcap and leg electrodes of other chairs, which could be employed alternatively as needed.

The first victim of this new chair was Charles McElvaine, a 19-year-old boy who had stabbed a grocery store proprietor to death during the course of a robbery. He was led into the execution room and strapped to the new chair. His face had the look of terror. He was muttering prayers to Jesus for help and clutching a crucifix in his left hand. As his right hand was immersed in the jar of salt water, the crucifix was removed from his left. And once his left hand had likewise been immersed in its jar, the switch was flipped, electrifying McElvaine from hand to hand.

McElvaine's torso arched and strained against the straps as his body muscles violently contracted and his facial muscles contorted. The current was allowed to flow for 50 seconds and then turned off. An attending physician put his hand into one of the jars to feel for a pulse. Remarkably, McElvaine still had a pulse. At this point, MacDonald quickly ordered the skull and leg electrodes to be attached, and McElvaine was then jolted again, this time from head to leg. Again, his body went into extreme contortions, and a hissing sound came from his leg. After 36 seconds, the current was stopped. This time he was dead. [32]

Apparently, Edison had been wrong, and MacDonald had the actual data to prove it since the electrical parameters had been recorded. The hand electrodes had produced a current of 2 to 3.1 amps, driven by 1,600 volts, and left McElvaine alive, while the skull and leg electrodes that had killed him delivered 7 amps, driven by 1,500 volts. The data suggested skull and leg electrodes had only half the resistance of hand-to-hand electrodes, completely refuting Edison's contention that the skull's electrical resistance greatly impeded current flow during Kemmler's execution. In a statement obviously designed to throw Edison's words back in his face, MacDonald quipped to the press, "Edison probably reasoned all right from his standpoint as an electrician, but all wrong from the standpoint of a physician."[33] Henceforth, all future criminal electrocutions in the United States would use skullcap and leg electrodes.

Over the subsequent decades, the electric chair, with some further refinements, became a well-established and routine means of execution for capital crimes in the United States. Electrocution was institutionalized to the point where a new prison professional was specifically responsible for executing prisoners: the state electrocutioner. It's noteworthy that state electrocutioners were chosen entirely based on their expertise with electricity, not for any human electrophysiology knowledge. Doctors still attended the electrocutions, but their function was simply to pronounce the time of death and to conduct a subsequent autopsy. They had no control over the electrical parameters of the lethal shock.

Robert G. Elliott became New York's official state electrocutioner in 1926. He was the state's third, but ultimately became its most well known. During his career, he electrocuted 387 prisoners on death row, some of them nationally notorious, including Bruno Hauptmann, who was charged with kidnapping and killing the infant son of the famous aviator Charles A. Lindbergh.

Elliott didn't have any moral problem with being an electrocutioner. He saw it as a legitimate profession, and he tried to perform his duties in the highest professional manner. Prison officials admired his serious, no-nonsense approach to his work, and he ended up being regularly hired as a freelance electrocutioner by the neighboring states of Pennsylvania, New Jersey, Massachusetts, and Vermont.

Elliott kept a very low profile, partly because he was by nature a very quiet and private family man, but mostly because he was continually the target of death threats from people who disliked his line of work. In fact, his house was once bombed while he and his entire family were at home asleep. Fortunately, nobody was killed. Unfortunately, the perpetrator was never apprehended. This incident didn't cause Elliott to change professions, but he did change his behavior. From then on, he seldom spoke to anyone about how he made his living, and he slipped in and out of prisons in secret. His phantomlike appearances and disappearances at electrocutions tended to increase his mystique with the public.

Elliott kept detailed records of all the executions he performed, and as he approached his own death, he used his collection of notes as an original source to dictate his autobiography, finishing it while on his deathbed. The autobiography was entitled *Agent of Death: The Memoirs of an Executioner*.[34] It went to press in 1939, just days after Elliott died. The book was widely read due to the fascination people had for this shadowy figure and the gruesome job he performed so well. But many readers were surprised at the take-home message of Elliott's book: the death penalty should be abolished. The man who had executed so many was, in fact, opposed to capital punishment!

In his book, Elliott tells a highly detailed and seemingly credible story about exactly what he did, how he did it, and why he did it. He did not relish his work, but he saw it as a job that needed to be done with technical expertise as well as a lot of compassion. But he didn't feel personally responsible for any of the 387 deaths: "My job is to see that this is done as humanely as possible. Outside of that, my responsibility for this individual's untimely death is no greater than that of any other member of society that endorses or condones capital punishment."[35]

Elliott, who started out as an ordinary small-town electrician, eventually became the head electrician for Dannemora State Prison in New York. As such, he was assigned to assist the traveling state electrocutioner whenever he came to the prison for an execution. The state electrocutioner at the time was Edwin Davis, the very man who had electrocuted Kemmler. Davis had also electrocuted Leon Frank Czolgosz, President William McKinley's assassin, and dozens of others in the following years. He was credited with the modified design of the electric chairs that followed Kemmler's chair, in which the lower-back electrode was replaced with an electrode affixed to the lower leg. Davis had incrementally perfected his technique over the many electrocutions he conducted in the ensuing years, and had actually patented the head and leg electrodes he had designed, and used his patent rights to keep out potential competitors for his job. He was a true professional killer who performed his job very well.

Davis ended up training Elliott to serve as a substitute executioner if, for some reason, Davis himself was unavailable to carry out a sentence. Davis taught Elliott how to do a test run of an electric chair using a large piece of beef, up to 15 pounds. The idea was to use 1,700 volts to deliver 7 amps to the beef. Davis had learned from experience that an electric chair precalibrated with a hunk of beef would easily dispatch even a large man. But, on one occasion, a chair pretested with beef did not do the job. The man was discovered to still have a pulse while his body was lying on the autopsy table. He was quickly returned to the electric chair and shocked again, at which time he finally expired. No one could understand what had gone wrong. But it was noted at autopsy that the man had a very large heart. In fact, it was larger than that of any person who had previously been executed using the electric chair. It was surmised that his exceptionally large heart might have been better able to withstand the current. After this incident, Davis upped his voltage from 1,700 to 1,800 to help ensure this didn't happen again.

When Elliott ultimately took over as state electrocutioner,[36] he routinely used an amperage meter, or *ammeter*, to assess the amount of current going through people's bodies during electrocutions and noticed it was typically around 11 amps. He recognized that a body's electrical resistance level was the major determinant of the amperage. If body resistance was low, the current could get to levels as high as 16 amps. When resistance was high, the current could be as low as 7 amps. Elliott was surprised, however, that there seemed to be no correlation between body size or muscle mass and resistance. There did seem to be an apparent relationship, however, with resistance and tuberculosis, a chronic disease of the lungs that was common at that time. Specifically, he noticed that when he electrocuted men and women suffering from tuberculosis, they had very high electrical resistance. Someone had told him tuberculosis patients had low chloride-ion concentrations in their blood, and he postulated the low levels of this highly conductive electrolyte might be the reason for their bodies' higher resistance.

Elliott was wrong about tuberculosis patients being chloride deficient; they are not. So that cannot be the explanation for his anecdotal observations about tuberculosis victims having higher body resistance. In fact, he was also likely wrong about the alleged association between tuberculosis and resistance. He hadn't done a controlled study; he had only made anecdotal observations, which are often unreliable. But we now know a lot more about the body's electrical resistance, and we can make an educated guess as to why he saw more than a twofold difference in amperages between victims.

The overall body resistance is dependent on the resistances of the various tissues the current passes through, with the bulk of the current taking the path of least resistance. Resistance varies tremendously by tissue. Bones, tendons, and fat are among the most resistant tissues, while nerves, muscle, and blood have the least resistance. Skin is particularly noteworthy in that its resistance can vary 1,000-fold, depending upon whether it is dry or moist—dry skin having the highest resistance. As we've discussed earlier, moist skin contributes to the flashover effect of lightning, in which the current takes a route over the surface of the skin rather than penetrating into the body tissues. So the absence or presence of sweat on the skin could have been a major factor driving the variations in overall body resistance that Elliott measured.[37] He doesn't mention anything about sweat levels of the victims, but Elliott does say some victims met their fate calmly and stoically while others were in a frenzied state when strapped in the chair. One can imagine their perspiration levels varied similarly, and that alone could more than account for a twofold difference in current for any constant voltage.

Flashover also may be the reason why so many electrocuted prisoners, to everyone's amazement, have maintained a pulse after their initial jolt. But the initial flashover likely vaporized the skin moisture, making the skin drier, more electrically resistant, and less prone to another flashover when the second jolt was administered, thereby making the current's penetration into the body and subsequent death more likely. But this is mere speculation. No one has ever researched flashover occurrences during prisoner electrocutions. However, Franklin had observed something similar when he did some experiments electrocuting rats. He noted with surprise that "a wet rat can not be kill'd by the [Leyden jar], when a dry rat may."[38] Perhaps Franklin's wet rats were being spared by the flashover effect.

Elliott also reported that the amount of heat produced in the body by electrocution can be very high. He recorded postelectrocution core body temperatures and found an average body had a temperature of

138°F (59°C).[39] (Normal body temperature is 98.6°F, or 37°C.) But tissues in the vicinity of the electrodes could be much hotter. On one occasion, a copper electrode attached to a leg actually melted. On another occasion, a guard severely burned his hand, requiring him to seek medical attention, when he barehandedly grabbed a female victim's lifeless body to remove it from the chair.[40]

Despite all this gruesomeness, Elliott felt confident his victims felt no pain. He related that medical experts had determined that unconsciousness occurs in less than 1/240th of a second after the switch is flipped.[41] Too quickly, he reasoned, for them to feel anything even if their heart remained beating or they showed other apparent signs of life, like muscle contractions. (The measurement of 1/240th of a second is a very precise number, suggesting it came from a highly sophisticated study, but Elliott didn't cite his source for this information.)

If there were legitimate medical scientists researching the underlying physiology of electric chair executions, that research must have been happening elsewhere, because Elliott was never involved. He claimed that, other than one time when an electrocardiograph was used on a condemned prisoner, there was no other medical research into the electrophysiology of prisoner electrocutions during his tenure. No other, unless you consider the work of George M. Ogle, an electrical engineer with the National Electric Power Company, to be "research." Ogle had no medical training.[42] Elliott reported that Ogle, who lived just six miles from Sing Sing Prison, once asked for and received permission from the warden to run his rubber-gloved hands over the body of a victim while the lethal current passed through his body, in order to study the degree of muscle contractions caused by the current.[43] Odd research indeed.

For the most part, Elliott received no outside medical guidance and largely relied on his own intuition about how best to electrocute a human. He empirically modified his procedure based on his observations and his own unverified medical theories: "Lowering the current slowly following the final shock is my own idea. I believe that it effectively weakens the heart and stops the action. Perhaps this is fallacy, but once or twice when I have not done it, the heart has continued to beat after the current was turned off."[44]

After performing 387 electrocutions and witnessing all their various outcomes, he remained convinced the electric chair was "efficient and quick." If the state was going to kill someone, Elliott believed electrocution was the most humane way to do it.[45] Yet, he did think its killing efficiency could be improved. Elliott was permitted to modify only the voltage, current, and duration of the electrocutions, not the placement of the electrodes. Since McElvaine's electrocution, all states mandated the head electrode and a lower-leg electrode be used. Which leg typically depended upon the specific state—right leg in New York, New Jersey, and Connecticut; left leg in Pennsylvania, Massachusetts, and Vermont.[46] To stop the problem of the heart continuing to beat, a persistent issue that often necessitated a second jolt, he recommended the leg electrode be dropped and rather "that an electrode be placed over the prisoner's heart, thus stopping the action of this vital organ almost as soon as the electricity is applied."[47]

Elliott ended up having some very strong opinions about capital punishment, carried out by electrocution or any other means. He thought the death penalty should be abolished. He saw no good that came from it. It wasn't an effective deterrent to crime and simply amounted to societal revenge. He even had a suggestion on how to get rid of it: "So long as capital punishment exists . . . witnessing an electrocution [should] be made a civic duty, just as jury service has been. Thus would the repugnant horror of, and responsibility in, legal slaying be impressed on the average person. I venture to predict that if this were done, the abolishment of the death penalty would soon follow."[48] At the end of the day, that was the lifelong take of a man who had carried out court-ordered death sentences 387 times.

Nationwide there have been over 4,300 prison electrocutions in the United States since Kemmler was first executed in the electric chair.[49]

Many people today believe the electric chair is extinct, replaced by lethal injection or life sentences as more civilized ways of dispensing justice. But that isn't true. Although it is no longer the primary means of execution in any state, the states of Alabama, Florida, Kentucky, Mississippi, South Carolina, Tennessee, and Virginia maintain the option of using the electric chair for capital punishment, as of 2020. Also, Arkansas and Oklahoma currently have state laws to bring the electric chair back into use should lethal injection ever be declared unconstitutional by the US Supreme Court. The electric chair has not been declared unconstitutional because the Supreme Court hasn't deemed electrocution to be "cruel and unusual punishment"—the constitutional criteria to ban it.[50] The United States is the only country in the world where the electric chair is used for capital punishment.

The state of Tennessee has been using the electric chair regularly as of late. Within a recent 16-month period, five murderers have been electrocuted. Edmund Zagurski, who had been convicted of a double murder, was electrocuted on November 1, 2018. Zagurski actually asked for the electric chair, preferring it to lethal injection.[51] Zagurski said he chose the electric chair over lethal injection because be believed electrocution to be quick and painless. Despite Elliott's 80-year-old recommendation that an electrode be placed over the heart, the electrodes were attached to Zagurski's head and his lower leg in the traditional manner that's been in use for over a century. When asked if he had any last words, Zagurski smiled and replied, "Let's rock!" He was shocked with 1,750 volts of electricity for 20 seconds, then jolted with the same voltage for another 15 seconds—an electrocution regimen virtually unchanged from what Elliott used in the 1930s. Zagurski slumped in his chair between jolts, and after the last jolt he was declared dead. Witnesses saw no signs of pain or struggle, and there was no apparent smoke or burning flesh. All had gone well.[52]

On December 6, 2018—one month later—Zagurski's fellow inmate David Earl Miller took his seat. Miller had been convicted of stabbing a young woman to death while in a fit of rage fueled by alcohol and drugs. Miller too opted for the electric chair, and he too died similarly without incident.[53] His last words: "Beats being on death row!" Miller had been on death row for 37 years.

As of this writing, the last inmate to die in the Tennessee electric chair was Nicholas Todd Sutton, who was executed on February 20,

2020, giving no explanation as to why he had chosen electrocution as his means of death. Sutton was 58 at the time of his execution. He had murdered four people, including his own grandmother when she was 58 and he only 19. Sutton had been in prison 39 years and on death row for nearly 35. In his final words, he spoke of the "power of Jesus Christ to take impossible situations and correct them."[54]

If you've had to look away for some (or most) of this chapter, I understand. Electrocution is not for the squeamish. But don't worry; the important takeaway message is simply this. Both low-voltage and high-voltage shocks to the body can be deadly, particularly if the current goes through the heart—the prime target for death. The chief difference is that high-voltage electrocutions are also usually accompanied by tissue burning, because the higher voltages are typically pushing higher currents, which in turn dramatically increase heat. Low-voltage electrocutions, in contrast, produce little heat and consequently little to no burning and thus no telltale body markings, often frustrating coroners' attempts to determine the exact cause of an accidental death. Although sometimes electrocutions are intentional, as is the case for capital punishment, most often they are accidents. And the best protection against accidental electrocution is a well-grounded electrical system. There you have it, in a nutshell. You can open your eyes now.

Before leaving this chapter, I would be remiss not to mention that the apparently noble ideal of adopting electrocution as a more humane method to execute condemned prisoners was badly tainted by special interests. Those special interests came from the business world, and they revolved around two warring titans of industry: Thomas Edison and George Westinghouse. Each man had hitched his business wagon to opposite modes of electric current delivery—direct current (DC) and alternating current (AC), respectively. And each man had a huge vested interest in his particular mode of current delivery becoming the standard household current for the United States.[55]

One trumped-up issue was relative safety, and Edison was determined to convince the public that Westinghouse's AC was inherently unsafe; much too dangerous for Americans to let into their homes. What better way to do that than to convince prisons to choose AC to electrocute prisoners because of its supposed higher lethality? Edison thought a mental connection between the electric chair and AC would be all that he needed to motivate consumers to demand his allegedly "safer" DC current from their local electric utility. So he performed public electrocutions of dogs and other animals as a demonstration of AC's lethality, and lobbied prison authorities for AC to be used in electric chairs. He even went so far as to design his own electric chair, which, of course, ran on AC.

Edison was successful in getting AC current used for the electric chair. The electrocution of Kemmler—the very first electric chair victim—was done with AC, and every subsequent electric chair electrocution in America has used AC. But Edison wasn't successful in keeping AC out of homes. His safety threats fell on deaf ears, and AC ultimately became the standard current that we use to this day to power our home appliances. Why AC? Because it is much cheaper to deliver. The consumers followed the money.

But what exactly are AC and DC currents, precisely how do they differ, and is one really more life threatening than the other? As it turns out, that's a whole other story.

7

A FIELD DAY

ALTERNATING CURRENT AND LINES OF FORCE

[He] found a bridge for the lightning of thought.

—DANISH AUTHOR HANS CHRISTIAN ANDERSEN'S EULOGY OF HIS LATE FRIEND HANS CHRISTIAN ØRSTED

Jostled from my slumber by pounding on my bedroom door, I abruptly awaken to hear, "Get up, we're going to miss the tide!" My 15-year-old self groans at the prospect of rising at this perverse hour of 5:00 a.m., but my father is insistent. "We need to get our butts down to the dock right now!" The previous evening the two of us had agreed to go fishing in the morning, but I hadn't fully grasped the implications of what "morning" actually meant when I had committed. Regardless, the tide waits for no one. So, I drag my sorry butt out of bed and down to the dock. A-fishing we will go.

When I was a boy, my family vacationed on the New Jersey seashore, very close to Little Egg Harbor Inlet. An inlet is a very narrow waterway linking an enclosed body of water, like a bay or estuary, with the outside

ocean. Small boats regularly use inlets to access the open ocean from their inland harbors. My father regularly used our local inlet to get his recreational boat out to the ocean fishing grounds. We relied on that inlet as our passageway to and from the sea. And when you rely on inlets, timing is everything.

Boaters who regularly pass through coastal inlets are very familiar with the concept of alternating current. Driven by the combination of the rotation of the Earth and the gravitational force the moon exerts on the Earth's oceans, sea levels rise and fall on a regular cycle. These rising and falling tides, in turn, push seawater in and then out of coastal inlets. For just over six hours, the water current will run through the inlet in one direction and then reverse its flow, running for six hours the opposite way—an endless cycle as old as the moon.[1] The cycle of the tide takes slightly over 12 hours before things get back to where they started, so there are nearly two complete tidal cycles each 24-hour day. Boaters take note of these current cycles. They time their activities to make optimal use of the intensity and direction of the water flow through the inlet. Ignore the cycles and you risk running your boat aground.

To seafarers, current cycles seem quite natural. They don't question too much how it all works; they just accept that inlet currents alternate their direction in a predictable cyclical pattern as a consequence of a force of nature—gravity—acting on a moving object—the Earth. They needn't wonder which direction the current might be flowing when they get to the inlet. They know. They just time their travels around the predictable pattern of water movement. In fact, they are very thankful that the current cycle is, at least, predictable . . . unlike the weather.

Likewise, people who work with electricity are also familiar with the concept of alternating current. They know direct current (DC), the type of current that a battery produces, is electrical current flowing in one direction only. In contrast, alternating current (AC), the type of current that comes from an electrical wall outlet, flows in one direction and then switches its direction to flow the opposite way, just like the tidal currents—only faster, much faster. While current flowing through a coastal inlet has just 2 cycles per day, standard AC has 60 cycles per second. Very fast indeed.

Although seafarers are accepting of alternating currents as the norm for inlet navigation, not all electrical scientists and engineers were originally tolerant of AC becoming a norm for transmitting electrical power. Some thought it inferior to DC for technical reasons and some claimed it was very dangerous to work with. So a major controversy was spawned. In fact, there was a virtual "war" in the 1890s between two industrialists, Thomas A. Edison and George Westinghouse, to determine which type of current would become the standard for electrical power transmission in the United States. Edison favored DC largely because he already had a cache of patented electrical inventions based on DC, and Westinghouse favored AC because he was developing inventions based on AC with the help of Edison's nemesis, the great electrical engineer Nikola Tesla.[2] People familiar with this AC/DC war will tell you Edison ultimately lost the war and Westinghouse won it, which explains why AC now rules the day for electrical power transmission worldwide. End of story. All of this is true . . . except that the story is definitely not over.

If you are like most people, you probably have a cell phone charger or laptop computer charger in the room with you right now. You may even have one of your numerous electronic appliances being charged at this very moment. If so, go over and feel the main body casing of the charger. I'll wait until you get back.

It feels warm, doesn't it? It feels warm because the charger contains both a *transformer,* to step the incoming 110 volts (typical US household voltage) down to a 5-volt output, and a *rectifier* or *converter,* to convert the AC household current into DC, the only type of electrical current capable of recharging batteries and running electronics. The housing feels warm because this highly localized current conversion is inefficient, and it releases the wasted energy in the form of heat. Multiply the wasted energy from your charger by the total number of chargers in the world, and you can see charging up personal electronics wastes a great deal of energy worldwide. So converting AC to DC at the place of electricity's use is certainly not a "green" process. Since most electricity is being generated from combustion of fossil fuel, we might even say it has a large "carbon footprint." So what's the deal with alternating and direct current? Is all of that local current conversion really necessary? It certainly seems to come at the cost of appreciable energy loss.

The problem is that all truly "electronic" devices need DC to function. This is because, by definition, *electronic devices* are electrical appliances that rely on transistors—miniature electric switches—in order to operate.[3] Transistors are now in virtually every appliance, so almost everything that uses electricity has at least some electronic components. These transistors cannot utilize AC at all. Either they must use a battery as an electricity source (DC) or, alternatively, household AC current must be converted to DC in order to power them. In contrast, appliances that have electric motors in them, such as washing machines, kitchen blenders, and garage door openers, often run directly on AC. But even these appliances typically have internal electronic controllers, so at least some AC current needs to be internally transformed into DC in order to operate the appliance's controls. Even with household lighting there is an ever-increasing demand for DC because, while the older incandescent lightbulbs can use AC or DC, the modern light-emitting diode[4] (LED) lights, which are quickly replacing them, cannot. LED lights run only on DC.

What all this means is that, as the use of electronic devices grows, there is increasing usage of DC over AC in our everyday lives. Some have estimated over 50% of our total electricity usage may be in the form of DC within 20 years, and locally transforming all that AC current into DC will entail tremendous energy costs due to the low efficiency of local conversion. Yet, AC is the standard current used to transmit electricity through the nation's power grid, thanks to Westinghouse's winning the current war. So why was it we ever started generating electricity in the form of AC to begin with?

We have seen that electricity was first generated by rubbing two different materials together and harvesting the static electricity that was produced. Static electricity was the chief source of man-made electricity in Franklin's day. It produces current that runs in just one direction, making it DC by definition. Later, the invention of the electrochemical battery by Volta provided a new source of electricity. Likewise, the current from a battery runs continuously in one direction from the positive terminal of the battery to the negative terminal, and thus is also DC. It would take the invention of yet another means of generating electricity—*electromagnetic induction*—for AC to make the scene.

The acclaimed British scientist Michael Faraday discovered electromagnetic induction in 1831. Like Franklin, Faraday was an electrical genius. On paper, Faraday's background and training even looked like Franklin's; both came from nonaristocratic backgrounds and both had nearly no formal education. They were largely self-educated men who were able to scientifically train themselves through their access to a tremendous amount of reading material. In Franklin's case, it was because he worked in the printing trade. For Faraday, it was because he worked as a bookbinder.[5] Likewise, they both ended up among the scientific elite. Faraday might even be called the new Franklin. In fact, Faraday was born in 1791, which was one year after Franklin died. In many ways, his birth ushered in a new generation of electrical scientists and new ways of thinking about electricity.

Faraday and Franklin were also similar in their scientific approaches. Both men liked to "play" in the laboratory. They would spend hours hooking one thing up to another just to see what would happen. If nothing happened, fine. But if something interesting happened, they would try to explain it. For their electricity experiments, Franklin would try to explain his results with his single-fluid theory of electric charge. Faraday, in contrast, wasn't a big fan of electric "fluids." He preferred to explain his findings in terms of electric "fields"—a type of force field. He wasn't exactly sure what a force field was, but he was gaining insight through his many experiments.

Faraday was criticized for his force field ideas because they contradicted what scientists thought they knew about how forces acted at a distance. Forces such as gravity, according to the great scientist Isaac Newton, exist instantaneously and were supposed to exert their force only in straight lines. That is to say, unlike for light, which moves from place to place at a constant speed, no time supposedly was needed for gravitational forces to get from the sun to the Earth, or from the Earth to the moon. Faraday's alleged force fields violated these "rules." He contended electrical and magnetic (i.e., *electromagnetic*) forces acted along curved lines and they took some time to appear.[6] This concept directly contradicted Newtonian theory. Heresy!

Faraday credits his work on force fields to the foundational work of Hans Christian Ørsted, a Danish scientist who had made a remarkable discovery.[7] Ørsted discovered that if you took an ordinary compass—the navigational type with the suspended magnetic needle—and brought it near a wire that was conducting an electrical current, the compass needle would point at the wire. If you turned off the current, the compass would revert to its normal north/south orientation with the axis of the Earth. The current flow seemed to be producing some kind of force field around the wire that interacted with the needle's *magnetic field*.[8] Ørsted went on to further show that when an electric current flows through a wire, it produces its own cylindrically shaped magnetic field along the axis of the wire. One could even amplify the magnetic field by using a very long insulated wire coiled around itself. By coiling the wire on itself, like a garden hose, the magnetic field generated by each coil adds together, thus producing a very powerful magnetic field. If the wire was coiled around an iron bar, the bar took on all the characteristics of a permanent bar magnet, as long as the electrical current was flowing. When the current was interrupted, the magnetism disappeared. This was the first demonstration of an *electromagnet*.[9]

These observations by Ørsted got Faraday thinking about the exact relationship between electricity and magnetism. Ironically, this issue had been around for thousands of years but largely had been dismissed. Back when the electrical properties of rubbed amber were being investigated, the question had arisen whether amber's attractive and repulsive actions, and the attractive and repulsive actions of lodestones (which are naturally magnetic stones formed from an oxide of iron ore), were just different manifestations of the same natural force.[10] The fact that amber's forces were created by rubbing while lodestones needn't be rubbed, and the fact that lodestones, unlike amber, produced no shocks or sparks suggested they were entirely separate phenomena. But

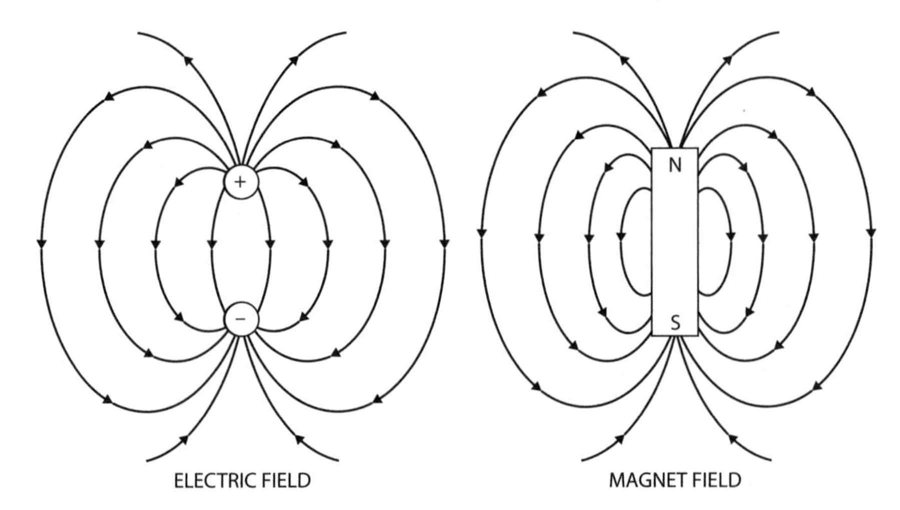

FIG. 7.1. Electric fields versus magnetic fields. Electric and magnetic fields are really the same phenomenon viewed from different perspectives. Consequently, these two types of fields are often discussed collectively as electromagnetic fields, particularly because they share so many characteristics. One major difference, however, is that while electric fields have endpoints—the positive and negative electrodes mark the ends of the fields—magnetic fields do not. The north and south poles of the bar magnet do not represent the two ends of its magnetic field. Rather than terminating at the bar magnet's north and south poles, the lines of force of a magnetic field loop back around lengthwise through the interior of the magnet, taking the form of continuous ellipsoids. Another difference is that, although a magnetic field must have both a north and a south pole, a positively charged locus need not be paired with a negatively charged locus to produce an electric field. An electric field will be present around a single charged locus of any type, but it will take on different shapes when in the proximity of another charged locus, the shape dependent upon whether that other locus has a positive or negative charge.

the compass needle and bar magnet results seemed to suggest that, although they may be different forces, these two forces seemed to interact with one another in peculiar ways.

One maxim of science is that many natural phenomena can run both forward and backward. In other words, natural processes that tend to run in one direction often can be compelled to run in reverse. As we have already seen, this is true for electrical batteries. Electrochemical reactions in the battery produce an electrical current output that persists until its chemicals have been depleted, but it is also possible to run electrical current back into the depleted battery and drive the chemical reaction in reverse, thus restoring the original chemicals. This is the principle behind a rechargeable battery.

So, given this mind-set, it was logical for Faraday to wonder if electromagnetism could be reversed. Since running a current through a wire wrapped around a metallic bar can magnetize that bar, Faraday thought that perhaps wrapping a wire around a permanent magnet might produce a current. That is, if the metal bar in Ørsted's experiment was already a magnet, would it have the effect of producing a current in the wire? Put another way: Could Ørsted's experiment be run in reverse? An interesting idea, but unfortunately wrong. To Faraday's dismay, he got no current out of wires wrapped around permanent magnets.

But then Faraday tried something else. He wrapped wire around a hollow cylinder, something like the cardboard cylinder from a toilet paper roll. Then he placed a permanent bar magnet inside the cylinder and measured the wire's current. Again, he found nothing. But he noticed when he removed the bar magnet from the roll, his instruments detected a faint current moving through the wire while the magnet was moving through the cylinder. As he withdrew the magnet, current moved in one direction, and as he put it in again, current flowed through the wire in the opposite direction. When he stopped the magnet's movement, no current. By repeating this action over and over, he thus produced a crude alternating current in the wire. He further showed it needn't be the magnet that moved. He could leave the magnet stationary and move the wire-wrapped cylinder back and forth and the current would likewise flow, altering its direction in synchrony with the back-and-forth movement. It was as though the moving magnet was pushing and pulling the metal wire's innate charge—what Franklin would call its electrical fluid—back and forth in the wire. Faraday was thrilled. In his own words:

> Ørsted showed how we were to convert electric into magnetic forces, and I had the delight of adding the other member to the full relation, by reacting back again and converting magnetic into electrical forces.[11]

What Faraday had discovered was a new way to generate electricity. He called it "electromagnetic induction," and it joined the list with static electricity generators and electrochemical batteries as yet another means for humans to produce electricity. Electrical induction is now the major way humans convert mechanical energy into electrical energy. Virtually all electricity in use today can trace its source to induction. This is true even for the batteries in our electronic devices, because we use electricity generated by induction to recharge them.

Faraday next tried to produce a similar induction effect by using a rotary motion. He took a copper disk revolving on an axle, just like an old vinyl record spinning on a turntable. At its edge, he mounted a classic horseshoe-shaped magnet so that the edge of the spinning copper disk passed through the gap between the magnet's two poles without actually touching the magnet. He found that when he hand-cranked the instrument, to get the copper disk spinning, he got an electrical current output from the spinning disk. This was the first example of a type of machine that would later come to be known as a *dynamo*, a mechanical device that generates electrical current. In this case, it generated DC from the rotary motion of a metallic conductor moving near a permanent magnet fixed in position.

As we've seen previously in the case with Volta's battery and its subsequent development, instrument makers often picked up and ran with scientific inventions, even when they really didn't understand the underlying science. The same was true for Faraday's discovery of electrical induction. Hippolyte Pixii, a young French instrument maker, accidentally built an early-model AC electrical generator, based on the principle of electromagnetic induction. His instrument was similar to Faraday's dynamo, except in Pixii's design the magnet wasn't fixed in position but was spinning around itself.[12] The AC it generated was a consequence of having the magnet spin, pole over pole, in the vicinity of the coiled wire. The changing polarity of the spinning magnet in relation to the coil caused the current to change directions in synchrony with revolutions of the magnet. But Pixii saw no practical use for such an alternating current. So he solved the "problem" of AC generation by adding a commutator to his instrument. A *commutator* is an electrical switch that periodically changes the route of the current in synchrony with each cycle of the generator, thus always keeping the current output moving in the same direction. Or more simply stated: It changes AC into DC.

Pixii was not the only one who didn't see any value in generating AC. No one did. Well, perhaps "no one" is an overstatement. In 1849, Guillaume Duchenne, the neurologist and electrophysiologist we met earlier, took an interest in electromagnetic induction machines as an alternative to batteries for electrotherapy, and he initially reported better patient outcomes for induction machines.[13] Yet, in his further experiments shocking frogs and rabbits and in treating patients, he saw no major differences in physiological responses between electricity produced from batteries (DC) or electricity produced from induction machines, either with direct or alternating currents. On a volt-to-volt and amp-to-amp basis, AC and DC seemed equivalent in their ability to stimulate muscle contractions. He concluded AC and DC were "perfectly comparable" for medical treatments.[14] Nevertheless, Duchenne reported he preferred using induction machines to batteries. He was won over by the utility and versatility of hand-operated electromagnetic induction technology, even if it didn't provide any particular physiological treatment advantages for his patients.

But exactly how did electromagnetic induction work? No one yet knew, but Faraday believed the secret lay in interacting "force fields" between the magnet and the electric current. The problem was that force fields were no easier to understand than electrical fluid, and they were equally invisible to humans. That is, force fields were usually invisible. But there was a way to visualize a magnet's force field, if you had some iron filings.

Magnets, of course, attract iron. If you file down a piece of iron, you can produce iron filings that look like a coarse black powder, which are

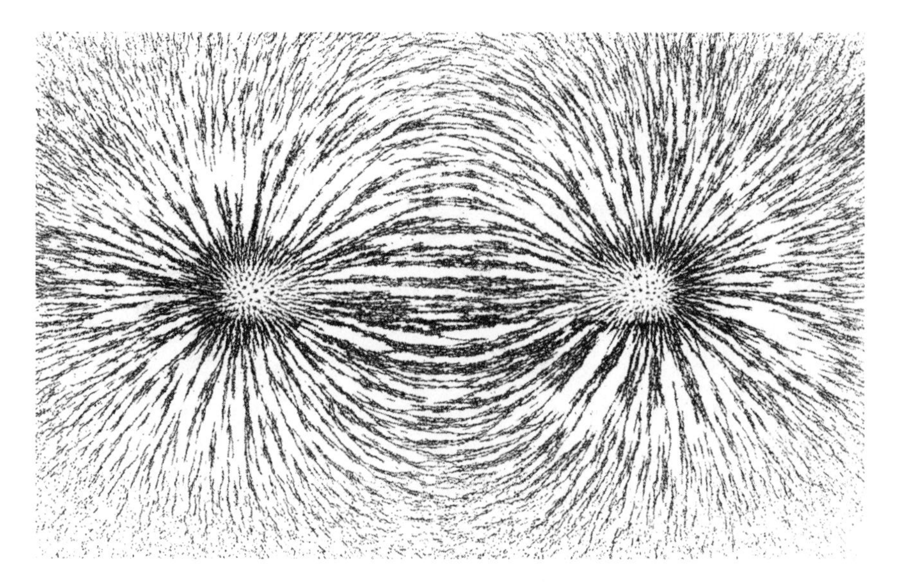

Fig. 7.2. Lines of force. Magnets attract iron. If you file down a piece of iron, you can produce iron filings that will be attracted and stick to a permanent bar magnet. But if you lay the magnet on its side, place a piece of paper over it, and sprinkle iron filings on top of the paper, the filings will spontaneously align themselves into a curious pattern (shown here). The filings form curved lines, looping away from the axis of the magnet and then back in, spanning from one pole of the magnet to the other. This was a long-known, but unexplained, physical phenomenon, until British scientist Michael Faraday introduced the concept of "lines of force" in 1846. We now know that lines of force define the shape and direction of a magnet's magnetic field and that electrical fields also exhibit lines of force. (Phil Degginger/Alamy Stock Photo)

likewise attracted to a magnet. But here's the interesting thing: If you lay the bar magnet on its side, place a piece of paper over it, and sprinkle the iron filings on top of the paper, a curious pattern will appear. The filings will form curved lines looping away from the axis of the magnet and then back in, from one pole of the magnet to the other. This was a well-known but unexplained phenomenon in Faraday's day.

Seeing such a phenomenon, you could interpret it in one of two ways. It could mean that iron filings, when subjected to a magnetic force field, assemble themselves single file into outward looping lines. In other words, forming curved lines is an intrinsic property of iron filings. Or you could interpret it to mean that there are preexisting force fields taking the form of invisible looping lines and the iron filings are simply aligning themselves along the preexisting lines of force, thus making the lines of force observable. This may sound like double-talk, but actually it's not. The different interpretations have very different implications for how magnetics exert their force.

Let's use an analogy of a bank lobby to explain this concept. When a bank opens in the morning and customers rush in to do their financial business, they always form straight parallel lines in front of the teller windows. Why?

Perhaps it's because the room previously has been set up with some mysterious "invisible" roped stanchions that only the bank customers can see. These invisible stanchions direct the customers into lines. In other words, the lines of people form because of a physical property *intrinsic to the bank.*

Alternatively, there may be no roped stanchions in the bank lobby at all, and the people just naturally assemble themselves into single-file lines when they enter a bank because they know it to be the most orderly and efficient way to get to a teller. In other words, line forming is just what people do; they don't need to be herded into lines by any outside influences like stanchion ropes. Under this scenario, forming lines is simply a natural behavior *intrinsic to people.*

Most scientists believed the visible lines of iron filings simply reflected the behavior of iron filings when in the proximity of a magnet; that is, the scientists contended *there are no stanchions*. Faraday believed, however, that actual magnetic lines existed (i.e., there are invisible lines of force that surround the magnet) even in the absence of iron filings. And these lines, taken together, constitute the magnet's *force field*. The iron filings just make the force field visible to us. Stated another way, Faraday believed magnets actually produce invisible stanchion-like lines of force that can be visualized by things, like iron filings, that can "feel" the forces and respond accordingly. The filings themselves have no way of forming their own lines; they are simply pushed and shoved into line formations by the force field they happen to be within.

The lines of force provided Faraday with an explanation of how a wire moving by a magnet generated an electric current. He postulated that energy was imparted into a wire every time the wire crossed one of these magnetic lines of force. Thus, a wire moving near a magnet, or a magnet moving near a wire, would result in the induction of an electric current in the wire proportional to the number of lines of force it crossed. This is a difficult thing to conceive and was intuitively obvious only to a mind as great as Faraday's, but we mortals might think of it in another way. To help us understand, let's try another analogy, this time using speed bumps to replace bank lines.

If you've ever driven on a highway with tolls, you probably have had an experience with "rumble strips" in the roadbed, which are usually positioned just before an upcoming tollbooth to shake your body to attention. Your car is riding along smoothly, and then you suddenly encounter a series of parallel speed bumps crossing the road. The car's tires absorb more and more force with each bump they cross, and soon the whole car is vibrating, alerting you to slow down for the upcoming toll. For Faraday, the moving wire is represented by the moving tires, and the lines of magnetic force are the speed bumps. The speed bumps each convey a lot of kinetic energy to the moving car. But when the car isn't moving, there is no energy conveyed because it crosses no speed bumps. Is everything perfectly clear now? Perhaps not. If you are still having trouble with the concept of lines of force, you aren't alone. In fact, Faraday took a lot of abuse when he first revealed his "silly" idea, largely because of the way he revealed it.

Faraday didn't intend to fully disclose his concept of lines of force until he had the experimental evidence to support it. The few times he had hinted at his idea to others resulted in criticism, so he was cautious. But one day he extemporaneously revealed what he was thinking to the world.

His disclosure occurred on April 3, 1846, during the Royal Institution's Friday Evening Discourse, a lecture series Faraday regularly attended. The speaker of the day, Charles Wheatstone, was to talk, at Faraday's invitation, about his latest invention: a device that measured extremely small time intervals. But Wheatstone suffered a bad case of stage fright just minutes before he was to talk and fled the building, leaving an audience of considerable size without any speaker. To get out of the awkward situation, Faraday decided that he, though unprepared, would address the audience as best he could. As he spoke, he took the audience on what amounted to a stream-of-consciousness adventure through the physical world as he saw it, a world filled with lines of force. These invisible lines of force crisscrossed space in all directions, and matter existed at the spots where the lines of force intersected. A strange world indeed. He even went so far as to say he thought gravity itself was likely the result of such lines of force. To many in the audience, this seemed like a direct assault on Newtonian theory. Faraday had gone too far and crossed a sacrosanct line: Newton's theory of gravity. Needless to say, his message was not well received by the scientific community at large. The criticism came fast and furious. By speaking off-the-cuff, Faraday had opened himself to ridicule.[15]

In an effort to walk back his lecture remarks on lines of force, he subsequently published an explanation in an article in *Philosophical Magazine*, in which he confesses:

> I think it likely that I have made many mistakes, . . . for even to myself my ideas on this point appear only as a shadow of speculation, or as one of those impressions of the mind which are allowable for a time as guides to thought and research. . . . [The experimental scientist has many such ideas, yet he] knows how often their apparent fitness and beauty vanish before the progress and development of the real natural truth.[16]

As it turned out, Faraday's ideas about lines of force would not vanish. They *are* the real natural truth!

Faraday's fundamental professional problem was that he was his own worst enemy when trying to convey his ideas to other physicists. This is because physics at the time was in the process of adopting a language that Faraday couldn't speak. Mathematical jargon and equations were fast becoming the dialect of all branches of physics, but Faraday, who had never attended school beyond the age of 14, had little understanding of mathematics. In fact, his publications are remarkable in that they contain virtually no mathematics. Every scientific principle he formulated was conveyed pictorially and by description. To make matters worse, Faraday's descriptions of physical phenomena typically employed his own idiosyncratic terminology (such as "lines of force") that had meaning only to him.

Not understanding the meaning of Faraday's novel terminology, it was hard for other physicists to decipher what Faraday was trying to tell them. In contrast, ideas conveyed in the universal language of mathematics could be easily understood between physicists. And the implications of those ideas could be readily tested because mathematical relationships and formulas could be definitively proved or disproved, but fantastical stories of lines of force crisscrossing empty space could not. Unfortunately, Faraday was speaking largely to an audience of one—himself.

Sadly, Faraday well understood he was becoming a dinosaur among physicists. He knew he was one of the last remaining from the prior generation of physicists, where science could be performed without an education grounded in mathematics. He was still working like Franklin had—conducting experiments and describing his experimental findings. But what worked for Franklin was not working for Faraday. These were different times.

But then a precocious young physicist, who happened to be mathematically gifted, took an interest in Faraday's work. He was a Scot named James Clerk Maxwell. Maxwell came from an aristocratic background and was highly educated, first at boarding school in Scotland and then at Cambridge University in England. He was particularly skilled in applying mathematical principles to physics problems. So, while Faraday may have envisioned lines of force crisscrossing space, Maxwell imagined numbers and equations. With some trepidation, the unknown Maxwell sent the famous Faraday a letter expressing admiration for his work, and he enclosed a manuscript that translated Faraday's ideas into mathematical equations. To Maxwell's relief, Faraday was overjoyed that someone was able to successfully achieve what he himself lacked the mathematical skills to do, and he responded to Maxwell's letter enthusiastically: "I hang on your words because they are so weighty and . . . give me great comfort."[17]

Maxwell's equations had provided the mathematical voice that Faraday lacked. In 1855, Maxwell published a paper, "On Faraday's Lines of Force."[18] In it, Maxwell showed that Faraday's lines of force were entirely analogous to the well-established lines of tension in fluids.[19] The mathematics of the two phenomena was the same, suggesting both were representations of some fundamental law of nature. Faraday must have grimaced at the analogy to a fluid, since he was trying so hard to guide electrical science away from the invisible "fluid" idea, but he was surely overjoyed that what he was describing in his pictorial model held up to the rigor of a mathematical analysis. And each approach reinforced the other. The experiments suggested a theory—a theory that could be mathematically modeled—and the model could be validated by further experimentation. It is exactly through such iterative approaches that science advances. Maxwell, by applying mathematics to Faraday's ideas about force fields, had legitimized those ideas in the minds of other physicists, and soon everyone had to accept the notion that a flowing electric current produced its own force field, which took its physical form as lines of force.

Still, some remained confused. They couldn't understand how the mechanics of a fluid was in any way related to electricity. But Maxwell chided them that a mathematical model is merely an aid to thinking about a scientific problem. He explained that his mathematical treatment of Faraday's ideas attempted to show how "the connection of the very different order of phenomena which [Faraday] has discovered may be clearly placed before the mathematical mind."[20] The "fluid" part simply allowed for the problem to be explored by way of a mathematical analogy. Electricity is *not* a fluid. The fluid is a useful mental ploy, not a physical reality.[21]

It would have been possible for Maxwell to have picked a different mathematical model, such as the movement of heat through a piece of metal, to support Faraday's lines-of-force ideas. But the fluid mechanical model was appealing in that people were already conceptualizing electricity as an invisible fluid, thanks to Franklin's work, so the shift in thought didn't need to be all that great. In Maxwell's model, the fluid movement was due to pressure differences along a tube, moving from a high-pressure sink to a low-pressure sink. The pressure difference was analogous to electrical voltage, and the rate of flow was analogous to *electrical flux*, the speed an electrical field moves through space. The pressure gradient, simply defined as the fall of pressure per unit length along the tube, Maxwell called the *field intensity*. The field intensity (or *field strength*) was equivalent to the force of the electric field, expressed as volts per meter.[22] This model gave mathematically minded physicists the access to explore Faraday's ideas further.[23]

With Maxwell's help, Faraday had won the day despite all of his critics. But Faraday wasn't dismayed at the rancor often displayed among physicists, each trying to promote his own ideas about the laws of nature. Faraday saw it as just part of the normal scientific process, and once remarked:

> Is it not wonderful that views differ at first? Time will gradually shift and shape them. And I believe that we will have little idea at present of the importance they may have in ten or twenty years hence.[24]

Such was exactly the case with his discovery of AC. Although Faraday first demonstrated the induction of AC in a wire when he moved a bar magnet in and out of a hollow cylinder coiled with wire, he saw no practical use for it. And Pixii had found AC to be a nuisance that needed to be corrected when his newly invented spinning-magnet electric generator started producing it. After that, the concept of actually using AC for something lay fallow for many years, until Nikola Tesla showed that high-voltage AC provided tremendous advantages for long-distance power distribution.

It is well established that the most efficient way to transmit electrical power over long distances is to use very high voltages at low amperages.

That may not be intuitively obvious, but based on what we've already discussed in the last chapter, we can deduce that it should be the case. Let's pause here to consider why.

Most people use the term *power* quite sloppily, but physicists and engineers specifically define power as the ability to produce work over a specified time period. The easiest way to think of power output is when it is expressed in terms of *horsepower* units. The modern definition of one horsepower is specifically defined as 550 foot-pounds per second, but the unit was originally chosen because it roughly equated to the power output of an average draft horse, a breed of large horses used to pull heavy wagons. The unit was frequently used to assess the power output of steam engines. For example, when someone said a steam locomotive had a 20-horsepower engine, people had a mental picture of the load size that the locomotive could pull and the number of draft horses it could replace. No wonder an early nickname for the steam locomotive was the "iron horse"!

Using horsepower, however, to measure electrical power coming from things like lightbulbs and handheld hair dryers isn't quite as intuitive, is it? For example, what would it mean to have a one-eighth horsepower hair dryer? It doesn't produce a very vivid picture. You could use horsepower units if you wanted to, but most people prefer to use the unit of *watts* to express electrical power output.[25] The watt is named after James Watt, a prominent eighteenth-century engineer.[26]

Many people are familiar with the watt unit, because the power output of classic incandescent lightbulbs is measured in watts, with 60- or 100-watt bulbs commonly preferred for normal indoor household lighting. A watt doesn't give any better mental picture than horsepower does when applied to a lightbulb, but the watt has a major advantage over horsepower when used in electrical power calculations. That's because a watt is simply calculated by multiplying the number of volts by the number of amps—two readily measurable electrical parameters. This makes watts very simple to employ when trying to measure electrical power output in real-life situations . . . or even real-death situations. As you may recall from our earlier discussion of the electric chair, New York State's first electrocutioner, Edwin Davis, typically delivered

1,700 volts and 7 amps to condemned prisoners. That would amount to a power output of 11,900 watts, which Davis found to be sufficient power to kill a man or, alternatively, enough power to light 119 100-watt lightbulbs, if you can picture that. Let's consider the implications of this simple multiplication for a moment:

$$1{,}700 \text{ volts} \times 7 \text{ amps} = 11{,}900 \text{ watts}$$

As you can see from the equation above, a defined level of power can be achieved either with high voltage at low amperage or low voltage at high amperage. It makes no difference in terms of the power output. But electrical resistance within the transmission wires causes a loss of power due to heating, and the heat loss, as we've already seen, is a function of the amperage and not the voltage. Therefore, if the amperage is very high, so is the amount of power lost as heat.

As we've seen before, the heat produced by a current moving through a conductor is proportional to the square of the current (amps × amps, or amps^2) multiplied by the overall resistance of the conductor. So, if you want to transmit a lot of electrical power over long lengths of wire, you can minimize the heat loss problem by keeping the amperage levels as low as possible. And the only way to do that, while maintaining the same power delivery level, is to greatly increase the voltage. Hence, we end up with high voltage/low amperage power lines as the best way to go. Such a strategy minimizes the power lost to heat. This is true regardless of the type of current, either AC or DC. High voltage is definitely preferable no matter what type of current you're transmitting.

In the days of Edison and Westinghouse, power stations could easily be designed to produce either high-voltage DC or AC. The problem was that bringing high voltage into homes was unacceptable. Apart from the dangers of high voltage, few appliances could use such high voltages. In fact, Edison's incandescent lightbulb could handle no more than 110 volts, which is largely the reason American household voltages are still standardized to 110 volts today. Consequently, if you were going to use high voltages to transmit electrical power, the voltage needed to be stepped down before it went into homes. For AC, there was no problem doing that. The Gaulard-Gibbs transformer, invented in 1882, could step down AC voltage before it entered the home.[27] But there was no equivalent voltage reduction device for DC.[28] And it wouldn't be until the mid-twentieth century that methods to step down DC voltage in commercial power transmission would be invented. Therefore, if you wanted low-cost electrical power transmission with low-voltage home delivery, high-voltage AC electricity generators coupled with step-down voltage transformers were the name of the game.

Westinghouse's AC power transmission lines ran at very high voltages but with lower amperages, to minimize the power lost as heat. Edison, in contrast, used lower-voltage and higher-amperage DC lines without transformers, but was finding it nearly impossible to transmit electrical power any great distance because of the great power losses. Thus, there is some truth to Edison's claim that the high-voltage AC transmission lines present an increased lethal hazard, notwithstanding that amperage is what kills. Even though the amperage is relatively low as power transmission goes, it is still high enough to kill people. And as we've already seen, it doesn't take much amperage to kill.

Another important hazard factor was that Westinghouse chose to run his AC transmission wires above the ground on poles, rather than to bury them as Edison did. It is much cheaper to erect poles than to dig trenches. But Westinghouse's electrical lines would sometimes fall to the ground and electrocute people, while Edison's lines never did, precisely because they were underground. Thus, whenever there was a power-related electrocution reported in the newspaper, it was typically one of Westinghouse's lines that was responsible. That was because his falling wires were more easily accessible to the public, not because AC was intrinsically more deadly than DC. If Edison had run his DC lines aboveground on poles, a similar number of accidental electrocutions would have occurred. But the public understood none of this. People came to associate accidental electrical deaths more with the type of current than with the proximity of the wires. Thus, it was no accident that they blamed AC for the deaths. The public had been brainwashed.

Edison didn't come up with the notion of AC being inherently more dangerous than DC, but he played a major role in promoting the idea. The person who actually originated the idea was Harold Pitney Brown. Brown was a young engineer, and a self-taught one at that. Nevertheless, his lack of medical education didn't prevent him from asserting in a letter to the *New York Evening Post* that AC was too much of a health hazard for residential use.[29] He claimed the rapid change in the direction of the current is what made AC lethal, and suggested it be banned. His letter caused quite a stir.

Critics claimed there was no evidence to support Brown's claim, and indeed Brown didn't seem to have any. So Brown contacted Edison and asked for the use of his laboratory facilities to demonstrate the lethal nature of AC. Edison was all too happy to oblige. Not only did he provide Brown with laboratory facilities, but Edison also assigned his chief electrician to assist Brown and help him acquire a supply of large dogs for his experiments.

But Brown's "experiments" turned out to be more public demonstrations of dog electrocutions than science, with some demonstrations attended by as many as 800 spectators. Typically, they took the form of shocking a single dog multiple times with DC current at various voltages, resulting in the dog exhibiting severe convulsions and apparent pain, but not killing the dog. When audience members called for him to stop the spectacle of torture, Brown would quickly put the dog out of its misery with a single jolt of AC, supposedly to show that even a single AC shock was lethal.

But one highly credentialed spectator to Brown's dog electrocution events decided to bring down Brown's charade. Dr. Peter H. Van der Weyde was a physician and medical professor at the New York Medical College, a professor of physics and chemistry at the Cooper Institute, and an expert on electricity. He claimed it wasn't surprising that the final AC jolt was fatal to the dogs, since they had been previously shocked to near death by multiple prior DC jolts. Van der Weyde further suggested Brown was guilty of deception, since he surely must have known that by only reporting the voltage he was delivering to the dogs, he was obscuring the fact that voltage without sufficient amperage would cause pain and suffering but not be lethal. He pointed out that, in Brown's experiments, two steam engines were used to drive two dynamos producing the AC, while just one steam engine with one dynamo was used to produce the DC.[30] So, for any given voltage level, the dogs were receiving twice as much AC amperage as DC amperage. No wonder the AC was killing the dogs. Current is the killer, and the dogs were receiving at least twofold more AC current.

Then in the July 13, 1889, issue of *Electrical World*, Ludwig Gutmann, a prominent electrical engineer, published a devastating critique of Brown's experiments, in which he enumerated the many ways Brown had stacked the deck within his experiments to make AC current appear to be more dangerous.[31]

Shortly after this, Brown's reputation was impeached further when someone gave the *New York Sun* a stash of Brown's private correspondence. The letters provided strong evidence that Brown was in cahoots with the Edison Electric Company and that his attacks on AC were financially motivated. Brown was in Edison's pocket and, therefore, was biased.[32] The newspaper published the letters in its Sunday edition, on August 25, 1889, under the headline

FOR SHAME, BROWN!
Disgraceful Facts About the Electric Killing Scheme

While this controversy was going on, Brown and Edison simultaneously pushed their anti-AC agenda. They had reasoned that the ultimate demonstration of AC's lethality was not killing dogs but killing people. So they lobbied prison officials to power the newly invented electric chairs with AC rather than DC, arguing AC was superior to DC in producing an instantaneous and painless death. If electric chairs were powered with AC, it would become inextricably linked with death in the minds of the public, just like the guillotine had been inextricably linked with the French Revolution. Edison and Brown reasoned that the mental link alone might be enough to persuade the public to reject AC power transmission as being far too lethal to let into their homes. As we've already seen, that's largely how the story played out. Kemmler, the first prisoner to be electrocuted, was killed with AC, due in part to Edison's and Brown's lobbying. The fact that Kemmler actually didn't die suddenly was initially blamed on the AC, but Edison quickly steered that conversation from current type to electrode placement, and AC has remained the standard for powering electric chairs to this day.

When Brown was asked why he thought AC was supposedly so lethal, he contended it kills by vibrating body fluids to the point that internal vital organs are destroyed. An interesting hypothesis; how unfortunate that he never tested it. As for Edison, he acknowledged he didn't really know the mechanism of AC's lethality, but he speculated it might have something to do with excessive excitement of the nerves and muscles—another interesting but uninvestigated claim.

So, what is the truth? Is AC inherently more dangerous to people than DC? At high power levels, there seems to be a consensus among medical experts that they are equally deadly. High power kills no matter how it's delivered. At lower power levels, however, it isn't as clear.

You can still find the debate raging on the internet, with some people claiming AC is more dangerous and others claiming DC is the one to worry about. Each side proposes various physiological mechanisms to support its position. For example, some DC critics claim it produces more tissue electrolysis, while some AC critics say it's more likely to affect the heart's natural pacemaker. But there seems to be a reticence to point to any hard data, likely because there are few to none. The problem with the AC versus DC hazard question is that it is always couched with "all else being equal," and all else is never equal.

The issue is that, in changing its direction, AC also alters a number of other current parameters. First, AC current must stop momentarily while it changes direction—once per cycle, or 60 times per second. What this means is that AC's current is intermittent in flow as well as alternating in direction. So, if there really are differences between AC and DC, are they due to the alternating direction of its current or to its intermittent flow? Duchenne realized this problem. To try to equalize everything and focus just on the alternating direction when he compared the two, Duchenne interrupted the DC current flow in synchrony with the AC current cycles, such that both AC and DC were equally intermittent, supposedly leaving only the current's direction as the difference between the two.

We can see that Duchenne demonstrated an impressive level of experimental sophistication. He understood how DC and AC currents worked, and he tried to control the variables. But even his experimental setup didn't capture all of the complexity. In addition to switching direction and being intermittent, AC's voltage and current also oscillate during its cycle.[33] For DC current, in contrast, the voltage and amperage levels remain constant over time. As you can see, AC and DC current differ in many ways, and it's unclear which of these many parameter differences may be relevant to the lethality question. But it is largely an apples-to-oranges comparison question and, therefore, very difficult to answer without qualification and equivocation. To some extent, AC versus DC safety will always be an apples-to-oranges comparison.

But in the end, it doesn't really matter. The AC versus DC safety issue never really was a scientific question. It was politicized from the very moment it was conceived, and it has remained so ever since. In practical terms, it makes little difference which current type is more dangerous. The question itself is, in many ways, nonsensical, and the measures taken to protect us from AC hazards protect us from DC hazards as well.

We are still, however, weighing the technological value of AC versus DC. As mentioned earlier, the increasing growth of electronic appliances keeps driving the need for DC, which requires that power transmitted in AC form be converted to DC at the site of use. Since the local conversion process is so inefficient, a significant amount of the energy savings afforded by long-distance transmission by AC is then lost by local AC-to-DC conversion.

Some believe that the solution to this dilemma is the DC *microgrid*.[34] Microgrid systems can be utilized in those industrial settings where electronics are in heavy usage, such as computer-based businesses.[35] The power savings are realized by replacing the inefficient individual converters with more efficient centralized converters along with small power grids to distribute the DC current throughout the facility. Others have suggested a more wholesale conversion from AC to DC power transmission is warranted; in effect, what Edison had wanted all along. Now that DC "transformers" are available, AC's former technological advantages are no longer what they were. Still others think it is much too late for a DC power transmission takeover. They note the AC power transmission has been the standard for so long that the nation's entire electrical power industry infrastructure would need to be replaced for a wholesale conversion from AC to DC power transmission to be achieved, requiring perhaps a century to realize any return on such investment. So, the AC/DC war may be long over, but there are plenty of skirmishes still going on.

Fortunately for Michael Faraday, he didn't have to suffer through the debacle of the AC/DC war and its highly sensationalized safety debate. He died in 1867, about 10 years before the war started. And Faraday had no electric power in his home that might electrocute him, either AC or DC, since there were no home appliances that required electricity in his day. All of that stuff was still a long way off. Faraday would have had little interest in it anyway. He always found the commercialization of science to be a bit unseemly.

Faraday was more interested in a scientific suggestion voiced in a letter he had received from another admirer, Alexander von Humboldt, the great German explorer and scientist who had encountered those horse-shocking electric eels in South America. Humboldt, like Maxwell, had been following Faraday's work very closely. In his letter, Humboldt praised Faraday on his work with electromagnetic forces and told him, "I am convinced that with the knowledge of electro-magnetism and physiology that we possess today, the study of the phenomena of the *Gymnotus* [an eel-shaped electric fish] should shed strong light on the function of nerves and muscular movements of man."[36] He further coached Faraday on how he might obtain some *Gymnotus* specimens and how to keep them alive in the laboratory. Faraday was intrigued.

After encountering many difficulties, Faraday ultimately was able to get a 28-inch live specimen that had been imported to England from South America. He set straight to work.

Having discovered the existence of electromagnetic lines of force, Faraday wanted to see if something similar was going on in electric eels. It was always important to him that electrical "laws," or principles, be universal, and not restricted to particular systems. Electrical principles should hold true for static electricity, battery electricity, electromagnetically induced electricity . . . and even electricity produced by electric eels. If they were actually the same phenomenon, they should all produce the same results. So the eel he obtained allowed him to test the universality of the electrical theory in his own laboratory.

Gymnotus was the same South American eel in which John Walsh first demonstrated, in 1775, that an electric fish could produce a visible spark just like amber, supporting the notion that electric fish actually produced the very same electricity as produced by amber. Faraday was able to replicate Walsh's sparking results. Now Faraday would use the eel to show an animal could also produce an electromagnetic field.

His experimental design was to put the eel in a pan-shaped glass container, having a diameter of about twice the fish's length, and to fill the pan with just enough water to cover the eel's body. A good mental image of his setup would be to envision a six-inch sausage (representing the eel) submerged in water an inch deep, in the middle of a 12-inch frying pan. Now hold that image of the sausage in the pan and further imagine that the sausage has been equipped with two little saddles, like western-style horse saddles, that straddle the sausage at each of its ends and hold it in place, so it can't roll around in the pan. This is exactly the type of setup you would need to determine whether sausages emit electric fields. Sausages don't, of course, but electric eels might. So Faraday substituted the sausage with an electric eel and tried to detect an electric field in the pan.

Lacking suitable devices to measure electric current distribution within water in the pan, Faraday used his own finger as a probe of the eel's force field. He stuck his finger in water at various positions in relation to the eel's body. Even though the eel wasn't actively releasing a jolt of electricity at the time, Faraday still could feel a tingle from its electricity in the surrounding water. The closer he put his finger near the midpoint of the eel's body the stronger the tingling sensation. As he moved his finger away, the feeling became less intense but was still present. By moving his finger to various positions in the pan and recording the tingling level at each location, he was able to construct a very crude map of the electric field around the eel's body. And yes, there seemed to be lines of force, from the fish's head to its tail, and these lines seemed to loop out and back in toward the ends of the fish, similar to what happens with iron filings and a magnet. But then Faraday went further.

Recall those little western-style eel saddles again. Now imagine that the top knob typical of western saddles has been replaced by antennae-like rods that protrude above the surface of the water and to which electrical wires can be attached. Faraday decided to use the rods as though they were battery terminals and the eel was the battery. He took a bird's wing feather and cut away the feathery parts, leaving only the hollow tube of the quill itself (something we might achieve today by simply substituting the quill with a plastic drinking straw). He then cut off a length of about an inch, producing a 1-inch-long hollow tube. He wrapped the little tube with tightly coiled fine wire and attached the ends of the wire to each of the saddle electrodes. Next, he stuck a nail through the little hollow tube and tested it for magnetism by assessing the nail's ability to attract small metal objects. When placed inside the tube, the nail did indeed become magnetized. Although the magnet was tiny and its magnetic force weak, it was still the first demonstration of an electric eel–powered electromagnet.

Faraday was thrilled with his electric eel findings. These eel experiments and others like them convinced him that the electricity emanating from a battery and the electricity emanating from an electric eel were exactly the same force of nature. They both behaved the same way, in everything from making sparks to forming force fields to powering electromagnets. And he came to agree with Humboldt that electrical experiments had the ability to reveal much about the human body and how its nervous system works. When he published his eel findings—yet another of his papers containing no mathematics—Faraday gushed:

"Wonderful as are the laws and phenomena of electricity when made evident to us in inorganic and dead matter, their interest can bear scarcely any comparison with that which attaches to the same force when connected with the nervous system and with life."[37]

Yes, Faraday had learned much about the electricity and lines of force emitted from electric fish. But he began to wonder, as was his way, whether the whole process could be run in reverse. What would happen if a man-made electric field were applied to a fish? An interesting question indeed.

8

ZOMBIE FISH

ELECTRIC FORCE FIELDS

> And the sportsmen were mentally on tiptoe to out-think the fish—which they sometimes do.
>
> —JOHN STEINBECK

I lean out over the bow of the boat with net in hand, peering into the water and waiting to see what swims to the surface. I'm *electrofishing*.[1] In case you've never heard of it, electrofishing is a method of catching fish that involves stunning them by running an electrical current through the water and scooping the stunned fish up with a hand net as they come up to the surface in an electricity-induced stupor.

I've been at it for only three minutes when I spy my first fish. A big yellow common carp (*Cyprinus carpio*), a distant wild relative of the domesticated goldfish (*Carassius auratus*), swims up from the murky bottom and heads straight toward the positively charged electrode. The electrode dangles just below the water surface, suspended by a chain from the end of an 8-foot (2.5 m) pole projecting horizontally from the front of the boat, about a foot (0.3 m) above the water surface. As the carp approaches the electrode, it turns on its side and appears paralyzed, which is a sure sign it has succumbed to *electronarcosis*, a fancy word for being stunned senseless by electricity. I snatch the fish from the water with my net and transfer it into the boat's live well, a water-filled box designed to keep the caught fish cool and well oxygenated while they recover. I resume fishing, but keep an eye on my carp to ensure it's all right. It is. Within just a few minutes, it is finning around normally, like nothing happened.

Electrofishing is not a sportfishing technique. It is a research tool used by fisheries scientists to assess the demographics of fish populations in lakes.[2] The scientists electrify an area of water with submerged electrodes and collect all the fish that succumb to the electricity. They record the numbers of fish caught, by species and size, in representative habitat areas and then extrapolate their findings to the whole body of water. In this way, they get a rough estimate of the population sizes and age distributions for the various species of fish that inhabit the lake.[3] When done correctly, the fish suffer no long-term damage, and they are quickly returned to the water to resume their business once the data have been logged.

Electrofishing works very well. As a lifelong devotee of fly fishing—the sportfishing technique notorious for its inefficiency—I am simply astounded at how extremely effective electrofishing actually is. But don't even think about trading in your fishing rod for electrofishing equipment. Recreational electrofishing is illegal almost everywhere. Even fisheries scientists must obtain a permit from state and federal fish and game agencies when they want to use it, and they are closely monitored by those same agencies to assure no fish are abused. Typically, there are regulations limiting the maximum voltages and shock durations that can be used for the particular species being shocked, with the goal of zero fish casualties. Salmonids, a group including the various salmon and trout species, receive special protection because they have very fine bones, including their vertebrae (backbones). If electric stimulation is too high, the induced muscle contractions can be strong enough to break their fragile vertebrae. As a further protection for the fish, the electricity is delivered in short bursts, called *pulses*, each lasting no more than 5 milliseconds (1/200th of a second), and the frequency of the bursts is usually limited to under 30 per second (30 hertz).[4]

Although properly conducted electrofishing may be safe for fish, electrofishing done incorrectly is dangerous to both fish and humans. That's why people need to be expertly trained before they use electrofishing equipment. My little electrofishing excursion on the lake actually is part of a training exercise being administered by Patrick Cooney, a professional fisheries scientist and the director of Electrofishing Science at Smith-Root, Inc., a leading manufacturer of electrofishing equipment based in Vancouver, Washington. I recently had been picking Cooney's brain over the phone about the details of electrofishing and how fish respond to it. He graciously invited me out to Vancouver to take the electrofishing training course he periodically teaches to users of his company's equipment so that I could get some firsthand electrofishing experience. I enthusiastically accepted his offer, and that's how I ended up in this electrofishing boat on Lacamas Lake in Washington.

I've been curious about electrofishing ever since I had learned something that electrofishers have known since 1885.[5] When fish are within an electric field in the water, they will swim toward the positive electrode until they come close enough to it to become electrically stunned. In fact, the fish-attracting property of the positive electrode significantly contributes to electrofishing's effectiveness. Rather than swimming away from a source of electricity, as one might expect, fish seem to orient their bodies along the lines of force of the electric field and swim directly into the electric current, ultimately ending up at the positive electrode. Unfortunately for the fish, the electrofisher has slyly positioned himself near the positive electrode, waiting to snatch them from the water as soon as they come within reach. The fish attraction capabilities of the positive electrode make the electrofisher somewhat of a Pied

Piper, but rather than rats being lured by flute music, it's fish being lured by electricity.

The positive electrode can even be used to draw fish out from their hiding places beneath rock overhangs and underneath docks. The exact mechanism of this electro-attractive phenomenon, called *electrotaxis*—the movement of an animal in a particular direction in response to the stimulus from direct current (DC)[6]—has been the subject of debate for over a century. Each generation of fisheries biologists has looked to the latest developments in neuroscience to come up with a new explanation. But so far, all of the proposed explanations have had their shortcomings. Nevertheless, one thing is clear: fish respond very oddly when they are subjected to an electric field in the water.

We need to take a break here and introduce a little electrical physics into our electrofishing discussion, just to keep us all on the same page with regard to what's meant by the "electric field in the water."

As we've touched upon already, electric and magnetic force fields have much in common. One might even say they actually represent the same physical phenomenon seen from different perspectives. Consequently, these two types of fields are often discussed collectively as *electromagnetic fields*. One major difference, however, is that while electric fields have endpoints—the positive and negative electrodes—magnetic fields do not. That's because magnets themselves don't have discrete ends. Let's take the example of a permanent magnet in the form of a bar of magnetized iron. Rather than terminating at the bar magnet's "north" and "south" poles, the lines of force of the magnetic field loop back around lengthwise through the interior of the magnet, so that, collectively, the lines of force take the form of doughnut-shaped ellipsoids. When you cut a bar magnet in half, you produce two smaller magnets, each with its own north and south poles and its own smaller ellipsoid lines of force. You can keep cutting the smaller magnets in half, over and over again, making the magnets smaller and smaller yet, until you get down to the level of just one iron atom, and it too will have its own magnetic field.

In contrast, an electric field is produced around any place where electric charge—positive or negative—is localized. A positive charge locus need not be paired with a negative charge locus to produce an electric field. But when they are paired, the two force fields interact with each other and produce connecting lines of force that look very similar to the magnetic lines of force you see when you sprinkle iron filings on a piece of paper over a bar magnet. If you pair two positively charged (or two negatively charged) loci together, the interacting fields will repel each other. The lines of force won't connect and rather will stream away from each other because of the repulsive forces between like charges.

Just as iron filings tend to line themselves up along the lines of force of a magnetic field, electric current tends to flow along the lines of force of an electric field. This sometimes leads to the confusion that the electric current flow is the electric force field. It is not. An electric force field exists between two electrodes even when there is no flow of current between them. So when electrofishing electrodes are in dry air—an electrical insulator—there is still an electrical force field between the electrodes. But when the electrodes are immersed in water, the conductivity of the water allows electric current to flow between the electrodes, and the current follows along the path of the electric field's lines of force.[7] Thus, the current's path and intensity match the electric field in shape and size. We cannot visualize an electric force field as easily as a magnetic force field, but it is present nonetheless.[8] It is precisely because the electric force field defines the path of the electric current flow that the term *electric field* is often used to mean both electric force field and electric current flow. You'll even catch me slipping into this more convenient unifying terminology as we progress through our story. But it's best to keep in mind the distinction between the *electric force field* and the *electric current.*

The reason to make the distinction is this: When fish enter an "electric field" and are shocked, they are not shocked by the electric force field; the electric force field is something their body tissues cannot sense. *They are shocked by the electric current,* which is running through the water along the fixed pathways defined by the electric force field—Faraday's lines of force. This is another subtle but very important distinction, because the biological effects of electric current are what interest us, not any alleged biological effects of electric force fields.[9] Physics interlude over. Time to get back to our fish story.

Cooney drives the boat while seated behind the steering console in the boat's stern, which gives him an unobstructed view of me and my fish-scooping technique up at the bow. He instructs me to scoop quickly and transfer the fish to the live well as soon as possible, to limit the time the fish is in the electrified water. "Make one sweeping motion with the net, lifting it up from the water and swinging it around to the live well, by twisting your torso while being careful to maintain your balance," Cooney advises. Grace of movement has never been one of my strong points. But Cooney is a kind and patient instructor, seeing my every failing as a teachable moment and encouraging me to incrementally move to ever-higher levels of electrofishing skill. He compliments me on every fish scooped, no matter how awkward my movements, and assures me I just need minor adjustments to my technique before I become a master electrofisher. I'm not so sure.

We continue down the lake's shoreline, shocking and scooping. Soon we have caught ourselves a sizable number of carp, as well as largemouth bass, smallmouth bass, largescale suckers, and yellow perch. Unfortunately, all but the largescale suckers are nonindigenous invasive species.[10] Cooney laments that the species native to the lake, predominantly salmon and trout, have largely been driven out, surrendering the water body to these unwanted transplant species. The only trout found in the lake are those stocked by the state for recreational angling. Electrofishing has revealed the extent of the invasive fish problem in the lake, although it hasn't provided a solution. It just isn't possible to electrofish all the invasive fish species out of a lake.

As an angler who mostly fly fishes for trout in streams, I've always been more interested in what's going on with the fish populations in the streams rather than in the lakes. Do the streams I fish have large populations of trout, or am I wasting my time on fishless waters? Fortunately, trout streams can also be electrofished to get that kind of information, but the electrofishing technique requires some modifications for use in a stream.

When electrofishing a stream, a boat isn't needed. Electrofishers just wade through the water with a battery-pack power source strapped to their backs. They carry a positive electrode mounted on a long pole, which they sweep back and forth in the water in front of them as they wade upstream. They resemble the folks you sometimes see at beach resorts who sweep metal detectors back and forth over the sand, hoping the device will reveal some type of metallic treasure buried there. However, in this case, the treasure is fish, not coins, and they don't have to be dug up; they swim right up to the positive electrode, where they must be scooped up quickly before the flow of water washes them downstream. Scoop too slowly and you'll miss the fish, resulting in a poor population sampling.

This is why electrofishers surveying a stream usually work in multiple teams of three. The trio comprising each team is typically (1) a shocker with the battery backpack, (2) a fish scooper with the net, and (3) a person trailing shortly behind the other two carrying a water-filled bucket to hold the catch. All of them wear rubber waders and gloves—rubber is a good electrical insulator—so they aren't shocked by the electric current flowing through the electric field in which they are literally knee-deep. In the event of a fall into the electrified water, the backpack has an automatic safety switch that instantly shuts off the current coming from the power source. With multiple teams spanning the width of the stream and moving upstream as a cohort, few fish escape, and a very accurate assessment of the fish population in the stream can be achieved.[11]

The fish densities of streams are usually reported as the average number of fish per mile (FPM), which is a fairly good indication of the overall health of the stream's fish population. However, anglers, like me, need to keep in mind that averages are just averages; it is still possible that specific miles of the stream are fish ghost towns. (I seem to have a

FIG. 8.1. Fishing with electricity. Electrofishers surveying a stream usually work in multiple teams of three. A team typically includes (1) a shocker wearing a battery backpack and carrying a long wand with a hooped positive electrode at its end, which the shocker keeps submerged midway down the water column; (2) a fish scooper with the net and bucket, who stands next to the shocker; and (3) a backup scooper trailing shortly behind, who nets any shocked fish the upstream scooper misses. Stream electrofishers also trail behind them a long, thick wire known as a "rattail," which serves as the negative electrode (seen here trailing behind the right leg of the shocker). The team progresses upstream at a slow pace, carefully probing all areas where fish may be hiding and capturing any fish immobilized by the electricity. (Photo courtesy of Jake Ponce)

particular talent in finding those fishless miles.) Departments of fish and game take ongoing FPM measurements as a fisheries management tool. It tells them which streams are healthy and which have problems that need their attention. Some even categorize their state's streams into Class A to F trout waters, based at least partially on FPM values. And most states post their streams' FPM values online, for the interested public to see. World-class trout streams can have greater than 20,000 FPM, while the worst trout streams can be as low as . . . well . . . 0 FPM, of course! (This is why I have trouble outsmarting fish.)

The basics of electrofishing are deceptively simple. Put electrodes in the water, turn on the current, and scoop up any fish that surface, all the while being careful not to shock yourself in the process. But the details are a bit more complex, and much more interesting.

With regard to the biology of electrofishing, electrotaxis has attracted the most interest. This odd behavior of the fish—swimming along lines of force toward the positive electrode and thus moving in a direction opposite to the electrical current flow—is crucial to the high catching efficiency of electrofishing. If you reverse the polarity of the electrodes, the fish will turn themselves around and move in the opposite direction. Simply amazing!

The German physiologist L. Hermann first reported this phenomenon in 1885. His studies were specifically done with larvae—juvenile fish that have recently hatched from eggs. Many other investigators soon confirmed his results with larvae and then extended those findings to adult fish of various species. Scientists have been trying to explain the phenomenon ever since.[12]

Theories about electrotaxis evolved in parallel with theories in neuroscience. Early electrotaxis theories centered on pain avoidance, since it was considered one of the main behavioral mechanisms underlying *stimulus-response theory*—an animal behavior theory very much in vogue at the time.[13] According to stimulus-response theory proponents, fish swim parallel to the current lines because it is the least painful orientation. The thought was that, if the fish's body were perpendicular to the electric field, it would feel more pain. But swimming parallel to the lines supposedly would minimize the number of electrical lines of force the fish's body would cross, thereby reducing its pain level. Central to this theory is the idea that fish consciously choose to swim parallel to current lines because it reduces the pain of the shock. We've actually seen such pain avoidance behavior before; it's the reason why bears move away from electric fences and not toward them. If electrical shocks produced a pleasant sensation, we'd have to find another way to repel bears. According to this theory, the fish are behaving like bears; they are avoiding being shocked as best they can. You're probably having a hard time understanding how crossing multiple lines of force would increase the fish's electrical pain experience. I know I am. But never mind. This theory doesn't hold water anyway.

The trouble with the pain avoidance theory for electrofishing is that it requires a functional brain to register pain and make the decision to avoid it. Yet, subsequent experiments showed that even when the connection between the fish's brain and spinal cord is severed, it behaves the same way.[14] So electrotaxis cannot be due to any conscious pain avoidance behavior.

Soon scientists began to think that electrotaxis might be some kind of reflex. A *reflex* is a response to a nerve stimulus that doesn't involve the brain. In a classic reflex response, signals coming from the *sensory neurons*, the nerve cells that are triggered by one of our bodily senses, go only as far as the spinal cord. There they are rerouted back to the muscles via *motor neurons*, the nerve cells that control muscle contractions. The result is the production of a muscle contraction independent of conscious control. Most of us have had firsthand experience with our own reflexes. For example, if you touch a hot frying pan, your hand recoils away even before your brain is conscious of exactly what's happening. If your hand needed to await instructions from your brain, the delay would increase the severity of the burn. The pain signal simply does not register with your brain soon enough to prevent a bad burn. But, fortunately, your hand doesn't wait around to get your brain's instructions. It acts by reflex while your brain is still trying to figure things out.

Seventeenth-century philosopher René Descartes, who had a strong interest in the underlying science of the brain and nervous system, first proposed the possible existence of reflex actions in the nervous system. (Descartes is better known as the originator of the frequently quoted assertion "I think, therefore I am.") But reflexes first got serious, widespread attention from neuroscientists with the 1906 publication of *The Interactive Actions of the Nervous System*, by Sir Charles Sherrington, a professor of physiology at Oxford University.[15] By that time, research had confirmed the existence of a *reflex pathway* (sometimes called a *reflex arc*) that goes from sensory nerve to spinal cord to motor nerve. Although a reflex arc doesn't require a functioning brain, it does require a functional spinal cord.

Sure enough, destruction of the fish's spinal cord abolishes the electrotaxis. This suggests electrotaxis in fish is indeed a reflex. The fish seem to have no cognitive control over electrotaxis. In effect, they are swimming like zombies in an automated fashion toward an attracting positive electrode, independent of any cerebral thoughts they might have.

But a reflex response, by definition, requires sensory neurons to pick up the signal that triggers the motor neurons to then act. That means the sensory neuron should be the initiator of the electrotaxis response. Incongruously, that doesn't seem to be the case. If you eliminate the sensory neurons, leaving just the spinal cord and the motor neurons intact, you still get electrotaxis in the fish.[16] So electrotaxis doesn't require either a brain or any sensory neurons. The only thing required is an intact spinal cord and motor neurons.[17]

The ability to remove the sensory neurons from the alleged reflex arc suggests the response is something other than a classic reflex arc. Also, it really doesn't make much sense for electrotaxis to be a true reflex. Reflexes are believed to be inherited behavioral responses that provide some type of evolutionary survival advantage. That being the case, why would swimming toward an electrode be an appropriate reflex response to being shocked by it? That type of reflex behavior simply would increase the likelihood of being killed by the electricity.[18] So electrotaxis would seem to be an evolutionary liability rather than a benefit, which raises further questions about it being a true reflex.

When it became generally accepted that fish electrotaxis was neither a conscious action nor a reflex action, *local action theories* came into play. Local action theories attribute biological effects of electricity to their direct effects on nearby nerves and muscles. For electrotaxis, direct electrical effects within the fish's swimming muscles were proposed to be the driving force.

In the 1960s, four French scientists—M. Blanchetau, P. Lamarque, G. Mousset, and R. Vibert—proposed a local action theory for fish electrotaxis, in which electricity acts directly on motor neurons to stimulate them into firing, thereby causing the contraction of swimming muscles.[19] They hypothesized that the side-to-side swimming motion was due to a differential response between the left-side and right-side motor neurons in the swimming muscles of the fish, depending upon which side was closest to the positive electrode. They further hypothesized the effect to be dependent upon the fish facing the positive electrode rather than the negative electrode, because the motor neurons in the fish themselves are allegedly electrically polarized, with the cell bodies relatively positively charged compared with their axons.

There is no doubt that a fish's motor neurons are structurally and functionally polarized, but the electrical polarization explanation of electrotaxis was claiming something more.[20] Motor neurons are structurally "polarized" in the sense that they have two distinct ends.[21] To illustrate, let's think of the shape of a motor neuron as being similar to a lollipop with a very long stick. The part you lick would be the *cell body* and the long stick is the *axon*. Furthermore, nerve signals always move through neurons in a single direction—from the cell body down the axon—thus making the flow of information through motor nerve fibers functionally polarized; that is, information moves only in one direction. Internal structures in the neuron are also polarized to keep nerve signals moving in a single direction. You can't simply hook the nerve up backwards as you might a copper electrical wire and expect it to work the same. Neurons are polarized such that they only work when oriented correctly; copper wires are not.

In addition, all the motor neurons in the swimming muscles of a fish are facing the same way. They are all oriented with their cell bodies anterior to their axons. In other words, their cell bodies are closer to the fish's head than the axons, which all point in the direction of the tail. Taken together, the identical orientation of all the polar motor neurons makes the fish's entire neuroanatomy polarized, and nerve signals moving through the motor neurons are likewise polarized because they flow exclusively in a head-to-tail direction. All of this is consistent with the local action model of electrotaxis. But in order to explain swimming toward the positive electrode rather than the negative one, the model goes one step further, onto very shaky ground.

The model claims that, when fish face the positive electrode, the electric polarity of their motor neurons is enhanced by the external electric field of matching polarity, and they are thus hyperstimulated to send a nerve signal to the fish's swim muscles. In contrast, if the fish faces the negative electrode, their motor neurons' alleged electric polarity would be the opposite of the polarity of the external electric field, and the neurons would be inhibited from sending a signal to the swim muscles. The result is a supposed swimming bias in favor of swimming in just one direction within an electric field; that is, in the direction of the positive electrode.

The longevity of the local action theory of electrotaxis is impressive—it has been around for over 50 years. Although there are precious little data to support it, it remains the most frequently cited explanation for fish electrotaxis in fisheries textbooks to this day. Nevertheless, frequent repetition doesn't make something true, and my gut tells me it is probably not true for at least one important reason. The theory rests on the supposition that motor neurons themselves have positive and negative ends, and we now know that isn't the case. Don't get me wrong; there is charge segregation in neurons, as we will soon see. But the charge segregation is not from one end of the neuron to the other; rather, it's between inside and outside. The inside of a neuron is relatively negatively charged compared to its outside. In this respect, neurons are like tiny Leyden jars with stored electrical energy inside, just waiting for their internal negative charge to be discharged in the form of an electrical signal to neighboring neurons and muscle cells. The interior of the neuron is electronegative relative to the outside environment along its whole length, and that electronegativity doesn't vary from end to end. Neuroscientists say a resting neuron is electrically "polarized" *across its cell membrane,* not from end to end. (We will later see how this is the key to its function in sending electrical signals to other neurons.) Since end-to-end charge separation is the cornerstone of the electrotaxis-through-local-action hypothesis, the fact that no such charge segregation apparently exists is a serious blow to supporters of the local action theory of electrotaxis.

About 20 years ago, an entirely different electrofishing theory made the scene, a theory not based on local action. This theory proposes electrofishing produces seizures in fish similar to what happens in the human disease epilepsy.[22] This theory's proponents note that electricity can be used to induce an epileptic seizure in otherwise normal people. (This is the basis for shock therapy for depression.[23]) They equate all the symptoms of a human epileptic seizure with fish behaviors seen in electrofishing. For example, electrofishing's forced swimming is equated with an epileptic person's *tonic-clonic seizure,* an initial contraction of the musculature followed by rhythmic and uncoordinated muscle contractions. Is there any validity to this theory? Perhaps for some aspects of electrofishing. But it still provides no explanation as to why the fish's tonic-clonic seizure orients it to the positive electrode, which is the most curious aspect of electrofishing. It also doesn't explain how a fish experiences electrotaxis even when its brain is severed from its spinal cord, since epileptic seizures in humans are known to originate in the brain. For these and other reasons, the epilepsy theory hasn't gained much traction in electrofishing's scientific community despite being around for two decades.

Thus, all extant theories of electrotaxis in fish over the last 150 years seem to have their flaws, and we are left with no single theory that explains exactly why fish swim along lines of force toward the positive electrode. But Cooney is unperturbed by the neuroscientists' disagreements over the mechanism of electrotaxis. He finds it inconsequential to the practical aspects of electrofishing. Besides, he says, the neurological aspects of electrotaxis are just one part of the reason that fish show up at an electrofisher's positive electrode. The other reason is that electrofishing equipment is designed to "stack the deck" to ensure they do so. He thinks these electrical engineering factors are underappreciated.

Cooney points out that a fish's entire anatomy, from fin orientation to body scale direction to tapered torso to body musculature, is designed to efficiently propel it in a forward direction only. Most fish can back up only with great difficulty. So, when a fish's body muscles contract due to electrical stimulation, the contractions cause tail movements that naturally propel the fish forward. Thus, an electrically shocked fish must move forward whether it wants to or not.

Cooney further points out that the electricity would be expected to cause the most uniform contractions on both the left and right sides of the body when the fish's body is aligned with the electric field's lines of force. Any misalignment with the field line would cause stronger muscle contractions on one side of the body, and thereby put the fish's torso back in line with the electrical field lines. Additionally, the fish's body experiences the greatest voltage drop, and thus the greatest muscle stimulation from head to tail, when it is oriented along a field line.[24] And those intense contractions quickly propel the fish forward along the electric field line.

But I still don't understand why they swim along field lines specifically in the direction of the positive electrode, so I press the question with Cooney. He has an answer to that, too. He says it partially has to do with the design of the negative electrodes. He tells me the negative electrodes used for boat electrofishing are specifically designed not to stun fish, and he says that is achieved by making the negative electrode much larger than the positive electrode. Having a disproportionately large negative electrode has the effect of spreading out the lines of force on the negative end of the field, so the electric field is less intense near the negative electrode. This means the amount of current flowing through the fish's body becomes progressively less as it nears the negative cathode, while it becomes progressively more intense when a fish approaches the positive electrode, resulting in stronger forward swimming contractions and, ultimately, stunning. So it's good to have a relatively small positive electrode positioned off the bow of the boat, where stunned fish can be easily scooped up, and have a very large nonstunning negative electrode trailing somewhere behind it. And the easiest way to make the trailing negative electrode very large is to make it the size of the whole boat!

Cooney explains that while the positive terminal of the DC power source is connected by a wire directly to the positive electrode dangling in the water at the bow and well in front of the hull, the negative terminal of the DC typically is connected directly to the boat's metallic hull. What this means is that the lines of force at the positive electrode are all converging at a single, relatively small spot in the water in front of the boat, but the other end of those lines of force terminates at various points on the large surface area of the hull. In fact, they are electrically engineered to spread the lines of force out over the boat's hull precisely so they don't converge near any single point. So fish swimming toward the hull do not experience converging lines of force of the electric field, and hence no increasing current flow within their bodies. But fish approaching the much smaller positive electrode experience increasingly more lines of force crossing their bodies as the lines of force converge near the positive electrode. The result is that fish approaching the positive electrode (dangling in the water in front of the boat) are stunned and fish approaching the negative electrode (the boat's hull) are not. Under this explanation, the swimming of fish specifically toward the positive electrode can be thought of as observer bias on the part of the electrofisher. She sees only the fish being attracted and stunned at the positive electrode, while any fish moving toward the negative electrode are actually under the boat and thus not visible to her.

Cooney doesn't deny there is an underlying neurophysiological basis for electrotaxis toward the positive electrode; he just believes electrical engineering factors, such as the relative sizes of the electrodes, are also

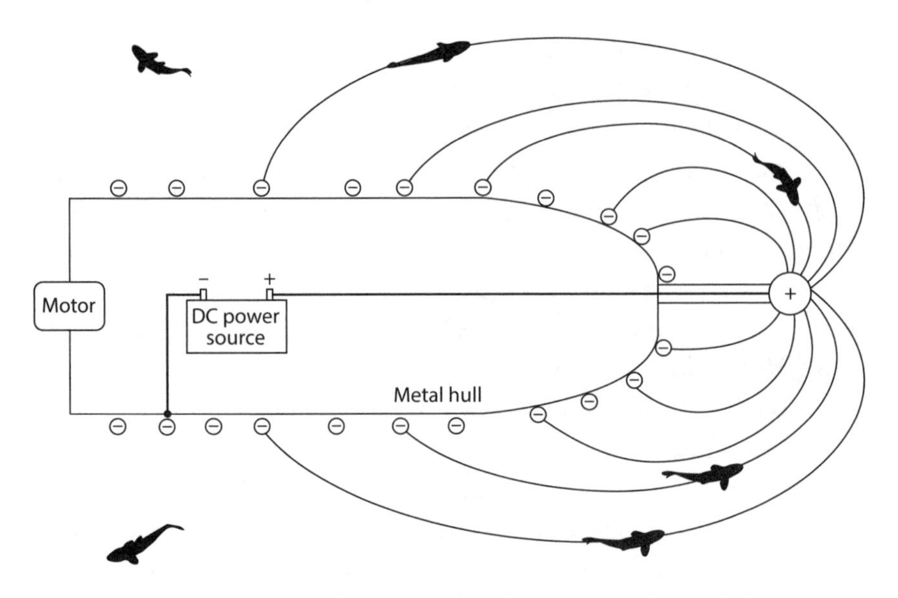

FIG. 8.2. Electric field surrounding an electrofishing boat. This diagram depicts the basic design of boats used for electrofishing. The negative terminal of a direct current power source is connected to the metal hull, thereby making the entire hull a negative electrode. The positive terminal is connected to a submerged electrode that dangles into the water from the end of a long pole that protrudes horizontally across the water from the bow (front) of the boat. This produces a large electric field that envelopes the water near the bow. As the boat slowly motors through the water, fish that enter the field are attracted to the positive electrode, which they swim toward parallel to the field's lines of force. Once a fish nears the positive electrode, it experiences a stunning electrical shock and becomes immobilized. Workers standing at the bow scoop up the stunned fish with hand nets, measure them, and return them to the water off the stern (back) of the boat once the fish have recovered and regained their mobility.

at play. And he's happy they both contribute toward the phenomenon of fish showing up at the electrofisher's positive electrode, otherwise electrofishing might not be as effective a tool as it is, and he might be looking for another job. I see his point, and I also see why it isn't likely that the details of the electrotaxis mechanism will ever be fully worked out in field studies; there are just too many parameters at play. This is something that needs to be studied under controlled laboratory conditions, where the different parameters can be varied one at a time. And it just so happens there is a group of scientists working on fish electrotaxis under controlled laboratory conditions right now, and they're doing some very innovative things with zebrafish.

Zebrafish are not zebras that swim like fish; they are fish that look like zebras, at least in that they are striped like zebras. These inch-long fish are very popular among aquarium hobbyists, who prize their robustness even when the water quality of their tank may not be at its best. Their hardiness also makes them popular with scientists who want to use them for research; they do very well in a laboratory setting where they may receive a lot of handling.

But beyond their sturdiness, zebrafish have other attributes that make them the darling species for doing a variety of research that is more generally relevant to all vertebrates (i.e., animals with a spinal column). Zebrafish (*Danio rerio*) are the simplest vertebrate-model organism with genetics that can be easily manipulated. For this reason, zebrafish are used widely in disease modeling, behavior screening, and drug discovery. Moreover, at five days postfertilization (dpf), they have fully developed sensory and motor neurological functions, making their larvae highly amenable to neurological function studies at a very young age—an age where they are doing little more than just moving around.

As I mentioned earlier, fish electrotaxis was first discovered in fish larvae. Perhaps mechanistic insight might be gained by going back to the beginning and studying electrotaxis in fish larvae, but this time using a modern technological approach with zebrafish larvae as the fish model of choice. That's exactly what scientists at York University in Toronto recently started to do, and already they've gotten some new insight about electrotaxis.

The scientists constructed a microfluidic device specifically designed for work with zebrafish larvae.[25] *Microfluidics* is a technology that deals with precise control and manipulation of fluids constrained to small spaces, typically handling volumes amounting to just a small fraction of a water droplet. Their device amounts to a combination of a microscope, a small microscopic channel filled with water, and a lot of accompanying electronics. The electronics control the amount of water in the channel, regulate the electric current that flows through the water, and collect the data from the experiments.

The basic experimental design is this. First, you introduce a single 5 to 7 dpf larva into the channel.[26] You observe the larva through the microscope to ensure it is swimming around normally and apparently randomly. Then you run an electrical current through the water-filled channel and observe how it affects the larva's swimming direction. If the larva starts swimming toward the positive electrode, you have presumptive electrotaxis. But, of course, the larva could have swum in that direction randomly, so you turn off the electricity and allow the fish to rest for a few minutes. Then you reapply the electric current, this time coming from the reverse direction. If the larva again swims toward the positive electrode, which is now in the opposite direction, you have thus replicated the response with a second shock from another direction, and that larva is judged to be displaying an electrotaxis behavior. You repeat this with about 30 larvae, and compare it to 30 larvae handled the same way but without any electric current (i.e., the control group). Findings indicate nearly 100% of the larvae exhibited positive anode electrotaxis as defined by the "two-shock test," compared to just 10% of the controls.[27]

But the true value of this apparatus for probing the mechanism of fish electrotaxis is that it enables screening for potential chemical inhibitors of the particular neurotransmitters driving the electrotaxis response. A *neurotransmitter* is a chemical substance that is released by one neuron in order to cause the transfer of a nerve signal impulse to a second neighboring neuron. *Receptors* for the specific neurotransmitter are needed on the second neuron in order for the chemical signal to be received properly. The different types of highly specialized neurons in the body have specific matched pairs of neurotransmitters and receptors, such that the type of neuron can often be deduced based on its receptor profile. By using chemical inhibitors of specific receptors and assessing their effects on nerve responses, the different functions of the various receptors can be inferred. That information matched to a neuron's receptor profile can provide a pretty informative picture of the role of that neuron in the nervous system.

This is important because there are a large number of different types of neurons in the body, and probably only a few types are involved with electrotaxis. If one could deduce the types of receptors that control electrotaxis, it would go a long way toward revealing the specific types of neurons that are involved in the underlying mechanism of electrotaxis. And then we would have come a long way from probing the neuroscience of electrotaxis by severing fishes' brains from their spinal cords and seeing what happens. To this end, the researchers tested two different types of receptor inhibitors—the D1-type and D2-type of dopamine receptor inhibitors. They were able to show inhibitors of the D2-type significantly suppressed electrotaxis in the larvae, while the D1-type did not.

This is exciting news because it provides a link with human biology and health. Dopamine is a very important neurotransmitter, involved in many neurological functions. Five types of dopamine receptors are found on different types of neurons, and dysfunction of these receptors is associated with a wide variety of neurological diseases. The D2-types of receptors, in particular, have been most strongly implicated in schizophrenia and Parkinson's disease. The fact that D2-type inhibitors can interfere with electrotaxis in fish suggests the neurological pathways involved in the mechanism of electrotaxis may share some commonality with the neurological pathways involved in the mechanisms of schizophrenia and Parkinson's disease in humans—an interesting connection that warrants further investigation. Perhaps the studies of electrotaxis on zebrafish will suggest treatment options for schizophrenia or Parkinson's disease.

But are we getting ahead of ourselves here? We've seen that each generation of scientists over the last 150 years has tried to explain electrotaxis in fish by proposing a mechanism that conformed to the prevailing neurological theory of the day. Could the dopamine link be just another example of the same—a theory to be abandoned when further details of basic neurobiology emerge? Perhaps. But Volta thought he was emulating the neurobiology of the electric eel when he invented his voltaic pile, which we now know has nothing to do with the way an electric eel generates its electricity. We have to start somewhere despite our ignorance of the fundamentals. Volta started with the electric eel and ended up with a battery, so why not go with zebrafish and see where it may lead?

Even though a detailed understanding of the underlying mechanism of fish electrotaxis remains elusive, parallel research developments in the areas of the physics of electromagnetic force fields and the biology of the nervous system have been very insightful in showing us how electricity and biology can interact in a highly coordinated fashion. But it's also clear that a true mechanistic understanding of how electricity affects the nervous system, as well as how the nervous system exploits electricity, requires an understanding of the basic functional units that generate both nerve signals and electrical signals. The basic unit of the nervous system is the neuron, a unique and highly specialized body cell with vital communication and muscle control functions. We've had a brief introduction to neurons in this chapter, and we'll be hearing a lot more about them as we progress through our story. But before going there, we first must deal with the fundamental unit of electricity, which is the elephant in the room that we thus far have been ignoring. Ironically, this elephant, which exerts its influence in so many big ways, is the infinitesimally small *electron*.

9

CROOKES OF THE MATTER

THE ELECTRON

> The electron is a theory we use; it is so useful in understanding the way nature works that we can almost call it real.
>
> —RICHARD FEYNMAN

With all the electronic amusements available to children these days, I wonder if there are any modern toys that don't require electricity. A quick internet search of top-selling toys suggests there are, and reveals some old standbys. I'm happy to see Mr. Potato Head still makes the list. He's been popular since I was a little boy. But wait, this year's model is new and improved; it now features "Moving Lips!" And guess what; the lips require three AA batteries. (Fortunately, they are included.) I'm not, however, at all happy to see Mr. Potato Head's modern competitor. Called "Buttheads," these tiny dolls have rear ends where faces are supposed to be. They are advertised as making "20 different fart sounds" and releasing a corresponding foul odor at a child's command. And, of course, all of those many variations of flatulence are powered by batteries. More disturbing to me still is that Buttheads are recommended as gifts for children "ages three and up." Oh well, I guess I'm just getting old.

Young or old, everyone has a fondness for toys, including physicists. And one of the most popular "toys" of all time among physicists also happens to be electrically powered: the Crookes tube. Invented by British scientist William Crookes in 1875, it transformed DC electricity into a glowing light show that entertained scientists and lay people alike.[1] It was as popular as the static electricity machines of an earlier generation that had both enthralled scientists like Franklin and entertained nonscientists with electrical parlor shows, like the Flying Boy.

The inner workings of a Crookes tube are actually very simple. A Crookes tube, in its most typical design, looks like a large, pear-shaped lightbulb and is about the size of an American football. And its construction isn't much different from an Edison-style lightbulb. There are two electrodes that penetrate the sealed glass chamber, from which most of the air has been removed. The main difference between a Crookes tube and an incandescent lightbulb is that the Crookes tube has no filament connecting the two electrodes. So, if you try to run electric current through the tube, nothing happens, because there is nothing connecting the electrodes and thus no filament to produce an incandescent glow.

But if you keep increasing the Crookes tube's voltage, eventually you get to the point that the electrical pressure across the electrodes is so high that the electricity starts jumping through the evacuated space, from one electrode to the other, despite the absence of a conductor. When this happens, the glass near the positive electrode starts to glow a brilliant bluish-green, and within the glow you can see the shadow of the positive electrode on the surface of the glass.

Often, the positive electrode of a Crookes tube is made in the shape of a Maltese cross, so you typically see a distinct shadow of the cross at the wall of the tube. The glass glows not because it's hot, although it does get warm, but rather because glass often contains fluorescent chemicals that can be energized to give off light when they encounter electricity. (Sometimes the glass is even coated with fluorescent paint to enhance the glowing effect.) A glowing glass bulb might not seem like much to us now, but the eerie effect of the green glow illuminating a darkened room was quite an attraction in the 1890s, as most homes didn't even have electric lighting at the time. It also seemed very mysterious, particularly because no one yet understood exactly how it worked.

It appeared that a shadow of the positive electrode was being cast by something coming from the direction of the negative electrode, very similar to the way rays of light from the sun cast shadows of trees onto the ground. If other metal obstructions were put in the middle of the tube, they too would cast their shadows on the glass, specifically on the end of the tube near the positive electrode. This led to the idea that some type of invisible ray was streaming out from the negative electrode, headed in the direction of the positive one. It was as though the negative electrode was releasing a high-pressure, invisible aerosolized spray, pointed at the positive electrode.

Physicists called these alleged rays *cathode rays* because they originated from the "cathode," another name for a negative electrode. Many physicists were "playing" with Crookes tubes because they were fascinated by cathode rays. Nobody had any idea what a cathode ray was made of, but they knew it was very important scientifically, whatever it was. No known physics could explain the phenomenon. It represented something entirely new.[2]

Crookes himself played a good deal with his namesake tube, and he wondered how the cathode rays would respond to a magnetic field. He took a large horseshoe-shaped bar magnet and used it to straddle the tube, so that the north and south poles of the magnet were on opposite sides of the tube. This created a magnetic field inside the tube, perpendicular to the path of the rays. He found this caused the cathode rays to bend toward the north pole of the magnet and away from its south pole. Crooke speculated that this was exactly what you might expect if the beam was composed of small negatively charged particles. Many physicists thought Crookes's magnet experiments were incomplete, insufficient, and inconclusive. Nevertheless, his work stimulated others to look at cathode rays. What exactly where they? The question remained unanswered for a long time.

You are probably now wondering why these odd rays, emanating from the negative electrode and increasing with voltage, weren't immediately suspected to be a spray of that invisible electrical fluid Franklin had postulated to be responsible for the flow of electrical current through wires. No one thought so for one reason: the spray was flowing in the wrong direction. Every scientist knew electrical current flows from positive to negative; Franklin himself had said so. The spray of rays coming from the negative electrode had to be something different because it was going in a direction opposite to the electrical current. Isn't that true?

I know you are just as excited as those physicists were and cannot stand the suspense any longer, so I won't make you wait until the end of the chapter for the big reveal about the nature of cathode rays. Cathode rays are simply flying *electrons,* and it is the flow of electrons that constitutes an electric current. Franklin had been wrong about the direction in which electric current flows. Although we still say to this day that electrical "current" flows from the positive electrode to the negative electrode, in deference to the great Franklin, those of us in the know about electrons understand that the electrons are really moving in a direction opposite to the imagined "current." And it is the flow of electrons that behaves so much like the postulated invisible electrical fluid everyone was looking for. Moving electrons explain virtually all the mysteries formerly linked to Franklin's invisible electric fluid.

You might think it surprising that we got halfway through this book about electricity and only just now are discussing the electron. After all, the concept most people think they know about electricity is that it is a current of electrons. You might expect we would have started the book discussing the electron, since electrons are such a fundamental part of electricity's definition. But the telegraph, the electric motor, the electric generator, the electric transformer, the electric relay, the telephone, the photoelectric cell, the phonograph, the lightbulb, the radio, and the electric power industry all came into existence before the discovery of the electron. You simply don't need to know about electrons in order to make use of electricity. In fact, maybe it would be best to forget about electrons altogether. After all, they cause a lot of problems in understanding electricity.

In his tongue-in-cheek book *There Are No Electrons,* electrical engineer Kenn Amdahl makes this very point. His argument is that most people aren't used to thinking about abstract physical entities. He says if we start teaching students about electricity by introducing them to a concept as abstract as the electron, we will surely dishearten most of them almost immediately. They'll become discouraged because they'll think electrical principles are too abstract for them to understand. Amdahl's argument boils down to this: Why risk losing your audience by forcing them to first learn about something that most of them don't need to know anyway? If anything, the electron is an extremely advanced concept, not an elementary one. So it's probably best not to start with the electron when learning about electricity.

Amdahl illustrates his point by describing all of fundamental electrical theory as though electric current were the result of Little Greenies, invisible little green people that live in electrical wires, who behave according to defined rules of behavior. Sounds silly, doesn't it? But Amdahl would argue it's easier to believe electricity is the result of invisible little green people hiding out in wires than it is to accept that an electron can behave both like a particle and a wave at the same time—a fundamental tenet of electron particle physics. Amdahl goes beyond being just blasé about the alleged need to understand electrons in order to work with electricity. He's downright dismissive of those who claim they understand electrons and how electrons produce electricity: "No one really understands electricity. But no one wants to admit it."[3] Amdahl even asserts that people who say they understand electrons actually are demonstrating their ignorance. Electrons are just too complex for any single person to fully grasp. If you doubt that, just spend some time perusing the host of scientific papers that claim exactly such insight into the electron's gestalt.[4] I guarantee you'll quickly come running back to the Little Greenies for relief.

But never mind. There is no shame in admitting ignorance of the electron. The electron has repeatedly humbled even particle physicists. Theodore Arabatzis acknowledges as much in his book *Representing Electrons: A Biographical Approach to Theoretical Entities,* in which he chronicles a continuous evolution of thought among physicists as they pondered what an electron actually is. Every time the physicists began to think they had the irksome electron firmly in hand, it slipped through their fingers once again.

Given that a full understanding of the electron is elusive, and knowledge of the electron might very well be considered superfluous to understanding the practical aspects of electricity, is there any reason to deal with electrons here? After all, our interest is somewhat narrowly focused on how electricity interacts with the body.

Despite the challenges, yes. I think it is important that we have a basic understanding of what electrons are and how they behave. And the reason is this: the circumstances that led to the discovery of the electron are important not just because they provide an explanation for how electricity works. They are important because they provide an explanation for how chemistry works; and chemistry, as we shall see, serves as the interface between the body's electrical and biological states.[5]

Before the discovery of the electron, chemistry was a rather "soft" science. It was highly descriptive and nonquantitative, and no one had any idea what was actually going on. Chemists would mix chemicals together and observe that they converted some things into other things, but they had no basic understanding of the mechanisms involved. All work was trial and error. No reaction products could be predicted. Furthermore, chemists had no tradition in mathematics, tending instead to rely solely on basic arithmetic to express their experimental weights and volumes. Indeed, most chemists were completely ignorant of higher mathematical methods. True scientists tended to avoid chemistry, although the great physicist Isaac Newton couldn't resist dabbling in chemistry now and then, particularly *alchemy*, where the potential payoff might be making gold out of lead.[6] However, with a few exceptions, physicists pretty much ignored chemistry (and chemists) altogether.

But it was becoming increasingly clear in the late 1800s that many chemical phenomena had physical aspects, and many physical phenomena had chemical aspects. Some scientists rightly were beginning to think physics and chemistry were just opposite sides of the same coin. As one put it, "The work of chemists and physicists may be compared to that of two sets of engineers boring a tunnel from opposite ends—they have not yet met, but they have got so near that they can hear the sounds of each other's advances."[7] With the birth of *electrochemistry,* the branch of chemistry that studies the interactions between electricity and chemicals, the tunnel borers met. Henceforth, the physicists and the chemists would join forces.

The growing availability of electricity, from either electrical generators or battery power sources, enabled chemists as well as physicists to easily experiment with electricity. They soon discovered a chemical reaction called *electrolysis,* literally meaning "breaking with electricity." By running a strong electric current through a chemical solution, they got new chemicals to form at each electrode. Sometimes those chemicals were in the form of a gas, causing the electrode to appear to emit bubbles. Other times a chemical crust would form on the electrode. But the chemicals deposited at the electrodes were always different from the chemicals that were used to make the solution. Something had changed due to exposure to the electric current.

Over time, chemists began to surmise, correctly, that electricity causes the breakdown of the molecules in the solution into their constitutive parts. Even the water in the solution could be broken up with a strong enough current. It was found the electrolysis of water resulted in gases bubbling at both electrodes. If you collected the gases coming from each electrode, you found there was twice as much gas coming from one electrode as the other—*exactly* twice as much. If you then mixed the two gases together and exposed them to a spark, you'd hear a loud pop and water vapor would appear. We now know this is because a water molecule is actually H_2O—two hydrogen atoms bound to one oxygen atom—that can be broken apart with electricity. You get twice as much hydrogen gas forming at the one electrode as oxygen at the other because water molecules have exactly twice as many hydrogen atoms as oxygen. The spark triggers the gases to recombine and turn back into water.

As more electrolysis data rolled in, they suggested molecules are made of smaller individual atoms held together by the attractive forces of electrical charge. Electric current running through the solution disrupts those atoms and their respective charges, so the molecules fall apart. The molecular pieces—some negatively charged and some positively charged—travel along the electric field's lines of force toward the electrode of opposite charge, similar to the way fish travel along the lines of force of an underwater electrical field toward the positive electrode.

And unlike much of chemistry, which for the most part seemed to work without any set rules or scientific laws, electrolysis did appear to work according to strict scientific principles. The *laws of electrolysis* were reported by none other than Michael Faraday, a scientist quite comfortable working from either end of the "chemistry meets physics" tunnel.[8] And since it was Faraday prescribing the laws, they contained minimal mathematics, of course. Although he proposed a couple of laws, the first, stated below, is the most important for our purposes here:

The amount of chemical change produced by current at an electrode is in proportion to the amount of electricity used to produce it.

All right, a little electricity can break up little molecules. So what? Is that really surprising given that a lot of electricity, like lightning, can break up buildings? Probably not. But why do a broken molecule's parts separate from each other and migrate to different electrodes in specific proportions? Are the molecule's parts—the atoms—electrically charged themselves? Are the positively charged parts being attracted to the negative electrode and the negatively charged parts being attracted to the positive one? Is that why most molecules are usually electrically neutral—because they are the summation of an equal number of positively and negatively charged constituents? As it turns out, the answer to all of these questions is typically yes!

But Faraday's first law of electrolysis suggested something more. Not only were the chemical products formed in specific proportions, so was the amount of electricity (i.e., electrical charge) needed in specific proportions. That is, the *stoichiometry*—the calculation of relative amounts of reactants and products in chemical reactions—seemed to include an electric component behaving as though it were an atomic particle. This suggested to some people that it might be that electrically charged particles enter into chemical reactions in a way similar to how atoms participate. German physicist Hermann von Helmholtz put it this way:

> The most startling result of Faraday's law is perhaps this. If we accept the hypothesis that elementary substances are composed of atoms, we cannot avoid concluding that electricity also . . . is divided into portions, which behave like *atoms of electricity*.[9]

Thus, the search was on for the "atom of electricity."

As mentioned, Faraday's first law of electrolysis literally says the amount of chemical change produced by an electrical current is proportional to "the amount of electricity" used to produce it. But when Faraday talks about the amount of electricity, he doesn't mean the volts (pressure) or the amps (current flow). Rather, he is specifically talking about the charge. The measurement of electricity's charge is something we haven't yet discussed, primarily because it doesn't easily fit into our simple model of electricity being a flow of an invisible fluid through wires, analogous to water's movement through hoses. But it's time we talk about measuring charge or we're never going to appreciate electrons.

Charge is related to amps in that amperage is defined as the flow of charge per unit of time. The unit of time is the *second,* of course, but we still need a unit for charge. Charge manifests itself to us as either a pushing (repulsive) or pulling (attractive) force. We are used to measuring forces. For example, the force of gravity gives us units of weight, be they in pounds or some other unit. Consequently, if we can measure the force of attraction or repulsion, we have a workable measure of the charge.[10]

Charge can be static, like when it's captured in a Leyden jar; or it can be dynamic, as when it's flowing in an electrical current. But it would be hard to quantify the exact amount of charge in a Leyden jar by measuring the forces between Leyden jars. In contrast, it is relatively easy to run a constant amount of current through two parallel wires and measure the pushing or pulling force between the wires. This is exactly what French physicist Charles de Coulomb did. And he found that, for any specific distance between the wires, the force between the wires increased and decreased in exact proportion to the amount of current (i.e., amps) running through them. That being the case, it seems a fixed amount of charge is associated with every amp of current. So, why not just define one unit of charge as the amount of charge in one amp of current? Not a bad idea. If one amp of current pushes one unit of charge per second, then charge can simply be measured in amp-seconds. In other words, if

$$Amps = charge/seconds$$

then

$$Charge = amps \times seconds$$

Thus, charge is measured in terms of an *amp-second*. But "amp-second" is a rather awkward term. So an amp-second was renamed a *coulomb*, to honor the person who came up with the brilliant idea of how to measure charge.[11]

Bingo! We have a unit for charge. And it is charge, measured in coulombs, that Faraday was referring to when he said "the amount of electricity."

Let's not go too far down this rabbit hole. If you find the above logic a bit convoluted, never mind. (I warned you that understanding electrons could be problematic!) The point is simply this. Faraday's electrolysis work suggested chemical reactions might involve some type of charged particles that acted like atomic glue, holding atoms together in different proportions to form molecules. So the seed of an idea that there might be little charged particles responsible for what we call electric current was planted in the minds of at least some physicists because of Faraday's electrolysis experiments. But it would take 64 years before this notion was proved correct.

The scientist credited with discovering the electron is J. J. Thomson.[12] When Thomson started his career in the 1880s, his work in physics was entirely theoretical. Upon entering Cambridge University in 1876, he obtained his physics training via the Mathematical Tripos, which focused on mastering theoretical physics rather than conducting physics experiments, so Thomson graduated without doing even a single physics experiment.[13]

Thomson was a renowned Maxwellian scholar, interpreting the underlying implications of James Clerk Maxwell's often cryptic equations. This was necessary because, although Maxwell was fluent in the "language" of mathematics, his equations often seemed to have their own dialect—a dialect other physicists found hard to interpret.[14] Mathematics was just emerging as the vernacular of physics, and Thomson built his early career in theoretical physics by interpreting the equations in Maxwell's classic two-volume work, *A Treatise on Electricity and Magnetism*.

Even though Maxwell was on the Cambridge physics faculty while Thomson was studying there as an undergraduate, Thomson never met Maxwell. This was in part because Maxwell died just three years after Thomson began his studies, but also because Maxwell headed Cambridge's famed physics laboratory, known as the Cavendish. The Cavendish was the physics teaching laboratory but was destined to eventually become a tremendous physics research lab as well. The Cavendish was all about conducting experiments, and theoretical physicists thought such experimentation somewhat superfluous to serious theoretical work. Thomson's biographer, Jaume Navarro, says this attitude was widespread at the time: "[To theoretical physicists] the experimental sciences appeared as a kind of second-class knowledge. They were provisional and particular and they lacked the rigor of mathematical formulation."[15] Thus, there was little intersection between theoretical physicists and experimental physicists. As a result and considering Thomson's youth at the time, Thomson unfortunately never directly encountered the great Maxwell himself.

That theoretical physicists, with their exclusively mathematical approach to physics, found experimentation superfluous seems particularly ironic given that Maxwell—the icon of mathematical physics—made his first scientific splash by applying mathematics to Faraday's experimental work on lines of force. And Maxwell himself was obsessed with conducting experiments ever since he had made his own laboratory in the attic of his childhood home. Maxwell found experimental data to be stimulating, and a challenge to the intellect to explain exactly what was going on. In this way, he was following very much in the tradition of Franklin and Faraday. Hence, he was a good leader for the Cavendish. As for Thomson, he too soon came to see experimentation as Maxwell did—vital to scientific progress—and began doing his own groundbreaking experiments. Thomson even came to the point where he would criticize his mathematical physicist colleagues: "[They] regard as the normal process of investigation in this subject the manipulation of a large number of symbols in the hope that every now and then some valuable result may happen to drop out."[16]

In the context of experimentation with various vacuum tubes that Thomson and other physicists were conducting at the time, Thomson ultimately focused his scientific attention specifically on the Crookes tube—the toy that was the obsession of all experimental physicists—to see what new physics he could learn from it. Crookes's earlier experiments with the magnet implied that cathode rays emitted from the negative electrode of a Crookes tube were some type of electromagnetic phenomenon, which was the branch of mathematical physics in which Thomson excelled. He also came to believe that solving the cathode ray question could have a tremendous impact on the understanding of exactly how electricity worked, a phenomenon that still remained largely a mystery. He saw the physics of cathode rays to be an area where all his theoretical physics training could be applied to a question of tremendous practical importance. He told his colleagues, "There is no other branch of physics which affords us so promising an opportunity of penetrating the secret of electricity."[17]

Thomson found it odd that cathode rays could be bent by a magnetic field. You couldn't do that with other types of rays, such as rays of light. Yet, as Crookes had already shown, when you brought a magnet near the glass of the Crookes tube, you could get the beam to stream either to the left or right depending upon the orientation of the magnet. The negative (south) pole of the magnet would repel the beam, while the positive (north) pole attracted the beam. This indeed suggested the cathode ray beam itself has a negative electrical charge associated with it. If that was true, Thomson reasoned, the path of cathode rays should also be influenced by an electric field as well as a magnetic field. So Thomson conducted an experiment where he replaced the straddling magnet used by Crookes with a pair of electrodes positioned on either side of the tube, thus creating an electric field, perpendicular to the path through which the beam needed to pass. Thomson found the electric field also bent the cathode ray beam, so that the beam curved toward the positive electrode just as it had curved toward the north pole of the magnet. This result supported the notion that cathodes rays consisted of flying negatively charged particles.

Thomson also found that the higher the voltage across the field electrodes—in other words, the stronger the electric field—the more the beam would bend. This finding wasn't too surprising. But Thomson further reasoned that the ability to bend the beam also should be dependent upon both the particle's charge and the particle's size. If the particle had a greater negative charge, it would "feel" a greater pull from the positive electrode, so the amount of deflection should be proportional to the amount of charge of the particle. But the particle's deflection due to its charge would be countered by its mass. In other words, heavier particles would have a greater momentum than smaller particles, so it should take a stronger electrical field to deflect the larger

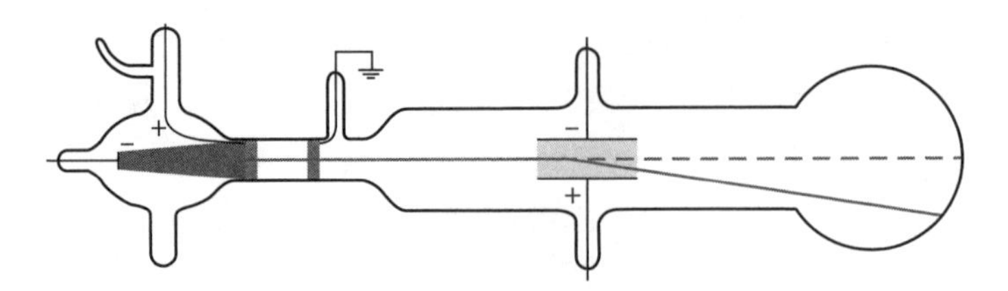

FIG. 9.1. The discovery of the electron. British scientist J. J. Thomson used the laboratory device diagrammed here to discover the electron. The device combines a cathode-ray-emitting glass bulb (left), known as a Crookes tube, with a longer glass tube containing two oppositely charged electrode plates (center). Thomson showed that a narrow beam of cathode rays passing between the plates (solid line) could be deflected from its normally straight path (dashed line) by maintaining an electrical charge across the plates. The direction of the deflection was always toward the positive electrode, and the magnitude of the deflection was dependent upon the amount of charge placed on the plates. The direction of the deflection suggested that cathode rays were composed of flying particles that were negatively charged. The magnitude of the charge needed to deflect them suggested the particles were much smaller than even the smallest atom (hydrogen), an assertion that was thought to be impossible at that time. But Thomson's findings were later confirmed, and he was awarded the Nobel Prize in Physics (1906) for the discovery of what we now call the *electron*. (WikiMedia Commons)

particles from their normally straight paths.[18] All else being equal, particle deflection should be directly proportional to the particle's charge and inversely proportional to the particle's mass. Thus, for any specific pathway of deflection, the ratio of charge to mass—appropriately termed the *charge-to-mass ratio*—should be a constant.

Thomson set himself to measuring the charge-to-mass ratio of the alleged charged particles that made up cathode rays. He soon determined the charge-to-mass ratio of the cathode ray particles: 1.758820×10^{11} coulombs per kilogram. It was an extremely high charge-to-mass ratio. But charge-to-mass ratios, like all ratios, are determined by both their numerators and denominators. If a particle had a high charge-to-mass ratio, was it because it had a very high charge or because it had a very small mass? Hard to tell.

The smallest particle known to scientists at the time was the smallest atom—the hydrogen atom—which could have either a positive or neutral charge. Positively charged hydrogen atoms were said to have one unit of positive charge. If the particles of cathode rays were the negative equivalent of the positive hydrogen atoms, they should have one unit of negative charge. Under such an assumption, how did the charge-to-mass ratios of hydrogen atoms and cathode ray particles compare? Thomson conducted particle deflection experiments with positively charged hydrogen atoms and then used the data to calculate the relative masses of the particles. He got a very surprising result. Assuming both had the same level of charge, Thomson calculated the mass of the cathode ray particle to be 1/200th the mass of a hydrogen atom!

This mass calculation caused a considerable stir. If you accepted it, you also had to accept that particles smaller than the smallest atom exist. Such a novel idea was very hard to swallow since atoms were thought to represent the smallest possible units of matter and they were, by definition, indivisible. In fact, the word *atom* comes from the ancient Greek word for "indivisible." Could it really be possible there exist particles 1/200th the size of hydrogen—the smallest atom? That was shocking, yet that's exactly what Thomson's findings suggested.

But there was some wiggle room about the size of the cathode ray particles. The size Thomson calculated was based on the assumption that the magnitude of the charge of each individual particle was equivalent (but opposite) to the charge of a hydrogen atom. Perhaps the cathode ray particles each had much greater charge; say, 200 times as great. In that case, Thomson's charge-to-mass ratio could be explained by each cathode ray particle having very high charge rather than very small mass. Thus, Thomson's mass determination for the cathode ray particle remained an open question until the charge of each particle could be determined. Were cathode ray particles highly charged or extremely small? Time would tell.

The breakthrough came in 1913 when American physicist Robert A. Millikan produced a device where cathode ray particles could attach themselves to microscopically small oil droplets that fell individually through the sensor of a charge detector.[19] In theory, any particular oil

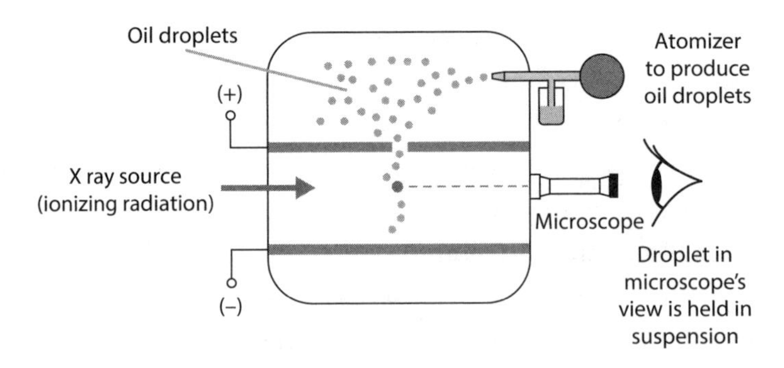

FIG. 9.2. The discovery of the electron's charge. American scientist Robert Millikan used the experimental device diagrammed here to measure the charge of an electron. He employed an atomizer—a spray-producing appliance similar to what is used to spray perfume from its bottle—to produce a mist of uniformly microscopic oil droplets into an empty chamber. Gravity caused the droplets to very gradually settle toward the bottom of the chamber, with some falling through a microscopic hole and entering a second lower chamber that was exposed to a source of X-rays. Since X-rays produce electrons, the second chamber was full of electrons, some of which got absorbed into the individual oil droplets. Most of the droplets continued to slowly fall, but those that had absorbed electrons could be impeded from falling by applying an electrical charge across the lower chamber by means of positively and negatively charged plates on the ceiling and floor of the chamber, respectively. The droplets that had absorbed electrons (negatively charged) experienced an upward electrical attraction toward the ceiling (positively charged), countering the downward force of gravity. By varying the charge across the plates, Millikan was able to find the minimum charge needed to suspend a droplet in a stationary position. Curiously, he observed other droplets suspended by applying exact multiples of that minimum charge. Millikan correctly deduced that, since electricity's charge is imparted by electrons, the minimum charge value he found ($1.60217662 \times 10^{-19}$ coulombs) must represent the charge of a single electron, and that droplets that were suspended at multiples of the minimum charge must have absorbed a corresponding number of electrons. Millikan was awarded the Nobel Prize in Physics (1923) for "his work of the elementary charge of electricity." (Image courtesy of A. Beléndez)

droplet could have anywhere from zero cathode ray particles attached to it to very many. When Millikan measured the charges on the individual droplets, he found the charges between droplets always varied by multiples of $1.60217662 \times 10^{-19}$ coulombs (less than a billion billionths of a coulomb) and the smallest negative charge measured for any individual droplet was also $1.60217662 \times 10^{-19}$ coulombs.[20] The most straightforward interpretation of these data was that the smallest negative charge on a droplet represented a droplet that carried one and only one cathode ray particle; the more highly charged droplets carried multiple numbers of particles. Furthermore, the $1.60217662 \times 10^{-19}$ coulombs of negative charge attributed to a single particle was exactly the same as the positive charge of a single hydrogen atom.

After Millikan's experiments, there was no other plausible conclusion. Hydrogen atoms and cathode ray particles have equal but opposite charges, but they were vastly different in their masses, as first suspected by Thomson. A cathode ray particle was but 1/200th the mass of a hydrogen atom, making it the smallest-known particle, much smaller than any atom. Cathode ray particles were, in fact, *subatomic*. And this meant the physics of atoms needed to be completely rethought. Atoms are not the tiniest units of matter and, therefore, indivisible; there exist even smaller things inside of atoms, the electron being one of them.[21]

Given that the cathode ray particle seemed to represent the smallest possible quantity of negative electrical charge, it led to the conclusion that it was the aggregation of these negatively charged particles that was responsible whenever negative charge accumulated, be it in a Leyden jar or thunderclouds. It followed that the cathode ray particle likewise represented the fundamental unit of electricity. Therefore, the coordinated flow of millions of these particles represented the underlying mechanism of electrical current, whether the current flowed through copper wires or human bodies. As such, cathode ray particles were renamed to acknowledge their fundamental role in electricity. Henceforth, they were called *electrons*.[22]

The discovery of the electron also spawned the completely new field of *nuclear physics*. It became clear that, within an atom, charge is segregated between negatively charged electrons and positively charged protons. (The nucleus of a hydrogen atom is actually just a single positively charged proton; it becomes a neutral hydrogen atom when an electron associates with it.) The protons are bound together (frequently along with uncharged particles called *neutrons*) within a highly dense atomic core that we now call the *nucleus* of the atom. The electrons, on the other hand, exist in a cloud-like state around the nucleus.

This atomic structure has been depicted in many ways. The "solar system" model was one of the first.[23] It makes the nucleus the equivalent of the sun, and the electrons are depicted as tiny, fast-moving planets, circling the nucleus in discrete orbitals at various distances from it. The ones closest in are said to be tightly bound while the ones in the outer orbitals are only loosely bound, and it is these loosely bound electrons that are available to participate in chemical reactions. Let's take a minute to look at how they do this.

The outer orbital electrons are most stable when they are present as pairs. These *Lewis pairs,* named after Gilbert N. Lewis, who introduced the concept in 1916, confer stability because each electron of the pair has an opposite *spin*. Electron spin is a form of *angular momentum*.[24] We don't need to understand spin and angular momentum for our purposes. There are very few people (perhaps none) who actually do. But the importance is that the opposing electron spins within a Lewis pair balance attractive and repulsive forces and confer chemical stability to the atom as a whole. In contrast, if the outer orbital of an atom has an odd number of electrons, one of them is necessarily unpaired and the atom has overall unbalanced spin. By definition, an atom with unbalanced spin is electrically unstable.[25]

These electrically unstable atoms are chemically reactive, because their unpaired electron forces them to react with neighboring atoms in a quest to find another electron with which to partner and thus form a Lewis pair. But two different atoms with unpaired outer orbital electrons can satisfy their individual electron needs by sharing a couple of electrons. Thus, when two atoms, each with an odd number of outer electrons, come together to form a molecule, their combined complement of electrons is now even in number, which means all the electrons can be part of a Lewis pair and the molecule is stable.

The simplest example of this is the hydrogen atom, which is the smallest of all atomic elements. Each hydrogen atom is made up of just one proton and a single orbiting electron; it is, therefore, one electron shy of forming a Lewis pair. But two hydrogen atoms can come together to form a Lewis pair by sharing their electrons. The result is that most atoms of hydrogen (H) partner with another atom of hydrogen to form *molecular hydrogen* (H_2), which typically takes the form of a gas. Nitrogen gas (N_2) is likewise the consequence of two atoms of nitrogen (N), each with 5 outer orbital electrons, coming together to form a two-atom molecule of nitrogen (N_2), with a total complement of 10 outer orbital electrons (five Lewis pairs).

This sharing of outer orbital electrons between atoms produces strong links between the atoms, known as *covalent bonds*.[26] Covalent bonds are the strongest of all the known chemical bonds and represent the superglue that nature uses to link small atoms together to make large, complex biological *macromolecules*, like protein and DNA. In fact, covalent bonds are so strong that breaking them requires a lot of energy. Usually, energy in the form of heat isn't sufficient; it typically requires a strong electric current. And that's precisely what's happening when a solution of chemicals is subjected to electrolysis. The electric current is breaking up the starting chemicals' covalent bonds and new chemicals are thereby being produced. And the amount of electricity (i.e., charge) needed to do this is proportional to the amount of chemical being changed. This explains the stoichiometry of electrolysis. The more electrons (electrical charge) you put in, the more changed molecules come out. This means electrons actually do participate in chemical reactions in a quantitative manner, just as Faraday's laws of electrolysis suggest. The explanation has to do with the formation of Lewis pairs and covalent bonds.

It's interesting to ponder the consequences if atoms could not satisfy their need for paired electrons by sharing outer electrons among themselves. If that were the case, there would be no covalent bonding, no conversion of atoms into molecules, and no ability to produce the large macromolecules that make up the substance of cells. In short, there would be no life on this planet. One might say, it is this chemical property of electrons that enables life to exist.

But enough with all the chemistry. I hope I've said just enough here to convince you that the connection between chemistry and physics is mediated through the lowly electron. As such, it was inevitable that the physicists and chemists, each working on their own end of the tunnel, would ultimately converge on the electron, which resides squarely in the middle. Whether the electron is a chemical entity or a physics entity depends upon your perspective.

But the perspective of this book is on the role of large numbers of electrons moving together in synchrony, a phenomenon we call *electricity*. Let's now return to the implications of the discovery of the electron to the theory of electricity.

Thomson himself immediately recognized the significance of his newly discovered electrons to Franklin's single-fluid theory of static electricity:

> The "electric fluid" corresponds to the assemblage of corpuscles [electrons],[27] negative electrification consisting of a collection of these corpuscles: the transference of electrification from place to place being the movement of corpuscles from the place where there is a gain of positive [charge] to a place where there is a gain of negative [charge].[28]

These outer orbital electrons are also the electrons that are so loosely bound that mere mechanical actions, such as rubbing two materials together, can sometimes dislodge them. Take, for example, the static electricity generated by rubbing two dissimilar materials together, like the amber and wool we talked about in chapter 1. The outer electrons of amber are readily "scraped" onto the wool, which gives the wool an electron surplus and the amber an electron deficit.[29] The electrons that the wool has sequestered make the wool negatively charged overall, while the protons left behind in the amber leave the amber with an overall positive charge. The amber becomes as much positively charged as the wool becomes negatively charged because the overall combined charges must always balance themselves out.[30] At some point, however, the charge imbalance between the two materials is so great that the electrons jump back as a group to the amber, in the form of a visible spark, and both materials thus are returned to neutral charge. Now doesn't this explanation sound an awful lot like what Franklin described for his invisible fluid? In fact, the electrons actually are Franklin's invisible fluid, and electrons are the illusive moving substance that drives electrical currents. It's just that Franklin had no way of knowing which direction the "fluid" was moving. It was a 50/50 guess, and he guessed wrong. It happens.

The discovery of the electron was undoubtedly one of the transformative events of the twentieth century. Thomson and Millikan are inextricably tied to its discovery and were acknowledged for their efforts; each man won a Nobel Prize. But, in fact, the realization that the electron was the fundamental unit of electricity came as a very slowly developing epiphany. You could say it started when Faraday noticed that chemical reactions seemed to involve discrete units of electrical charge, or you might say it actually started when Franklin noted that electricity behaved like an invisible fluid. But you could equally argue it all started when the first human rubbed a piece of amber, generated a spark, and pondered how that happened.

Millikan himself said it best back in 1917:

> In popular writing, it seems to be necessary to link every great discovery, every new theory, every important principle, with the name of a single individual. But it is almost a universal rule that developments in physics actually come in a very different way. . . . Each research is usually a modification of a preceding one; each new theory is built like a cathedral through the addition by many builders of many different elements. This is preeminently true of the electron theory.[31]

And the evolution of electron theory didn't end with Thomson and Millikan. As classical physicists, these two scientists would have been hard-pressed to foresee what electron theory would look like now, 100 years down the road.

Electrons behave in many strange ways. These odd behaviors were not captured adequately by the principles of classical physics and thus spawned the modern field of *quantum mechanics,* which amounts to mathematical physics on steroids. Understanding quantum mechanics requires a change in mind-set. Einstein himself had problems getting his head around quantum mechanics. For that matter, he even had issues with electrons themselves. In the advent of the splitting of the atom, when Einstein was asked what he thought about all the new subatomic particles regularly being discovered, he dryly quipped, "You know, it would be sufficient to really understand the electron."[32] But, unfortunately, that kind of insight was going to require quantum mechanics and a whole lot of mathematical physics.

If you're not a mathematician, don't sweat it. Neither am I. Fortunately, we don't need to employ any quantum mechanics to understand how electricity interacts with the nervous system. We just need to understand that the electron is the most fundamental unit of electricity and that its physical and chemical properties allow it to interact with *neurons,* which are the most fundamental units of the nervous system. Remember neurons? We already touched on them a bit as the possible targets for electrotaxis in fish. Neurons are to the nervous system what electrons are to electricity; both constitute the basic components of signal transmission. And transmitting signals is critical to everything. No signals, no communication, no life.

It may be possible to understand electronics without knowing anything about electrons, but it is impossible to understand the nervous system without knowledge of neurons. It's time we discuss neurons in a bit more detail.

10

REDBUD

THE NEURON

> Life imitates art far more than art imitates life.
>
> —OSCAR WILDE

Winter disturbs me. It's not so much the cold I'm bothered by, for like the Scandinavians say, "There is no such thing as bad weather, just bad clothing." I've stocked my wardrobe with good warm clothing, so I can deal with the cold. It's the gloominess of winter, with its long dark nights and overcast daytime skies, that starts to wear me down and gets me longing for spring.

But an overnight snowfall can perk up even the dreariest of winters. Snow-covered trees, in particular, shimmer in the morning's sunlight, brightening our spirits as we start our day. Unfortunately, I live in Maryland, a mid-Atlantic American state where snowfalls are infrequent and forests are dominated by deciduous trees—species of trees that drop their leaves in the fall. A Maryland forest in winter is a tangled mess of leafless trunks and branches, all colored a monotonous gray that blends in well with an overcast sky. This likely explains why landscape painters and photographers prefer pine forests, which remain "evergreen" year-round, as the subjects of their winter art.

It is insulting to say someone "cannot see the forest for the trees." The expression implies the person is small-minded; so bogged down in subunits of a problem that he cannot see it as a whole. But for a Maryland forest in winter, the problem is the reverse: you cannot see the trees for the forest. The crisscross of similar-looking branches devoid of any distinguishing leaves or coloration makes it hard to pick out individual trees within the branching chaos that comprises the forest flora. Everywhere you look, things appear to be the same: a mess.

But come spring, all that changes. Perhaps to recompense for the cold and gray winters, Maryland forests are blessed with an early-blooming tree with beautiful flowers: the eastern redbud (*Cercis canadensis*). I have no idea why it's called a "red" bud because the color of its little pea-size budding flowers is decidedly hot pink. Be that as it may, a redbud tree in full bloom is visually striking apart from its vivid color because of the very unusual places its flowers sprout. Rather than producing flowers only at the terminal ends of their branches, as is the case with most flowering trees and bushes, the redbud's flowers sprout all over the tree's barky surfaces, including its trunk and branches. The result is a tree completely painted in hot pink, from top to bottom. And all this happens before the leaves fully return to the other trees in the forest. This means that in early spring any sighted person can clearly pick out the structural features of an individual redbud tree, even if that tree happens to be immersed within an entanglement of twisting gray branches of its still dormant neighbors.

Neurons, with their long and delicate branching projections, are the "trees" that make up the "forest" of our brains. Unfortunately, there are no "redbud neurons" that stand out from the rest to assist in identifying an individual neuron from its neighbors. Consequently, early neuroscientists believed the tangled mess of gray lines they saw when they inspected brain tissue under a microscope amounted to a matrix of filaments that were not necessarily connected to the cell bodies also seen dispersed throughout the matrix.

Some scientists believed the filaments to be little tubes that bring nutrients to the nerve cells. Others thought the matrix of filaments itself functions to transmit nerve signals.[1] The idea was that the matrix of filaments acts like a three-dimensional spider web. Vibrations, or other stimuli, at one end of the fibrous matrix are supposedly transmitted, wavelike, to distinct locations in the nervous system. This was thought to be similar to the way a struggling insect transmits vibrations across a spider web, thus alerting the attentive spider that it has prey to dispatch. The debate about what actually happens in the forest of the brain went on for a long time. With no ability to pick out individual trees, one theory was as good as another. But this too would change. And the change would come as much through art as through science.

Do you remember Santiago Ramón y Cajal, the schoolboy who described for us the 1860 lightning strike on the priest in the church belfry? Well, that same boy grew up to be a physician and scientist, and he made great advances in the field of neuroscience. He was the one who first saw, and recognized, the trees that constituted the forest of the human nervous system, which, in turn, allowed him to see the forest for what it actually was. Remarkably, he arrived at these scientific insights through his art.

The young Cajal wasn't much of a student, but he had a passion for drawing and painting. He painted primarily outdoor scenes, with a keen eye for the beauty of nature. Certainly, he would have painted redbud trees in early spring if he had ever had the opportunity to see them. But redbud trees don't grow in his home country of Spain.

Cajal was a highly skilled, self-taught artist. Not everyone, however, appreciated his art, particularly his father. Cajal's father was a country physician, and he didn't approve of his son's wasting his time drawing trees, or anything else for that matter. Cajal's father wanted his son to abandon art, which he thought was the certain route to financial doom, and to follow in his footsteps in the more financially secure career of medicine. After resisting for many years, Cajal finally gave in to his father's wishes and entered medical school.

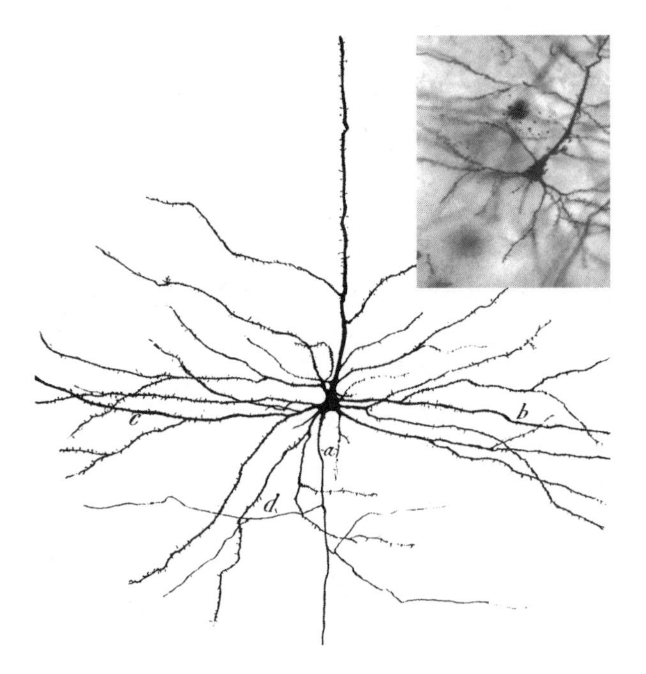

FIG. 10.1. Cajal's drawing of a brain neuron. Nobel laureate Santiago Ramón y Cajal is the author of the neuron doctrine, which says the nervous system is composed of a conglomeration of just one single fundamental element—the neuron—that can exist in various forms dictated by its specialized function and its anatomical location. Cajal worked at a time before microscopic photography. To record his scientific findings, he had to painstakingly draw everything he discovered from his countless hours of peering through a microscope. The richness and accuracy of his drawings can be appreciated by comparing his 1905 drawing of a pyramidal neuron of the brain with a modern photomicrograph of the same type of neuron (inset). (Cajal drawing from Science History Images/Alamy Stock Photo; light micrograph from Jose Calvo/Science Source)

Fatefully, in medical school, Cajal found a legitimate outlet for his artistic interests. In an age where photography was still in its infancy, there was a strong demand for accurate depictions of the human anatomy for medical books and journal articles. And that meant drawings, lots of drawings. Particularly needed were precise images of *microscopic anatomy*, the structures of various tissues as seen through the microscope.

Cajal was at first fascinated by what he saw through the microscope and drew the various cells of the body with enthusiasm. However, he soon became jaded. He described his state of mind at the time this way: "Now examine a drop of saliva, a little of the epithelium which covers your tongue, a drop of your blood, [etc.]—and always the same architecture appears: cells and more cells, more or less modified, repeating themselves monotonously, with wearisome uniformity."[2]

Nervous tissue, however, was very different. It had cells as well, but those cells were very beautiful. In fact, so beautiful they inspired Cajal to pen poetic soliloquies: "The nerve cell, the highest caste of organic elements, with its giant arms stretched out, like the tentacles of an octopus, to the provinces on the frontiers of the external world, to watch for the constant ambushes of physico-chemical forces."[3] Whew, that's a mouthful! I'm not sure this jumble of similes and metaphors makes for the best poetry, or even passible prose, but one thing is certain: neurons had made a big impression on Cajal.

But drawing neurons accurately was no bed of roses. (Please excuse the worn-out metaphor; Cajal got me going.) For one thing, there was the classic forest versus tree problem to deal with. But, in the case of brain tissue, it was the forest problem in the extreme. As Cajal lamented, for brain tissue, the problem was "finding out how the roots and branches of these trees in the gray matter terminated, in a forest so dense that . . . there were no spaces in it, so that trunks, branches, and leaves touch everywhere."[4] This forest of neurons that made up brain tissue needed some redbud trees. And Camillo Golgi would provide them.

Italian physician Camillo Golgi was a scientific contemporary of Cajal, and he too was interested in neurons. Golgi's work focused on better ways to stain cells, particularly neurons. This was very important work. For the most part, cells viewed under the microscope are invisible unless they are stained with contrasting agents that bind to and "paint" the cells. Sometimes these stains were nothing more than dyes, but often they were variations of the techniques used in photography to develop emulsions into prints, further adapted for delicate microscopy work.[5] Golgi employed such a photographic development approach for his staining. He first soaked slices of brain tissue with a solution of potassium bichromate for several days, and then dipped the saturated tissue into a dilute solution of silver nitrate. The result was the formation of a very dark precipitate of silver bichromate within the neurons. He called his technique the "chromate of silver method," but today it's just called *Golgi staining*.

A defining feature of Golgi staining is that it doesn't work very well. That is, it doesn't work very well in terms of its efficiency of staining neurons. Most neurons are not stained at all. For this reason, the method was largely ignored by many microscopists of the day. It is even described with disdain in some early neuroscience books. It took some practice to get Golgi staining to work, and few neurological microscopists had the time, patience, or desire to master it. So they largely criticized the method, sight unseen.

But the perceived weakness of Golgi staining turned out to be its biggest strength. Although Golgi staining fails to stain most cells, for some mysterious reason, a neuron sometimes fully takes up the stain. When that happens, the neuron's entire structure, from top to bottom, is completely visible under the microscope.[6] The result is that an individual neuron can be clearly seen as an apparently isolated cell, even though in reality the neuron is within the midst of a forest of unstained, and therefore nearly transparent and invisible, neighboring neurons. These Golgi-stained neurons are, in effect, the redbud trees within the forest of our brain.

Golgi himself realized the value of his own technique, and he made major contributions toward describing the fundamental morphology of individual neurons of all types. He described the neuron's *cell body* (or *soma*)—the part we earlier likened to the candy part of a lollipop. And he identified the single, long-stemmed projection, now known as the *axon*—the part we compared to the lollipop's stick. Although it previously served our limited purpose in describing how neurons are structurally polarized, comparing a neuron to a lollipop is laughably simplistic. And just like Franklin's invisible fluid model of electricity, a lollipop model of neurons gets us only so far in understanding how neurons actually work.

For neurons, the devil is in the details. The cell body comes in various shapes and sizes and contains the *nucleus*, the egg-shaped compartment that holds the cell's DNA. Coming out from the surface on the cell body are fibrous, branching projections that extend into the neuron's surroundings. One of these projections is much longer than the rest and extends some distance from the cell body. This long projection is the axon, while the other, typically much shorter projections are called *dendrites*. (Our lollipop model lacked dendrites.) In fact, some axons of human neurons can be over 3 feet (1 meter) long, although most are much shorter. The far end of the axon branches out like the frayed end of a long rope. Every neuron has these three basic components: a cell body, many dendrites, and a single axon. Simple enough. But after that, things quickly get complicated because cell bodies, dendrites, and axons come in all variations of shapes and sizes. Mixing and matching these multivariate components results in a host of different types of neurons, each presumably having its own unique role in the human nervous system. Golgi staining provided a means to sort out these morphological details and perhaps make some educated guesses regarding the function of the individual classes of neurons.[7]

Architects like to say "form follows function," meaning the function of the components of a building should be the primary determinants of their design. If this concept is a universal truth, shouldn't it be possible to deduce the function of neurons by studying their structural design? Cajal thought so. When he learned of Golgi staining, he adopted it wholeheartedly, honing and perfecting it for the various different tissues of the brain and spinal cord. Cajal took the technique well beyond anything Golgi had achieved, and did it on a massive scale. Furthermore, he drew precisely and accurately every detail of what he saw. Cajal's drawings of the various types of neurons and, more importantly, the patterns of their interactions with each other, allowed him to see some universal commonality within all nervous tissue and to propose what we now know as the *neuron doctrine*.[8]

The neuron doctrine says that the nervous system as a whole is composed of a conglomeration of single, fundamental elements—neurons—that can exist in various forms dictated by the individual neuron's specialized function and anatomical location. And these various neurons transmit signals from one area of the body to another just by interacting with each other through their dendrites and axons. Most importantly, the doctrine holds that neurons are distinct individual cells that never fuse together. The nervous system also includes nonneuronal cells that provide a supporting structure to the neurons, but the neurons rule when it comes to transmitting instructional signals from the brain to the various parts of the body. This, in brief, is the neuron doctrine. It claims the neuron is in full command as far as the nervous system is concerned. It is the fundamental unit of all nerve tissue, including the brain.

The neuron doctrine may seem self-evident to us now, more than a century later. But it wasn't so clear then. In fact, Golgi himself only slowly came to believe in the neuron doctrine. To Cajal, however, the neuron doctrine was unassailable. And he firmly believed it was specifically the axons of the neurons that "receive and propagate the nervous impulse, contrary to the opinion of Golgi, according to whom these parts of the cell perform a merely nutritive role."[9] Cajal thought Golgi's idea that axons provide nutrition was absolute nonsense. Alas, yet another scientific controversy was born. Just like the feud between Galvani and Volta, who disagreed over the source of animal electricity causing frog legs to twitch, Cajal and Golgi had looked at the same drawings of Golgi-stained brain tissue—the only available data—and had seen very different things. Cajal's art may have been imitating life, but art alone wasn't going to be able to settle the issue of what axons actually do.

The point where the tip of an axon or a dendrite communicates with another neuron is called a *synaptic junction*, or *synapse*. Due to the low resolution of the microscopic techniques of the day, it wasn't possible to definitively prove whether two neurons were merely interacting and communicating with each other at their synapses, similar to the way people do when they hold hands, or whether the two neurons were actually fused together, like conjoined (or "Siamese") twins. Therefore, the neuron doctrine proposed by Cajal required an act of faith from its adherents, just as most doctrines do. But believers in the neuron doctrine had a tremendous scientific advantage over those who were skeptics. The belief that the neuron is the fundamental and discrete unit of the nervous system provided an avenue toward understanding the underlying mechanism of the larger nervous system. If you could determine how an individual neuron relays the signal it receives from one neuron to the next, you could gain a fundamental understanding of the body's nervous system as a whole. In other words, it might be possible to better understand the forest if you had intimate knowledge of its trees. And as we already know, there were ample clues to suggest that the way neurons communicate with one another is by means of electricity. Unfortunately, drawing Golgi-stained neurons couldn't help decipher their electrical properties.

Cajal also suffered from the same problem as Faraday. He lacked the mathematical skills needed to take his studies to the next level. Biology, like physics before it, was becoming dominated by mathematics. Cajal regretted having neglected his studies as a boy: "[I] missed very much a knowledge of mathematics, which I ought to have learned when I had the chance."[10] But Cajal also believed mathematics, though essential to physics, unnecessary for the practicing biologist: "Being essentially descriptive, the biological sciences deal almost exclusively with the qualitative, which escapes quantitative determination."[11] That may have been true once, when Cajal was starting out, but it was no longer true later in his career, and it was particularly untrue for neurobiology.

Ironically, Cajal had once harbored a strong interest in physics: "How interesting I found physics. . . . Electricity and magnetism . . . with their marvelous phenomena, held me spellbound."[12] It's tempting to speculate that this interest might have originated from his unfortunate childhood experience with the lethal lightning strike at school. Regardless, he let his interest in physics wither, again being put off by the math. That too, was unfortunate. A better appreciation of electricity, in particular, would have served him well in his quest to better understand the brain.

The trouble with studying neurons is that they are small—very small. It might be possible to see them with the help of a microscope and staining, but even then, what you're looking at is actually dead. You are just seeing the corpse of a neuron, killed by the handling and staining process, and you're hoping what you see in the corpse holds true in life. But you aren't watching a living neuron doing its job in real time, so you can't know for sure.

Unfortunately, neurons aren't easily handled and manipulated, especially while still alive. This was the obstacle for *electrophysiologists*, the scientists who study the electrical properties of the body. They needed something like a Golgi stain to come along; something they could use as a tool to probe the electrical aspects of neurons. Enter the squid axon—the frog leg of the twentieth century.[13]

We have seen how valuable frogs were to Faraday, Galvani, and Volta in the study of a variety of electrical properties. Frogs may not be humans, but the way they kick their legs has parallels with the way humans kick theirs. Squid (*Loligo forbesii*) have tentacles, not legs. But they too have neurons. And according to the neuron doctrine, their neurons should work fundamentally the same way as any other neuron. Understand a squid neuron and you'll likely have gone a long way toward understanding a human one. But the same can be said of the neurons from any animal, so why study squid neurons? Because the axons of squid neurons are extraordinarily large.

When a frog wants to get someplace, it uses its legs to jump there. With each leg muscle contraction, the frog incrementally makes headway toward its goal. Likewise, when a squid wants to get somewhere, it squirts out a jet of water to propel itself, but it too moves incrementally, each squirt moving it ahead a little more. Both the frog jumps and the squid squirts are caused by the contraction of a muscle, a muscle under the control of a nerve fiber.

A nerve fiber is a bunch of neurons and their axons that are entwined, forming a string-like cord of nervous tissue. People sometimes call nerve fibers simply "nerves," but this can be problematic because we could confuse nerve fibers with the individual nerve cells, the neurons. A nerve fiber is composed of a bundle of neurons working together to transmit signals. The sciatic nerve fiber is the major nerve fiber in a frog's leg. (The ulna and the vagus nerve fibers are such examples in humans, and we will get to them presently.) But in the squid, what was once thought to be a nerve fiber that caused its mantle muscle to contract turned out not to be a bundle of neurons but rather just a single, giant axon of a lone neuron. The discovery was made at the Woods Hole Oceanographic Institution, on Cape Cod, Massachusetts, in 1936.[14] A fantastic discovery! What it meant was that scientists had found a neuron so large that it could be manipulated relatively easily for experimentation, particularly electrical experimentation.

If you're a cook or an angler, you may have had some occasion to cut up a squid either to make a calamari dish for dinner or to prepare bait for a fishing trip. If you sliced open the squid's body and spread out the *mantle*, the fleshy muscle portion that people and fish like to eat, you may have noticed on the interior surface of the mantle little "tubes." These tubes, which are about 0.5 millimeters (mm) in diameter, resemble the veins you see on the underside of a maple leaf. In fact, given its size, shape, and the array of tubes, the mantle does indeed look very similar to a maple leaf, albeit pink in color rather than green. But the tubes you see on the mantle are not "veins." They are gigantic individual axons of neurons, or, more specifically, *motor neurons*—the type of neurons that tell muscles to contract.

Now imagine exchanging your carving knife for a small probe with a delicate tip. Using a magnifying glass to better see, you slowly tease a delicate axon and its cell body away from the mantle it adheres to. Once you have done that, you have yourself a neuron—a complete and very big neuron. And if the squid is freshly killed, you have yourself a live neuron. Now what to do with it?

The Cambridge University physiologist Alan Lloyd Hodgkin had an idea as to how squid axons could be used, so he sent graduate student Andrew Field Huxley to Woods Hole to learn the technique of isolating squid axons for research. Upon Huxley's return to England, the two scientists first tried to determine the viscosity of the fluid within the axon by forcing a droplet of mercury through the axon and measuring the opposing force. I have no idea why they were interested in viscosity, but never mind; their approach didn't work and the project was abandoned. Nevertheless, in doing the work, Hodgkin and Huxley realized the squid axon's diameter was so large that it was possible to puncture it with a fine glass pipette. And if such puncturing was possible, it also should be possible to use the glass pipette to insert a tiny electrode into the axon. If they could do that, they had a method to detect any voltage difference across the axon's membrane. After all, it had been suspected since the time of Galvani and Volta that the nervous system had electrical properties of some kind. If that hypothesis was correct, it suggested that neurons—the fundamental unit of the nervous system—might also have intrinsic electrical properties. So why not stick an electrode in them and try to measure their electrical status?

When Hodgkin and Huxley did this, to their delight, they discovered a small but measurable voltage difference across the axon's membrane—70 millivolts (mV), or 0.07 volts. They also determined the inside of the axon to be relatively negative in charge with respect to the outside. So it's typically said that the membrane of a resting (i.e., unstimulated) axon has a negative 70 (or −70) mV *resting potential*. The implication of the finding was that a neuronal cell membrane evidently has very low electrical conductivity and acts as an electrical resistor, or barrier, to the flow of current across it. Otherwise, the charge between the inside and outside of the cell would quickly equalize and the voltage reading would be zero. Their nonzero reading had important implications for the biology of neurons.

By acting as a resistor, the neuron's cell membrane essentially is sequestering electrical charge within the cell, similar to the way the glass of a Leyden jar confines electrical charge within the jar. Of course, large Leyden jars can store a lot of electrical charge and reach very high voltages (as much as 70,000 volts). Neurons can only store less than a billionth of a coulomb of charge and attain a voltage of just a few thousandths of a volt.

The neuron's internal charge and the voltage level across its cell membrane is extremely small compared to that of a Leyden jar, but then again, the size of a neuron is also relatively small. But what does this mean in terms of the relative strengths of the electric fields for a Leyden jar and a neuron? Is the electrical field strength of a neuron, like its charge, voltage, and cell volume, also extremely small?

When we discussed electric fields earlier, we mentioned that their field strengths can be measured in terms of volts per meter—the voltage level between the electrodes divided by the distance between them.[15] It makes intuitive sense that electrical field strength drops with distance such that, if you double the distance between electrodes, you might expect to halve the field strength; and the physics shows this to be true. Using this principle, let's calculate the electrical field strengths for Leyden jars and neurons and see how the values compare.

As we've already mentioned, very large Leyden jars can reach voltages as high as 70,000 volts. Assuming a glass thickness of about 10 millimeters (0.010 meters), the electrical field strength of such a powerful Leyden jar is

$$70{,}000 \textit{ volts}/0.010 \textit{ meters} = 7 \text{ million volts per meter}$$

That's an impressive value, as Franklin and his turkeys could attest. Now let's apply this same field strength calculation to the neuron. As we said, the voltage across the membrane is a minuscule 70 millivolts (0.07 volts), but the thickness of the membrane is also small. Its thickness is no more than 10 nanometers (0.00000001 meters). Thus, the electrical field strength across the membrane is

0.07 volts/0.00000001 meters = 7 million volts per meter[16]

The two field strengths are the same!

This chapter is already drenched in metaphors, so what would it hurt to add one more? Let's do it. From our calculation above, we are led to conclude that a neuron is, in effect, an extremely small but very powerful Leyden jar. And just as discharging a powerful Leyden jar got the immediate attention of a long chain of hand-holding monks, so will discharging a neuron get the attention of its downstream neuron neighbors that are "hand-holding" by means of their axons and dendrites.[17] And, with this realization, we are getting very near to understanding the underlying mechanism by which the nervous system transmits its signals.

Earlier, when we were considering the structure of neurons, we used the image of a lollipop as a crude physical model of what neurons look like. But now that we appreciate that they act a lot like Leyden jars, it's time we upgraded our model. Replacing our lollipop model of the neuron with a Leyden jar model has some advantages. Lollipops might crudely model the shape of a neuron, but a Leyden jar model incorporates the neuron's electrical properties, and the electrical properties are what bring us closer to understanding the neuron's true function. Nevertheless, both models are limited in that they are only physical models of neurons. Often, the greatest insight comes from mathematical models.

In their essence, mathematical models are simply equations (although some of the equations themselves aren't very simple) that tell us what we might expect to see as a value of y given a value of x. For example, if x happens to be the compass direction an ocean hurricane is moving in, the mathematical model might predict a y value that equates to the location of the coastline where the hurricane likely will make landfall. For any situation, there can be multiple mathematical models proposed by scientists, some better than others. That's why weather forecasters usually show us the hurricane tracks from several models, each of which is based on a different equation. But a mathematical model proves its worth by consistently predicting actual events. If the hurricane doesn't make landfall where a particular model predicted it would, we lose confidence in that model. If it fails repeatedly, we would conclude the model provides no insight into what's actually going on with hurricanes, and abandon it in favor of a model where the real events better fit with the events predicted by the model.

An excellent model would be one where the actual outcomes turn out to be very close to the predicted outcomes (i.e., there is an excellent "fit" between the actual and the predicted data). When that's the case, we start to believe the model might be able to provide some basic insight into the underlying mechanism of what's going on. For example, if inputting barometric pressure data is a critical parameter in a good hurricane tracking model, we might start to believe barometric pressure could be fundamentally involved in the mechanism of storm progression.[18] With that clue, we can explore barometric pressure further, by other scientific means, to confirm or refute this hypothesis.

Hodgkin came from a mathematical modeling background, so his first inclination was to approach the axon as though it behaved mathematically like a simple electrical circuit, with the membrane providing both electrical resistance and an electrical switch, turning the flow of current across the membrane on and off, in the same way a wall switch controls room lighting.

The trouble was that this simplest of circuit models didn't match the squid axon time course data very well. That is, the measured changes in axon voltage over time suggested there was more going on than the opening and closing of a single switch. To get their squid axon voltage model to better fit the actual data, Hodgkin and Huxley tweaked the model. They added a second branch, in parallel, that had a different current level and its own switch. When Hodgkin and Huxley did this, they found their experimental data fit their circuit model exceptionally well.[19] Their new and improved circuit model became the foundation for a mathematical equation describing the change in membrane voltage with time at any point along the length of the axon. This mathematical equation is known as the *Hodgkin-Huxley model.*

The Hodgkin-Huxley model's scientific value is that it is able to predict the "shape" of *action potentials*—electrical pulses that move down an axon when the cell body of the neuron is stimulated.[20] These electrical "pulses" (sometimes called waves) behave similarly to pressure pulses that move through arteries when the heart contracts.[21] Who hasn't felt their own pulse by pressing a finger against their *radial artery,* the surface artery of the wrist? What you feel is the momentary rise and fall of pressure as blood surges by on its way from your heart to your fingertips. An action potential is like that. It is a pulse in voltage at a point along the axon as the electrical signal coming from the cell body passes by on its way to the dendrites.

When we say the shape of the action potential, what we actually mean is the shape of the line on a graph when we plot the voltage across the axon's membrane versus time (with time on the x axis and voltage on the y). The −70 millivolts resting potential I mentioned earlier is actually the steady state baseline voltage of the membrane before a pulse arrives. Graphing the action potential allows us to see how the voltage changes with time as the signal passes. Such a graph of an action potential has a very characteristic shape. Starting at time zero, an arriving signal causes the membrane voltage to rapidly shoot up from its initial −70 millivolts all the way into positive voltage values, reaching as high as about +40 millivolts—a 110-millivolt rise. Then, nearly as quickly, the voltage drops back down toward the resting potential. But, curiously, it overshoots −70 millivolts, going as low as −85 millivolts before slowly rising to its starting value of −70, where it remains until the next signal arrives. It was this unusual shape of the action potential that Hodgkin and Huxley were trying to simulate with their mathematical model of an electrical circuit with two parallel branches, each independently controlled by its own switch. It was a very simplistic model indeed, but it seemed to work. It nearly perfectly predicted the very curious shape of a squid axon's action potential.

At this point, you are probably asking yourself, So what? Why should anyone care about the shape of the action potential of a squid axon or the Hodgkin-Huxley model's remarkable ability to predict it? The answer is this. If we assume that there is nothing special about the squid axon (i.e., its electrical behavior is no different than the axon of any other neuron from any other animal, including humans), then what is true for squid neurons is also true of the neurons in a brain. And that should be a topic of great interest to everyone who has a brain. And it also means the Hodgkin-Huxley model might be the key to fundamental insight as to how the brain actually transmits its information at the electrical level. Furthermore, the fact that the Hodgkin-Huxley model is relatively simple mathematically, with regard to the very limited number of parameters it contains, could very well mean that the mechanism of electricity's passage through a neuron via its axon is also relatively simple, with just a limited number of biological components, thus providing a realistic hope that we might be able to figure out exactly how it works.

Of course, it also is possible to fit the action potential data using much more complicated circuit models, involving multiple branches and multiple switches within the circuit. But most scientists are adherents to the principle of *Occam's razor*—the belief that the simplest explanation for something is likely to be the correct one.[22] Applying Occam's razor to the electrical modeling of action potentials of the axon membrane suggests Hodgkin and Huxley might very well be correct. The squid axon must contain something equivalent to two electrical switches, or gates, each controlling independent branches of current flow, similar to a parallel electrical circuit. Each switch controls the flow of current through a different branch of the circuit, and the two branches have different currents. This was the simplest circuit model consistent with the time course of the axon potentials measured in squid axons, and it was the one Hodgkin and Huxley believed to be true. And the

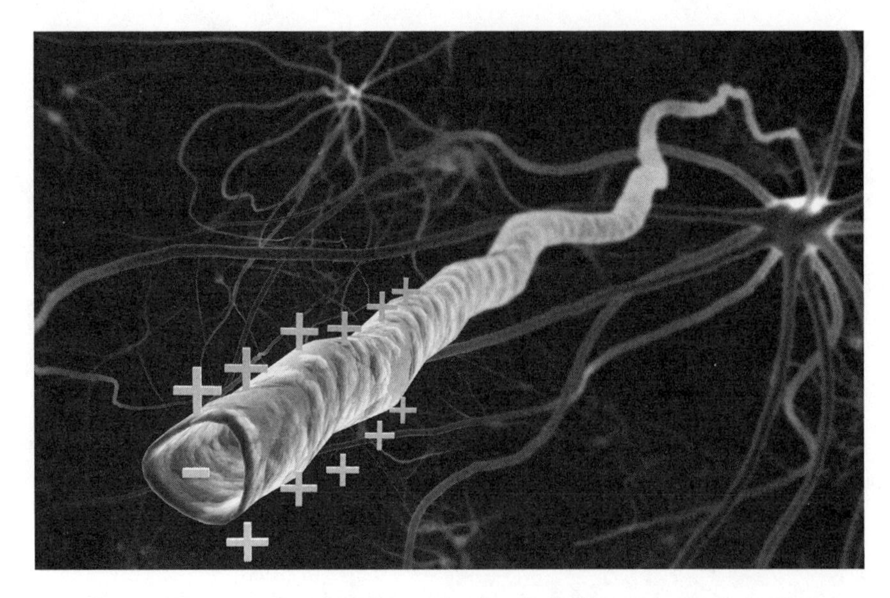

FIG. 10.2. Neurons transmit signals to each other through action potentials. Neurons connect to each other in sequence and pass along electrical signals from one neuron to the next by transferring a pulse of electrical charge (action potential), similar to how a bucket brigade transports water. The signals travel from neuron to neuron via long microscopic tubes called axons, which might be thought of as the nervous system's electrical wiring. But, unlike wires, axons are not electrical conductors. Rather, they transmit their signals by carefully managing the levels of specific ions (charged atoms) in their interior. By moving some types of ions in and others out, they are able to maintain a steady state of interior charge that is slightly negative with respect to the outside. Shown here is an artist's rendition of a neuron, its cell body in the background and its axon projecting forward. The axon is cut to show its hollow interior and the charge differential across its membrane. It is poised to receive the action potential (bright spot) that is being generated near the axon's connection to the neuron's cell body. When it arrives, the positive charges momentarily move in and the negative charges move out (the opposite of what is shown), until the axon's membrane restores its normal steady-state charge differential. (3D4 Medical/Science Source)

take-home message is that there are just *two* routes by which an electrical charge can pass through an axon's membrane, not three, four, or five.

The number two came as a big a relief. It meant that if you could figure out the physical identities of just these two "things" that allowed electricity to cross the axon's membrane, whatever they were, you would be closer to understanding the exact mechanism by which neurons transmit their messages. Once again, a model of a neuron—this time a mathematical model rather than a physical one—was pointing the way toward a better understanding of the underlying neurobiology.

However, the Hodgkin-Huxley model raised as many questions as it answered. Among the most important ones: What is the physical nature of the two electrical switches in the axon's membrane, and how is the neuron able to maintain a constant steady-state voltage across that membrane? But let's put these questions on the back burner. The answers will come in due time. For now, let's just be satisfied that, thus far, we have identified both the fundamental unit of electricity—the electron—and the fundamental unit of the nervous system—the neuron. Neither of these alone would be sufficient to understand electricity's role in in the human body; but, taken together, the possibilities for insight are limitless.

We often hear that people have different learning styles. Golgi and Cajal were visual thinkers; they deduced function from what they saw and drew. Hodgkin and Huxley thought mathematically; they calculated their way toward understanding. Although drawing skills and calculating skills are two distinct human talents, each can provide insight into a scientific mystery. Golgi and Cajal were never going to figure out how neurons transmit signals just by drawing them, but they did identify the aspects of a neuron's design that were amenable to the transmission of electrical signals, particularly the long axons that seemed analogous to electrical wires. Stimulated by Golgi's and Cajal's work, Hodgkin and Huxley investigated the electrical transmission properties of axons and deduced, through electrical measurements and mathematical modeling, a mechanism that somehow involved two parallel circuits of current flow with two independent switches to regulate that flow. But the actual physical identity of these circuits and switches was inaccessible by their methods. It would take future scientists, using still different approaches, to bring us closer to an understanding of the actual mechanism neurons use to transmit their electrical signals. Those future scientists would, in turn, build their work on the findings of their predecessors—Golgi,

Cajal, Hodgkin, and Huxley. So it is a very fitting tribute that all four scientists won Nobel Prizes in Physiology or Medicine for their pioneering laboratory research.

But I think this chapter has spent way too much time in the laboratory looking at very small things through microscopes. And all that talk in the beginning of the chapter about Maryland's bleak winters has gotten my spirits down. It's time we get out of the confines of the laboratory and head back out into the field, preferably someplace with a warm climate.

11

ZOMBIE PEOPLE

VOLTAGE-GATED CHANNELS

I had a dream last night I was piloting a plane,
And all the passengers were drunk and insane.
I crash landed in a Louisiana swamp,
Shot up a horde of zombies, but I come out on top.

—THE ROLLING STONES, LYRICS FROM THEIR SONG "DOOM AND GLOOM"

New Orleans, Louisiana, is the most eccentric and eclectic city in the United States. Mention New Orleans to people and various images flash into their minds, including Mardi Gras, jazz bands, and both Cajun and Creole cuisines. And most of these images are centered at a single mid-city location: the French Quarter. The historic French Quarter, with its frenetic nightlife and colorful buildings adorned with cast-iron balconies, is the heart of the city. And the nightclubs and restaurants of Bourbon Street—the French Quarter's main thoroughfare—keep that heart beating, day and night. Still, not all of the French Quarter's charms can be found on Bourbon Street.

My wife and I are in New Orleans to attend our son's wedding. The wedding is a few days away, so we have a couple of days free to do some touring. This isn't my first visit. I had been to New Orleans several times before on business trips. And on each of those trips, I made time for a brief visit to the French Quarter to experience its unique eateries and hear some live music, but I never strayed from Bourbon Street. This time, however, I am determined to see all the attractions the French Quarter has to offer, so we set off to explore some of the side streets. Turning a corner, we come upon a small shop: Marie Laveau's House of Voodoo. On seeing the shop's sign, I am reminded that New Orleans is also famous for its voodoo. In fact, New Orleans is the epicenter for the practice of voodoo in the United States. Curiosity about the shop gets the best of me. I go inside.

As I enter the shop, I am more pounced upon than greeted by a very aggressive saleswoman. "What can I help you with, sir? We have some nice decks of tarot cards on sale. We have all kinds of voodoo figurines, or perhaps a human skull candle will interest you?" My senses are assaulted by the sudden appearance of this hyperactive woman and the overflowing assortment of odd and scary-looking products for sale. I'm feeling a little overwhelmed and consider leaving. Then, suddenly, I'm able to elude the saleswoman when she's distracted by another potential customer who entered the shop just behind me. I'm left alone to explore the shop unpestered.

But the plethora of wares the shop offers goes well beyond anything related to voodoo. Take those tarot cards the saleswoman is hawking, for example; they have nothing to do with voodoo. They originated in fifteenth-century Europe as gaming cards and were later used in occult rituals to tell fortunes. And the "voodoo dolls" the shop is selling near the cash register aren't actually associated with voodoo either, despite their name. Such effigy dolls with pins in them were first used to ward off witches in Britain, when witches were thought to be the source of all bad luck. It wasn't until the twentieth century, when pop culture adopted the little dolls, that they were erroneously linked to voodoo. And what about the four-leaf-clover medallions the shop sells? Everyone knows they are Celtic charms. Yes indeed, this shop appears to be a fraud. Much of its wares have little to do with authentic voodoo.

Real voodoo is actually a religion, originating in Africa. Voodoo, as practiced in New Orleans, has evolved into an amalgam of African customs mixed with both Roman Catholic and Native American traditions, which likely were brought into the voodoo practices by converts. Voodoo practitioners believe in a visible and invisible duality of life and that both life-forms can alternately merge and separate from each other. Death marks a transition of the spirit from the visible to the invisible world, but it is also possible for the invisible existence to transition back to the visible.

You may be wondering how a guy hailing from New Jersey knows so much about voodoo. I first learned voodoo was a religion from a friend of mine of Haitian ancestry. If New Orleans is the epicenter of voodoo practice in the United States, Haiti is the epicenter of voodoo in the Americas. Voodoo, as practiced in Haiti, is closer to its African origins than what's practiced in New Orleans. In fact, voodoo probably arrived in New Orleans by way of Haiti, through the African slave trade, which was active in both locations. In Haiti, voodoo is still practiced primarily by the descendants of slaves, while most other Haitians dismiss it as superstitious nonsense. This much I learned from my friend. But apart from its ordinary religious traditions, voodoo also involves zombies—people who allegedly have been raised from the dead by voodoo priests. And what I know about zombies, I learned at Harvard.

In the mid-1980s, I was working as a postdoctoral fellow at Harvard Medical School. I was doing research in a biochemistry laboratory at the Dana-Farber Cancer Institute, trying to purify the various proteins that cells use to protect their DNA from damage caused by radiation. The idea was that such proteins might also protect against cancer (a contention that ultimately proved to be correct). The work was having its ups and down, and progress was slow. The setbacks were wearing me down. And it didn't help matters that, across campus, there was another protein purification project meeting with spectacular success. A graduate student was trying to identify the active ingredient in a Haitian zombie potion, and he seemingly had made an amazing discovery.[1]

The student was Wade Davis. He was working on his PhD in *ethnology*, the study of the characteristics of various peoples and how they differ from one another. For his research project, Davis's mentor had sent him to Haiti to find and bring back the rumored potion that voodoo priests use to make zombies. The mentor had been attracted to voodoo by a BBC News report of a man named Clairvius Narcisse, who apparently showed up in Port-au-Prince, Haiti, claiming to be a man for whom authorities actually had an official death certificate. Narcisse said he had been dug out of his grave, raised from the dead by a voodoo priest, and forced to work as a slave on a remote plantation until he ultimately escaped. His body scars and other known physical identifiers were consistent with the deceased's description. And he knew personal details of the dead person's life that only that person, or someone very close to that person, could have known. Even family members of the deceased were persuaded the man was their dead relative reincarnated. But Narcisse wasn't alone in his claim of being a zombie. Other Haitian people thought to be dead—both men and women—had sometimes reappeared in semicatatonic states after years of absence, with the locals attributing their return to "zombification" by voodoo priests. Attempts to provide a unifying medical explanation as an alternative to the multiple alleged zombifications were largely unsuccessful.[2]

Authorities were stumped. The death certificates appeared to be legitimate, having been written in reputable hospitals by licensed physicians. The exact causes of death were typically undetermined, but this particular man had arrived at the hospital suffering from some type of acute, sudden-onset nervous system disorder and died very soon thereafter.

Davis's mentor hypothesized that these alleged zombies were actually victims of a *neurotoxin*, a poison that attacks the nervous system.[3] Specifically, he believed it to be a poison derived from one of the poisonous plants known to grow in Haiti. His theory was supported by the fact that these poisonous plants weren't native to Haiti but invasive transplants from Africa. In fact, these invasive plants were indigenous to the same region of Africa where the voodoo religion originated. The idea was that both the voodoo religion and the poisonous plants were brought from Africa to Haiti via the slave trade, and that the toxic plants were important to voodoo rituals. Also consistent with the idea was the

FIG. 11.1. Cover of *Harvard Magazine*, January–February 1986 issue. A Haitian man named Clairvius Narcisse, pronounced dead by medical authorities in 1962 and buried, mysteriously reappeared in 1980. He claimed to have been the victim of a voodoo priest's zombie potion, putting him into a deathlike state. He allegedly was then removed from his grave and enslaved on a plantation until he finally escaped. His story led a Harvard researcher on a quest to find the ingredient in zombie potions that could induce a deathlike state. What the researcher found was a natural neurotoxin that interferes with the transmission of electrical signals through the nervous system. (Cover of the January–February 1986 issue of *Harvard Magazine* (88:3). Copyright © 1986 Harvard Magazine Inc. Original photograph taken by Wade Davis. Reprinted with permission. All rights reserved.)

fact that one of the plants (*Datura stramonium*) is a member of the nightshade family and has the local common name of "zombie cucumber," which might be an allusion to its possible use in a zombie ritual.

Davis's mentor theorized that these "resurrected" people had been intentionally poisoned by a voodoo priest with a dose of neurotoxin that suppressed the nervous system to the point where people appeared to be dead. They later revived, either because the poison's effects wore off or because the voodoo priest had administered some kind of antidote, thus simulating bringing the "dead" person back to life. An interesting theory, but it needed validation. That's where Davis came in. He was to go to Haiti, secure some of the zombie potion, and bring it back to Harvard for biochemical analysis.

To cut to the chase, Davis was successful in his mission of finding voodoo priests in Haiti and obtaining from them various samples of their zombie potions, which he brought back to Harvard and analyzed. As it turned out, there were no plant-based neurotoxins in them, but they did isolate tetrodotoxin from the various concoctions.[4]

Tetrodotoxin is a poison that gets its name from its presence in the organs of an order of fishes known as Tetraodontiformes. This order encompasses 349 different fish species, including 120 species of pufferfish, many of which are poisonous. Most people are familiar with pufferfish, otherwise known as blowfish, because they have the curious behavior of blowing themselves up into the shape of a ball when they feel threatened by potential predators. This behavior increases their apparent size and supposedly deters some of their smaller predators. But I would also think some of their larger predators would see the apparently larger fish as an even more tempting meal, so I'm somewhat skeptical of this predator protection theory. Regardless, if a predator does decide to eat the distended little fish, the tetrodotoxin it contains exerts its revenge, and the predator's prospects for survival are gloomy. And if the predator is lucky enough to survive such an unpleasant experience, I suspect it thinks twice before eating another pufferfish, whether inflated or not.

Tetrodotoxin has long been known to toxicologists because of human deaths associated with eating pufferfish. The problem is that pufferfish is a good eating fish as long as you just eat the muscles. But tetrodotoxin lurks in the fish's organs. If you eat the organs, or if the organs contaminate the meat during food preparation, death is a very possible consequence.

The localization of the toxin in the organs and not the muscles of the pufferfish was first suggested by a 1774 incident that happened to the crew of a ship sailing under the command of Captain James Cook, the British naval explorer often credited with the discovery of the Hawaiian Islands.[5] The sailors had prepared themselves a meal of pufferfish, saving the flesh of the fish for themselves, while feeding their onboard pigs the fish entrails. Afterward the crew experienced mild symptoms, such as numbness and breathing difficulties, but all the pigs died.

Tetrodotoxin is a problem in Japan, where pufferfish, known there as *fugu*, are considered a delicacy. But fugu is typically only eaten in≈restaurants where the chef has been licensed and is trained in its safe preparation. When eaten in such restaurants, fugu is completely safe. Virtually all of the annual pufferfish poisoning deaths in Japan are the result of home cooking. And the victims die from neurological distress.

Tetrodotoxin is toxic to the nervous system because it interferes with the flow of electrical signals through axons. If the electrical signals are totally blocked, every bodily organ that depends upon those signals fails, and the organism that consumed the tetrodotoxin is doomed to death. And it is not coming back to life.

This is our first encounter with a chemical that interacts with the body's electrical signals. Let's explore the details of how tetrodotoxin is able to do this.

I said earlier that an electric current is the rate of flow of electric charge past some particular point. Notice I did not specifically say "negative charge"; I just said "charge." Yes, I know; when we discussed the electron, my explanation left the impression that all electric currents are due to the movement of negatively charged electrons. I did that because for most electronic circuits that is exactly the case. But that's not so for electric currents in neurons. The electric current across the cell membranes of neurons is due to the flow of positive charge, not negative charge. If you feel I've deceived you, please accept my apology. I hope you aren't now skeptical of me. But I thought telling the half-truth was justified in order to illustrate a larger point: electrons represent the most fundamental unit of electrical charge. Now that we all appreciate that point, let's look at a situation where, remarkably, you can have an electric current without moving electrons.

As we discussed in chapter 9, the importance of the electron largely has to do with its chemistry. Electrons make up the "glue" that holds atoms together to form molecules. In the case of metallic elements, the atomic nuclei have only a loose grip on the electrons in their outer orbitals. This permits metals to act in a special way as far as electricity is concerned. The fact that the positively charged nuclei of metals don't have a solid grip on their outer negative electrons means that the atoms of metals readily *ionize* into positive ions when external forces dislodge one of their outer electrons. These charged atoms (i.e., *ions*) have a net positive charge because their vagabond outer electrons have gone missing, taking their negative charge along with them. What this means, in practical terms, is that for a solid piece of metal, like a copper wire, only the electrons are available to take a journey; the positively charged atomic nuclei must stay put. That's why, for electrical circuits made from metals, the movement of electrons, and electrons only, is producing the current. They move in synchrony among the immobilized metal ions, hopping from nucleus to nucleus, as they travel within the wire.

But for other materials, such as salts—compounds that spontaneously form ions when dissolved in water—and ions that are dispersed within a gas, electrons are only a part of the story. In such cases, all the ions, both positively and negatively charged, are free to move within the liquids and gases, and thus participate in generating the electric current, moving along the lines of force of an electric field toward the attracting electrode. However, the large size of most ions retards their movement. As a result, any race between ions and electrons in either a liquid or a gas is always won by the electrons. Their extremely small size allows electrons to travel amazingly fast. Ions, in contrast, are relative slowpokes.

You might wonder why neurons put their money on the slowpoke. Basically, it comes down to a matter of control. Electrons, because of their extremely small size, are very hard to manage; they tend to go wherever they want, like little birds flitting about. But ions, being much larger and slower, can be herded like sheep and moved from one pasture to another by means of one-way gates. You cannot herd hummingbirds.

Also, these gates in neurons are so discriminating that they will allow a sheep to pass through but not a goat. In fact, there are two types of gates—one type recognizes and admits only sheep and another is just for goats. And, oh, I almost forgot—there is one last feature of these gates: they are controlled electrically.

So what happens when you stuff a lot of sheep into a pasture and then open the sheep gate? As you might guess, a lot of sheep are going to quickly run through the gate to freedom. And if each of those sheep were to carry a single positive charge, you would have a flow rate of charge (i.e., running sheep) passing a fixed location (i.e., the open gate). And that, my friends, is the definition of an electric current. Just count the sheep as they run through the gate over a fixed period (being careful not to fall asleep while you're at it), and you then will have a measure of the electrical current. No electrons required.

The axons of neurons actually have many such molecular gates embedded within their membranes.[6] And each gate controls the passage of one specific type of ion or the other—either the sheep or the goats—through the axon's membrane, allowing the specified type to enter or exit the interior of the axon. These gates, or "channels," are extremely small. In fact, the channels are so tiny that the ions can pass through only one at a time. The direction of flow—toward inside or out—and the specific type of ion allowed to pass are dependent on the exact type of molecular gate. And whether the gate opens or closes is determined by the local voltage immediately around the gate. As such, they are called *voltage-gated ion channels,* and for neurons they come in two main flavors—one type controls membrane passage of sodium ions (the sheep) and the other potassium ions (the goats).[7] Sodium's gated channel lets sodium ions flow into the axon, while potassium's gated channel lets potassium ions out of the axon.[8] That is, these two types of positively charged ions flow in opposite directions.

By now, you're probably asking yourself what all of this has to do with tetrodotoxin. After all, weren't we talking about tetrodotoxin? Yes, we were, and here's you answer: Tetrodotoxin locks the sodium gate. And locking that gate is not good.

I really should now explain how scientists came to know all about voltage-gated membrane channels, their two different types, and that tetrodotoxin specifically affects just one of the types. After all, if we can understand this, it should mean we're homing in on exactly how neurons work their electrical magic at the molecular level. And that's central to understanding how electricity and neurobiology interact. But as you've likely guessed by now, in order to understand things at such a fundamental level, we're going to need to delve into a little bit of chemistry—specifically, biochemistry. If you are not happy that a book you thought to be about the physics and biology of electricity now seems to have veered off course and headed toward chemistry, don't worry. I promise our foray into the world of chemistry will be brief but necessary. We'll be back on track in no time.

As we've already seen, compared to physics and biology, chemistry got absolutely no respect among mainstream scientists. In the words of Thomas Thomson, a Scottish scientist of the early nineteenth century who was among the first to study chemical phenomena in a scientific manner, "Chemistry, unlike other sciences, sprang originally from delusions and superstitions, and was at its commencement exactly on a par with magic and astrology." (You can easily see how people might have gotten that impression, considering we're still talking, two centuries later, about things like chemical potions that can create zombies!) But largely because of Faraday's work with electrolysis—the splitting of molecules with electricity—and the scientific conclusions resulting from that work, chemistry finally acquired its legitimacy. As Faraday put it: "Chemistry is necessarily an experimental science; its conclusions are drawn from data, and its principles supported by evidence from facts."[9] I totally agree. So let's push aside the superstitions and bring on the facts.

You'll remember that Hodgkin and Huxley were able to show that, when an action potential passes, the axon's membrane momentarily switches electrical polarity, going from negative inside (relative to outside) to positive. And they were able to mathematically model the time course of the action potential by envisioning the axon's membrane as a simple electrical circuit where current can flow by one of two separate routes, each with its own resistance and switch. The question was whether this theoretical model based on electrical circuitry represented any physical reality, or was just some fortuitous mathematical result that mimicked the actual data. Were there actually two different types of current pathways in an axon's membrane, as the mathematical model implied? And, if so, what exactly were these pathways physically?

Another pair of scientists—Erwin Neher and Bert Sakmann—decided they'd work together to find these two theorized electrical pathways in the axon's membrane. They hypothesized that the pathways, or "gates," would likely take the form of electrical channels floating within the membrane. So they invented a technique to grab hold of a small patch of the membrane of a squid axon and measure the flow of electricity across it. They reasoned that, if you grabbed a small enough patch, you could conceivably isolate a portion of membrane that contains one, and only one, of the alleged channels. And then you could measure the current through that one isolated channel. Accidentally trapping two or three channels in a patch should result in exactly twofold or threefold multiples, respectively, of the single-channel current flow. Therefore, it would be easy to deduce the current due to a single channel, based on the higher-level currents resulting from multiple channels being trapped.

This approach is directly analogous to the way Millikan had deduced the charge of an electron. As you'll recall, he had deduced the electron's charge from his measurements of the negative charge of microscopic oil droplets. The measured charges among different droplets always varied by exact multiples of $1.60217662 \times 10^{-19}$ coulombs and were never less than that charge value, providing strong evidence that the charge of a single electron was specifically $1.60217662 \times 10^{-19}$ coulombs and higher droplet charges were due to their carrying multiple electrons.

To achieve their goal, Neher and Sakmann developed what is now known as the "patch clamp" method. And as you can imagine, they immediately encountered some technical challenges. One was mechanical and the other electrical.

The mechanical challenge was how to grab a small patch of membrane. Neher and Sakmann ended up using extremely small pipettes with ultrafine tips. They touched the tips to the surface of the membrane and applied slight suction, which drew the membrane to the tip and produced a very tight seal around a very small area of membrane—an area they hoped would contain just a single channel. You can do something similar by applying one end of a drinking straw to your skin and then sucking on the other end of the straw so you form an airtight seal with the patch of skin. Believe it or not, as difficult as it sounds, they ultimately perfected this membrane-grabbing technique and were able to isolate a single channel in an axon's membrane.

But then Neher and Sakmann were immediately faced with an electrical challenge. If there was actually an electrical current flowing through a single channel, it would certainly be tiny—on the order of picoamps (a picoamp is one-trillionth of an amp, or 0.000000000001 amps). Electrical technology at the time was unable to measure such small currents, so Neher and Sakmann decided to invent the needed technology. They end up creating their own, extremely sensitive, *electrical amplifier.*

Nobel laureate William Shockley, the inventor of the transistor, had a colorful way of describing what an amplifier does:

> If you take a bale of hay and tie it to the tail of a mule and then strike a match and set the bale of hay on fire, and if you then compare the

> energy expended shortly thereafter by the mule with the energy expended by yourself in the striking of the match, you will understand the concept of amplification.[10]

So that's the concept. I think we all get it. But if you're trying to amplify current through a tiny patch of an axon's membrane to the extent you can actually measure it by an ammeter, a mule isn't going to help you.

The standard solution to the amplification problem is to use the small electric current flowing through the tiny circuit as a variable control on the flow of a much larger current through a much larger circuit. Since you know the relative magnitude difference between the two circuits (i.e., the *gain factor*), you can easily calculate the amperage of the current through the small circuit by simply dividing the amperage of the large circuit by the gain factor.[11] But currents at the picoamp level are much too low to use this technique. Another approach would be needed to measure the extremely tiny currents in which Neher and Sakmann were interested.

The researchers ultimately developed another, more sensitive and sophisticated, method to amplify the current enough to measure it. For our purposes, we don't need to know exactly how they did it. But the important point is that it took a long time, and in the interim the study of biochemical mechanisms of putative ion channels in the axon membrane had to be put on hold to allow the electrical science to catch up.[12]

This situation was similar to Volta's predicament in trying to develop an electricity meter to measure the current used in his frog leg experiments. As we discussed earlier, Volta was frustrated because no electricity meter he came up with was as sensitive as the leg of a frog. He pressed on, trying to make a highly sensitive electricity meter, with the frog leg itself being the gold standard for electrical sensitivity. These two situations illustrate that biologically significant levels of electric current are very far below the levels of normal electricity sensors and circuitry.

It took Neher and Sakmann many years to perfect their technique, combining their highly sophisticated electrical amplifier and their delicate patch clamp suction device. Finally, in 1974, they were able to detect and accurately measure a mere 1 picoamp of transmembrane current lasting less than a tenth of a second. But that was enough. They had confirmed the existence of ion channels in the axon membrane, an existence first suggested by a mathematical model. The achievement earned them the Nobel Prize in Medicine or Physiology in 1991. Never had such a minuscule level of electrical current made a more significant impact on science.

The nervous system's electrical signals are certainly weak in terms of their individual components, but taken together they can sure do some powerful things. The question is how exactly they can do that. It turns out that an important part of the answer lies in the momentary reverse in charge polarity across the membrane, which Hodgkin and Huxley had previously reported. In fact, the membranes had developed a part of their structure that could "sense" this polarity change and alter their structure in response to it.

The channels in these membranes are made of proteins, as are most of the structural features of cells. Proteins are very large biological molecules formed by linked chains of much smaller molecules—various types of *amino acids*. When a protein is stretched out, it resembles beads on a string, with 20 different types of beads, each differing in size, shape, and, of course, charge. Under normal physiological conditions, most of these different bead types are neutral—neither positively nor negatively charged—but three types of amino acids naturally have a positive charge and two types have a negative charge.[13] The membrane channels use the positively charged amino acids to create a voltage sensor.[14]

At one end of the protein is a structural feature, called a *domain* in biochemistry parlance, that is rich in the positively charged amino acids. The protein is embedded in the membrane, with its positively charged sensor protruding above the inner membrane surface into the interior of the cell. When the cell's interior is at resting potential, and thus negatively charged (−70 millivolts), the interaction of the positively charged domain with the negatively charged cell interior leaves the protein in a "relaxed" state and its channel remains closed. But if the interior suddenly becomes positively charged, as when an approaching action potential reverses the charge polarization of the membrane, the positive domain is then immersed in a positively charged cell interior. Since like charges repel each other, the positive domain is now in a repulsive rather than attractive environment. The repulsive force introduces *torsional strain* to the protein—something analogous to the twisting force you apply to the cap of a soda bottle to open it—and the molecule twists itself into a new shape that now has a channel, or hole, through its middle. As a result, positively charged sodium atoms that were being excluded from the cell interior by the impermeable axon membrane now rush inside through the hole, making the interior even more positive relative to the outside. Thus, a spike in voltage (an action potential) progresses further along the axon. This moving wave of voltage encounters more protein channels in the membrane, which respond in exactly the same way, so the action potential keeps moving along the length of the axon.

As we have discussed, the rapid upward spike in voltage as the action potential passes by is immediately followed by a decrease in voltage, returning the voltage distribution across the membrane to its original state of −70 millivolts. This reduction in voltage is caused by potassium voltage-gated membrane channels, which also sit in the axon membrane but in the opposite orientation to the sodium voltage-gated channels.[15] Instead of admitting positively charged sodium ions into the cell, the open potassium channel allows positively charged potassium ions to escape to the outside. Consequently, the rise in charge due to the entry of the positive sodium ions is counterbalanced by an equal and opposite exit of positively charged potassium, restoring the equilibrium voltage of −70 millivolts and leaving the axon ready to transmit its next signal.[16]

Yes, I know, the details are complicated and perhaps, for some, a little too much information. But the interpretation is not complicated. In essence, voltage-gated ion channels are simply pores in a neuron's axonal membrane that open and close in response to a flip in charge polarization caused by an approaching action potential. In other words, they are the two electrical switches that the Hodgkin-Huxley model predicted. They first admit positively charged ions and then immediately expel positively charged ions, which has the effect of propagating the pulse of the electrical signal further along the axon. Most importantly, these ion channels are selective in that they restrict which types of ions can pass through the membrane—a sodium gate will not admit potassium and a potassium gate will not admit sodium.

As when we were counting the rate of charged sheep running through a gate, we similarly have here a movement of charge (positively charged ions) through an opening (a membrane channel) at a specific rate (ions per second). Thus, we have an electric current. The current isn't large—even considering all the channels in an axon together, it collectively amounts to only a few nanoamps (about a billionth of an amp)—but that doesn't matter. It's sufficient to keep the electrical signal moving down the axon.

A novel feature of this current is that it doesn't involve electrons and the charge movement is in the same direction as the current movement. For negatively charged electrons, you'll recall, the charge flow is in the opposite direction to the way we define current direction (i.e., current is defined as *always* moving from positive to negative). But that was in a situation where the negative charge is mobile. When the positive charge is mobile, as in the case of sodium and potassium ions, the direction of the charge and current flow are not at odds. So, in the particular case of axons, Franklin appears to have been correct about the flow of charge. As far as axons go, the flow of the charge (i.e., positive charge) and the "current" do move in the same direction.

Now let's consider exactly how tetrodotoxin screws this up.

As I mentioned, people have known about tetrodotoxin for a very long time. It is more lethal than cyanide, and there have been many reports of its lethal effects on people over the centuries. But tetrodotoxin also has been used therapeutically for almost as long. A Chinese pharmacology book estimated to date from 2700 BC notes that eating a meal of pufferfish eggs is lethal but suggests that regularly consuming just a few eggs at a time, as one would swallow pills, could be a health tonic. A much later Chinese publication suggested that pufferfish eggs too toxic to be used for therapy could be presoaked in fresh water to reduce their toxicity. This likely is correct, given that tetrodotoxin is water soluble.

Tetrodotoxin was first isolated and its mechanism of toxicity determined in the mid-1960s. As it turns out, the toxin actually acts like a cork to plug the entryway to the voltage-gated sodium channel and thus block the sodium ions from entering the channel. This prevents the axon from transmitting action potentials.

If all nervous systems work by transmitting action potentials using voltage-gated sodium channels, the question might be asked how pufferfish keep from poisoning themselves. We had earlier asked a similar question about how electric eels keep from electrically stunning themselves, and the answer there was that no one yet knows. In this case, however, we know exactly why the pufferfish is unfazed by its own poison. It's because the entryway of the pufferfish's voltage-gated sodium channel has a slightly different shape than all other sodium channels, so the tetrodotoxin "cork" does not fit the neck of the bottle. Isn't that a nifty trick?

We still might wonder why this poisonous fish uses a neurotoxin as opposed to some other type of toxin. The answer to that is likely similar to the reason many other plants and animals harbor neurotoxins: to kill and deter a variety of animal predators, and do it quickly. Since all their animal predators have nervous systems, using poisons that attack the fundamental features of all nervous systems is likely to provide widespread protection regardless of the specific predator species. Animals deploying defensive strategies generally use poisons, while venoms are used for offense. So it is worth noting that tetrodotoxin is also used effectively as a venom in some species of venomous octopuses.[17] No sense wasting a good idea.[18]

Speaking of venom, we now know snakes, scorpions, and spiders have developed a host of venoms that target the other ion channel: the voltage-gated potassium channel.[19] And these venoms seem to be as effective as tetrodotoxin. Although the venom from these species seems to be a cocktail of many noxious agents, the components that target the potassium channels have been isolated and characterized.

The neurotoxins that specifically target one ion channel or another have been very useful to neuroscientists. By using them individually to specifically block one type of ion channel, and that type alone, they can focus on the function of the other types for detailed study of their mechanisms. Much of what we know about how ion channels work has been learned through the use of neurotoxins that can target the axon's Achilles' heel—the voltage-gated membrane channels—and selectively inactivate them. What we now know about how these neurotoxins work comes not from studies motivated by a public health need to find therapies for pufferfish poisoning, as such poisonings are pretty rare. Rather, the studies are specifically motivated by the need to learn the details of how normal neurons transmit electrical charge. Gaining an understanding of these details could lead to even broader therapies for a host of neurological diseases and not just poisonings.

Nearly 35 years have passed since Wade Davis first proposed tetrodotoxin was the active ingredient of zombie potion. During those 35 years, a lot of tetrodotoxin research has taken place. Do the latest data support or refute Davis's 1984 zombie theory? I decide to find out.

The first source I turn to is the *Skeptical Inquirer*, a magazine devoted to putting questionable scientific claims under intense scrutiny. It amounts to a more intellectual version of the television series *MythBusters*, a show that subjects questionable urban myths to on-camera semiscientific experimental validations.[20] The *Skeptical Inquirer* has been around for 40 years. The magazine has its critics, but I think it does a credible job of trying to set the scientific record straight on dubious scientific claims, leaving its readers to make a final ruling for themselves.

Sure enough, the magazine has evaluated Davis's tetrodotoxin and zombies theory in an article published in 2008 by Terence Hines, a professor of psychology at Pace University and author of the book *Pseudoscience and the Paranormal*. In his article, Hines states that Davis's hypothesis stretches credulity to the brink. He bases his evidence on inconsistencies in Davis's zombie reports with the known pharmacological characteristics of tetrodotoxin, differences in symptoms between the alleged zombies and bona fide tetrodotoxin victims, and discrepancies in finding significant levels of tetrodotoxin among various zombie potion samples.[21] My conclusion, after reading the article, is that he makes a very strong case against the likelihood of Davis's hypothesis being correct. But I must say, he doesn't provide any convincing evidence that actually disproves Davis's theory either. He certainly has strongly assaulted Davis's theory, but he doesn't put the final nail in the zombie's coffin, so to speak.

I, for one, don't much care whether Davis's decades-old theory about how voodoo priests make zombies is correct or not. Davis's tale is certainly entertaining in its own right. And his zombification story incorporates within it the story of tetrodotoxin—a story that is unquestionably true and, in my opinion at least, even more interesting. It's a story that chemically bridges the fields of electricity and neurobiology. How can it not be interesting?

Davis is now a professor of anthropology at the University of British Columbia and an explorer-in-residence at the National Geographic Society. He is still actively working in ethnobotany. Over the years, he has studied how native cultures interact with their local flora in various countries throughout the world. But he has long moved on from zombies, pufferfish, and tetrodotoxin. And so should we.

My curiosity satisfied, I make my way to leave the voodoo shop when one last gift item catches my eye: blessed chicken feet. Sold as matched pairs—left and right feet supposedly from the same chicken—they are advertised to bring their owner good luck. I'm familiar with a rabbit's foot bringing good luck. In fact, I owned a rabbit's foot strung on a key chain as a kid. I had actually won it playing a wheel of chance at a carnival. That proves it brings good luck, doesn't it? But blessed chicken feet are new to me and I'm not sure they represent authentic voodoo. Should I just accept the risk that they may be fraudulent and go ahead and buy myself a pair? My son's wedding is in two days; maybe they will ward off bad weather. Decisions, decisions.

While I'm pondering this, my wife, who has remained patiently outside the shop this whole time, suddenly becomes impatient and calls for me to speed it up. There are other things she wants to see. I comply and leave the shop empty-handed. Two days later, we have torrential rains precisely at the time the wedding begins. I tell no one about the blessed chicken feet I neglected to purchase when I had the chance, and I silently reassure myself that the rain probably isn't my fault.

In any event, I think we've learned as much as we can from stories of the living dead. Let's now spend some time with the dead dead, and see what tales they can tell us.

12

THE WANDERER

NERVE FIBERS

> Exploration is in our nature. We began as wanderers, and we are wanderers still.
>
> —CARL SAGAN

I'm no pathologist, but I am quite certain this man is dead. If nothing else, the stench of embalming fluid is a strong clue as to his condition. As we unzip the bright blue body bag to expose his pallid upper body, I'm thankful to see his face is draped with a cloth. I started out a bit queasy this morning and his blank stare might be a little too much for my stomach. So I'm quite happy we're not facing each other eye-to-eye.

I'm out of my depth in this endeavor and happy to have an expert guide along. Dr. Carlos Suárez-Quian is a professor at the Georgetown University School of Medicine. He knows his way around the human body, having directed Georgetown's human gross anatomy course for 15 years. When I told him of my interest in the *vagus nerve,* he graciously offered to give me a private tour of one of the most famous and mysterious of all the human nerve fibers. "Meet me at the anatomy lab at noon," he instructed me. "And don't worry. We have a gurney for rookies!" I'm actually not a complete rookie to human dissections, having once witnessed a pathologist perform an autopsy. Nevertheless, I'm thankful there's a gurney available where I can retreat, should I start to feel faint.

It's now noontime and here I am in the anatomy lab with Suárez. Both of us are garbed in aprons and rubber gloves, exploring this man's neck and torso to find his once vital vagus nerve. But why, you may ask, is the vagus nerve so important?

There are 12 pairs of nerve fibers in the human body, collectively called the *cranial nerves*. As *cranium* is the Latin word for "skull," you might expect to find cranial nerves localized to the head area. And they all are found there . . . except for one. The tenth (or Xth) cranial nerve—the *vagus nerve*—is a wanderer.[1] Not content to stay close to the head like its counterparts, the vagus nerve emerges near the base of the skull and heads south, far south.

The cranial nerves are fundamentally no different than the *spinal nerves*, which emerge from the spine, except they come from the cranium rather than the spinal cord. Most cranial nerves emerge from the *brain stem*, the neurological structure that joins the brain with the spinal cord, as first observed by the ancient Greek physician Galen.[2] However, two of the cranial nerves (I and II) are actually outpockets of the brain itself and, therefore, technically not "cranial" nerves. But it's not polite to contradict the great Galen, so we continue to call them cranial nerves to this day. (We should probably cut Galen some slack anyway, since he made these observations nearly 2,000 years ago using heads severed from gladiators.)

The various cranial nerves can be composed of either *sensory neurons*—the neurons that transmit messages from our sense organs toward the brain—or *motor neurons*—those that transmit signals away from the brain to move our muscles. For example, cranial nerves I and II transmit smell signals from the nose and sight signals from the eyes, respectively. And cranial nerves III and XII control movement of the eyes and movement of the tongue, respectively. But some cranial nerves incorporate both sensory and motor neurons.[3] The vagus nerve is one of the four cranial nerves (V, VII, IX, and X) that contain both types of neurons. This means the vagus nerve transmits action potentials in both directions—toward and away from the brain—just as a telephone cable transmits its voice signals both to and from a landline telephone.[4]

The anatomy lab currently has eight cadavers, both male and female, each of which has been prosected to a greater or lesser degree. A *prosection* is the dissection of a cadaver by an experienced anatomist so that beginning medical students can examine its anatomic structures. The cadavers in this room have had various parts of their bodies prosected in order to reveal different organ systems and body structures. Suárez and I are examining one where much of the upper body has been prosected and the lungs detached for easy removal. The lungs have been neatly covered in wet gauze to keep them moist, and carefully replaced at their original locations inside the body cavity. The result is a fully dissected chest, with easily removable parts, that allows students to examine actual human organs. They can see and handle in 3-D what they've previously only viewed as 2-D images in their medical textbooks. This gives the students a much better understanding of the anatomy of the organs they one day will be treating—a valuable learning experience for an aspiring physician.

In order for Suárez and me to get a complete look at the vagus nerve, we're going to have to look deep into the chest cavity. To do that, we must make some room and move the overlying chest structures out of the way. Suárez grabs the edge of a skin incision that runs the length of the chest and lifts back the flap of skin and muscles covering the chest cavity, just as though he were opening a book. This exposes the man's ribs. He carefully raises the previously detached rib cage and places it aside. He then gently lifts out the left and right lungs and puts them to the side.

Suárez explains to me that the best approach for tracking the vagus nerve is to intercept it as it passes through the neck, where the nerve fiber is a single relatively thick bundle of axons about the diameter of a shoelace, and then follow it down to its entry into the chest cavity. This particular cadaver has a very well-prosected neck and chest cavity; that's why we chose him to begin our tour.

In the upper neck, the vagus nerve is relatively easy to locate because it tracks alongside the *carotid artery*—the major blood vessel that carries oxygenated blood from the heart into the face and brain. You can easily locate your own carotid artery by pressing on the side of your neck to feel your pulse. What you're feeling is blood pulsing through the carotid artery. The vagus nerve doesn't pulsate, but when you feel your pulse in this way, you are also pressing on your vagus nerve. In the upper neck, the vagus nerve is superficial and thick, and it has the relatively large carotid artery as its traveling partner. So I think I could have found the main trunk of the vagus nerve in the neck on my own. But once the nerve gets to the level of the Adam's apple (cartilage of the larynx), things aren't so obvious; and that's when Suárez's expertise comes into play.

Suárez soon finds the first branching of the vagus nerve, which occurs even before it leaves the neck. Around midthroat, a nerve fiber containing both sensory and motor neurons splits away from the main nerve. It wends its way into the muscles of the larynx (voice box) and pharynx (the entrance to the esophagus), where it both controls speech and detects throat sensations. A little further down, the vagus nerve leaves the neck and enters the chest cavity.

When the vagus nerve enters the chest cavity, it largely leaves conscious nerve action behind and becomes the major nerve fiber of the *autonomic nervous system*, which is the part of the nervous system that works below the level of awareness to control the automatic functions of the internal organs, like heart rate and breathing.[5] The first instance of this happens when the vagus nerve shoots off separate branches to the heart and lungs. As mentioned, this particular cadaver has had its lungs removed, but Suárez points out the branch of the vagus nerve going to the still-intact heart. He notes that the vagus nerve enters the heart at the right atrium, and remarks that this is why surgeons performing some of the original heart transplant operations tried to keep as much of the atrial tissue from the old heart as possible. They wanted to

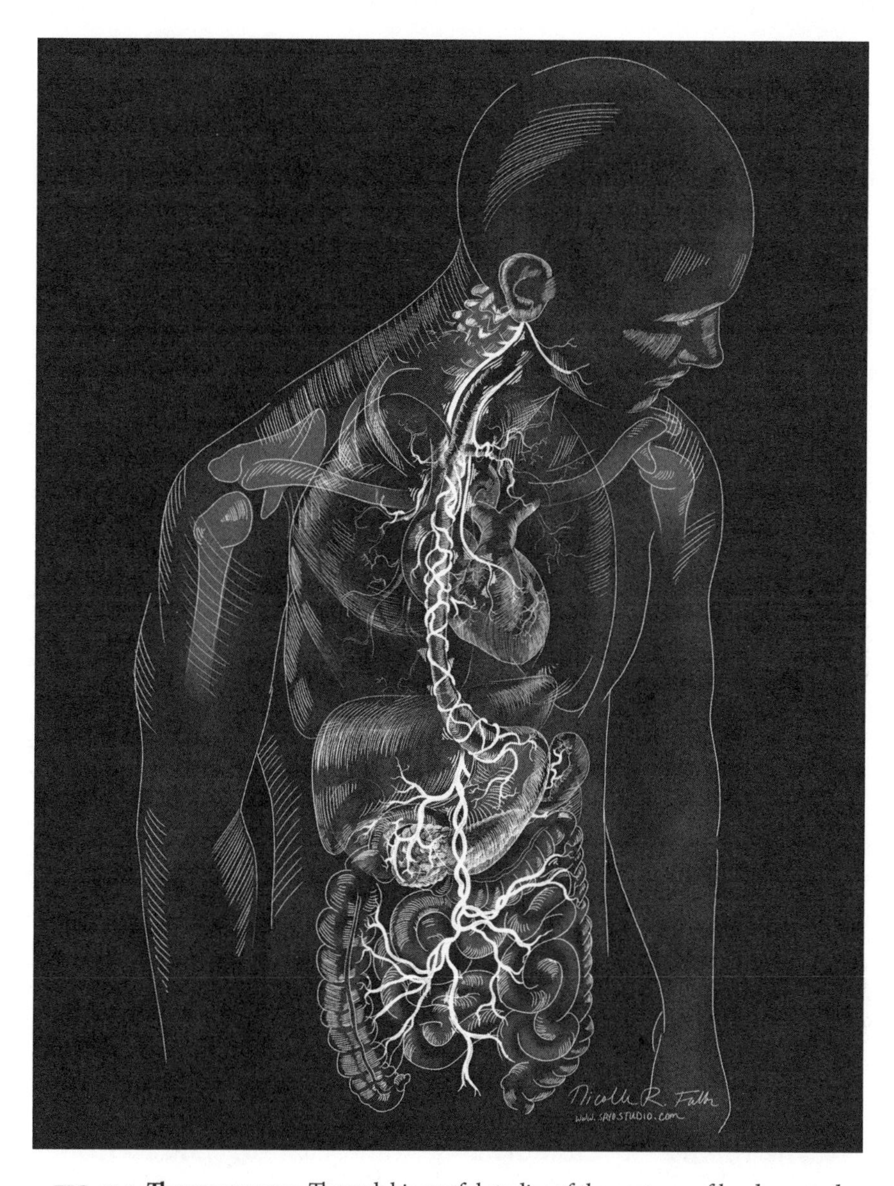

FIG. 12.1. The vagus nerve. Through his careful studies of the anatomy of heads severed from gladiators, Roman physician Galen (129–210 AD) discovered a small group of nerves that enter the body from the base of the skull rather than from the spinal cord, as most other nerves do. These nerves are classified as *cranial nerves*. The tenth (Xth) cranial nerve, known as the *vagus nerve*, is unique in that it emerges from the skull and extends far down into the torso. When the vagus nerve enters the chest cavity, it becomes the major nerve of the autonomic nervous system, which works below our level of awareness to regulate the subconscious functions of our internal organs. After leaving the heart and lung area of the chest, the vagus nerve passes through the diaphragm and into the abdomen. Once in the abdominal cavity, the main trunk of the vagus nerve sends smaller branches into the liver, stomach, pancreas, spleen, kidneys, small intestines, and upper part of the large intestines to relay information to the brain. (Image copyright Nicolle R. Fuller/Science Source)

preserve vagus nerve function as much as they could, so that it would be available to help regulate the heartbeat of the newly implanted heart.[6]

After leaving the heart and lung area of the chest, the vagus nerve continues to move down and passes through the diaphragm—the muscle that spans the chest cavity and separates the chest from the abdomen. To better see this, Suárez and I move to another cadaver, which has a very well-prosected abdominal cavity.

In the abdominal cavity, the main trunk of the vagus nerve sends smaller branches into a multitude of organs, including the liver, stomach, pancreas, spleen, kidneys, small intestines, and upper part of the large intestines (or *colon*). It's apparent that, as far as the internal organs go, the vagus nerve has things very well covered. With the exception of the adrenal glands, which sit atop the kidneys, and the lower colon and rectum, few organs escape the grasp of the vagus nerve. Thus, our vagus nerve keeps our brains well informed of what's going on down below, even if we aren't actually conscious of it.

Given the vagus nerve's widespread regulation of internal organ activity, one might infer that ailments of any organ innervated by the vagus nerve could be caused by dysfunction of the nerve. And sure enough, although strong evidence may be lacking, alternative medicine practitioners have implicated the vagus nerve as the source of a host of common physical illnesses of unknown or complex origin. They've given the vagus nerve an almost mystical quality, claiming that a dysfunctional vagus nerve is the root cause of most diseases. Consequently, alternative medicine practitioners have designed treatments that target the vagus nerve through massage, physical therapy, yoga, or acupuncture. Alternative medical therapies that are focused on the vagus nerve often are based on the *polyvagal theory*—supposedly a "radically different model of the autonomic [nervous] system"—and are frequently promoted in popular self-help books.[7]

Mainstream medicine remains skeptical that vagus nerve therapy is the panacea for a broad spectrum of common ailments. Nevertheless, very recent research performed at the Johns Hopkins School of Medicine in Baltimore strongly suggests Parkinson's disease, a brain disorder that causes tremors and slows normal movements, might originate in the gut and travel from the gut to the brain via the vagus nerve.[8] When a neurotoxic protein that can induce Parkinson's disease was injected into the gut of mice, the mice soon developed Parkinson's symptoms. But when the vagus nerve was cut prior to the injection, no such symptoms occurred. This suggests the toxin is gaining access to the mouse's brain by entering one of the vagus nerve's branches in the gut, traveling up the nerve through the chest, then through the neck, and ultimately gaining access to the brain through the vagus nerve's connection to the brain stem. Before these mouse experiments, autopsies of Parkinson's disease patients had suggested the brain problems might be associated with gut problems, but no one had direct evidence of how the brain and the gut ailments could be connected.[9] It now appears they may be literally connected by the vagus nerve.

And despite the lack of evidence that manipulation of the vagus nerve with alternative therapies has any medical benefit for common diseases, it is known that electrical stimulation of the vagus nerve provides some benefit for people suffering from epilepsy, especially for those patients who develop a resistance to antiepileptic drugs.[10] This is consistent with what little we know about epilepsy: it is an electrical disorder of the brain. Perhaps stimulating the vagus nerve with electricity somehow gets the brain's electrical activity back in synchrony. But I'm getting ahead of our story by bringing up the brain's electrical activity. We're not yet ready to discuss the electrical input and output of the brain. In fact, we aren't even finished exploring the vagus nerve.

Suárez shows me how the individual branches of the vagus nerve progressively become finer and tend to spread along the same routes as the finer blood vessels of the gut. Each branch eventually terminates in a *plexus*—a tangle of nerves, blood vessels, and lymphatic tissue that cannot be dissected any further.

But Suárez is quick to point out that the vagus nerve is not alone in its regulation of activities in the gut. There also exists, embedded in the lining of the gastrointestinal system, an *enteric nervous system* (*enteric* means "intestinal" in medical lingo) that lines the entire gastrointestinal system. The enteric nervous system operates independently of the brain and spinal cord, but interestingly enough, it can communicate with the autonomic nervous system by means of the vagus nerve. Clearly, there is little abdominal cavity activity that escapes the oversight of the vagus nerve.

At this point, my tour of the vagus nerve ends, and I'm appreciative of having been afforded the opportunity to see this important nerve up close and personal. I thank Suárez for acting as my guide, but I'm also thankful to these men and women who have donated their bodies to medical education. They may have died, but they haven't stopped giving. In fact, these cadavers will have more inquisitive visitors—first-year medical students—later today. And all of these students will become better physicians for having had access to these bodies.[11]

Examining the vagus nerve has been interesting and informative. But just as there were limits to what Cajal could learn from drawing the dead neurons he saw through his microscope, there is only so much a dead, and therefore nonfunctioning, nerve fiber can tell us about how nerve fibers work. Let's now explore a living and functioning nerve fiber in people who are very much alive and see what it can teach us. But, to do that, we are going to need to switch nerves. The vagus nerve lies deep within the body cavity and is much too complicated for our purposes. We need something more readily accessible and simpler to work with. And that would be the ulnar nerve.

The *ulnar nerve* is a nerve fiber of the arm consisting of motor neurons that branch off and attach to two different muscles in the forearm.[12] When you turn a doorknob to open a door, you send those instructional signals from your brain down to your arm via your ulnar nerve. The nerve receives its signals from the brain and then stimulates the appropriate muscle contractions in response to the signal. The ulnar nerve is not one of the cranial nerves, which means it gets to the arm by way of the spinal cord rather than directly from the brain stem. It originates from a *nerve root*, the spot where a nerve emerges from the spinal cord, located in the cervical (neck) region of the spine.[13] But instead of moving from the neck into the torso, like the vagus nerve does, the ulnar nerve heads straight to the forearm by way of the axilla (armpit). The ulnar nerve also differs from the vagus nerve in that it is not part of the autonomic nervous system; rather, it is very much under conscious control. Usually, it's under the conscious control of the person who owns it. But as we'll soon see, it is also possible to relinquish that control to someone else's thoughts.

We are all familiar with our own ulnar nerve, but we typically refer to it as our "funny bone." In reality, that strange tingling sensation in our forearm that we've all experienced from bumping the inside of our elbows has nothing to do with any bone; it's the result of impacting the ulnar nerve, which is located there, just below the skin. From the elbow, the nerve moves down the forearm and enters the hand.

Once in the hand, the nerve splits into branches. Two superficial branches enter the ring and pinky fingers. If you are a bicyclist who rides a bike with drop-down handlebars, like I do, you've probably experienced those two fingers becoming temporarily numb after long rides, a condition known as *bicyclist's palsy*. It's caused by the compression you are putting on your superficial ulnar nerve where it enters your hand when you press down on the handlebars, and it is an indication that it's time for a new pair of padded bicycling gloves. A deeper branch of the ulnar nerve spreads into most of the hand muscles, allowing us to spread our fingers.[14]

The ulnar nerve, as I've already mentioned, controls some movement of the human forearm and most of the muscles of the hand. As such, it can be considered the human experimental equivalent of the sciatic nerve in a frog's leg. Both the sciatic and ulnar nerves contain motor neurons, and motor neurons tell muscles what to do. Electrically stimulate a frog's sciatic nerve and its leg muscles will twitch; electrically stimulate a human's ulnar nerve and her forearm and hand muscles will twitch. So it's possible to reproduce Galvani's frog leg experiments in human beings using their ulnar nerves. And because most humans are better than frogs at complying with instructions, you don't need to decapitate them to get them to stay put during an experiment. (Although, as we've already seen, Giovanni Aldini found that freshly decapitated humans work just as well for electrical experiments.)

Because it runs just beneath the skin, the ulnar nerve can be electrically stimulated using simple, stick-on adhesive skin electrodes; no need to stick needles through the skin to impale the nerve. The skin electrodes function as electrochemical probes, able to convert the ion-based currents that axons use to send action potentials through nerve fibers into the electron-based currents needed by electronic devices, or vice versa. So, very conveniently, skin electrodes can be used to both detect nerve action potentials and generate an electron output in response; or they can receive electrons as input from an electronic device and then initiate action potentials in the nerve in response. (We'll talk about how skin electrodes are able to do this in chapter 13.)

If a pair of skin electrodes is attached over the ulnar nerve near the elbow and the nerve is electrically stimulated, the forearm of the subject will jerk inward in what neurologists call an *ulnar deviation*.[15] Since the external electrical stimulus is stronger than normal nerve signals coming from the brain, the subject won't be able to willfully stop the jerking because the external electrical signal overrides the brain's relatively weak signal.

Likewise, using the same type of skin electrodes, it is possible to detect the much weaker electrical signals coming from normal action potentials, sent from the brain, that pass through the ulnar nerve from elbow to wrist.[16] It's even possible to measure the time it takes for the action potential to travel through the ulnar nerve. If the person's brain sends a command to move the forearm, that action potential's speed can be measured and correlated with the movement of the forearm. When this is done, the measured speed of action potential transmission through the ulnar nerve is 80 meters per second, which is very close to the value Hermann von Helmholtz measured nearly 200 years ago for the speed that nerve signals travel through a frog's sciatic nerve—27 meters per second.[17]

You might wonder how it is possible to detect the ulnar nerve's passing electrical signal so easily when it was so difficult to detect the very weak action potentials of a squid axon. First of all, it isn't very easy. Just as with the squid axon, detection of the ulnar nerve's electrical signal requires an electronic amplifier, albeit a much simpler one. Having said that, what one actually detects in the ulnar nerve are the combined electrical signals from thousands of motor neuron axons firing in synchrony. The combined signal running through the nerve fiber is much stronger than the signal from a lone axon, so it is more easily detected through the skin. Nevertheless, it is still a very weak signal, and its detection amid ambient *electrical noise*—the extraneous electrical signals outside of the desired neural signals being measured—is a major challenge in our typical electricity-filled everyday environment.

It is possible, however, to use a rather clever tactic to readily detect the ulnar nerve's relatively weak signal just with skin electrodes. If the electrodes are placed over the ulnar nerve's target muscle rather than the nerve itself, the contracting muscle will spread the signal over a greater skin surface area and effectively increase it enough for the skin electrodes to detect it more easily. So the contracting muscle tissue acts somewhat like a biological "preamplifier" to boost the signal of the nerve action potential just enough to be picked up through the skin and transferred to an electronic amplifier for further amplification and analysis. This strategy of measuring signals from the nerve's target muscles, rather than the nerve itself, makes signal detection a whole lot easier.[18]

I wouldn't say putting the detecting electrodes on the target muscle makes the measurement of the ulnar nerve's electrical activity child's play . . . but almost. In fact, to further investigate how the ulnar nerve works, we need to leave the medical school and go back to high school. Fortunately, we don't have to go far; the closest high school to Georgetown's medical school is just around the corner.

Georgetown Visitation Preparatory School is a private Catholic high school for girls located in the historic Georgetown neighborhood of

Washington, DC.[19] The school is actively developing a *STEM* program, a curriculum with a focus on science, technology, engineering, and mathematics training. A STEM education helps ensure the school's graduates are on an equal scientific footing with their male counterparts when they go to college. There has been an ongoing underrepresentation of women in the technical fields, and the school is trying to address that problem head-on by encouraging more of its students to pursue science-related careers. Bravo! The school also requires *all* of their freshmen (or perhaps freshwomen?) to take a course called Conceptual Physics. Why? Because the faculty believes "physics gives girls a strong foundation for chemistry and biology." Bravo again!

It might not be too surprising that such a school also offers a number of advanced placement (AP) science courses, nor that these courses include some study of both electricity and neuroscience. It also may not be surprising that the science courses include a laboratory component, nor that the laboratory work includes both some electricity work as well as some neuroscience experiments. What may be a little surprising, however, is that some of those laboratory sessions combine hands-on electrical work with hands-on neuroscience in an integrated manner. This connects the two topics in the minds of the students and demonstrates their relevance to the students' lives.

My contact at the school is science teacher Nancy Cowdin, actually Dr. Nancy Cowdin—she has a PhD in neuroscience. I tell her I've heard she teaches a laboratory where the students do experiments on their own ulnar nerves, with electronic signal amplifiers built by students. Could this be true? She confirms it is indeed true, and she invites me to come see for myself the next time she teaches the class. I heartily agree. But I decide I need to do my homework before I go.

Cowdin teaches some of her neuroscience laboratory classes using kits purchased from Backyard Brains, a Detroit-based company started by two former neuroscience graduate students from the University of Michigan. This company has an unusual mission. It wants to bring the teaching of neuroscience down to the undergraduate and even high school levels.[20] As cofounders Timothy Marzullo and Gregory Gage see it, "Neuroscience is uniquely positioned to serve as a 'model discipline' for engaging students, subsequently improving performance in STEM-related fields, by combining biology, physics, electronics, health, and mathematical modeling in a single compelling field."[21] But neuroscience has some obstacles to overcome.

To be specific, the cofounders of the company believe that the high cost of the electronic equipment has been a major barrier to teaching neuroscience to students earlier in their educational years. Marzullo and Gage think neuroscience concepts are intelligible and potentially exciting to grade school and high school students, but the obstacle of paying for and operating sophisticated and expensive experimental equipment has resulted in the teaching of neuroscience being delayed until the students are in college. That is not good for the students nor for the field of neuroscience. But lowering the cost barrier isn't the only thing that motivates Marzullo and Gage. They say it's "also because hands-on-activities enhance learning." And that's not just their opinion; there is scientific data to back it up.[22]

The company's solution? Have the students build the equipment needed for the neuroscience experiments from do-it-yourself (DIY) electronics kits.[23] In this way, they kill two birds with one stone. The students learn about electricity and neuroscience simultaneously, and they also learn about how the two are connected. After all, isn't that how Galvani and Volta got started?

Fortunately, the neuroscience students of today have an advantage Galvani and Volta never had: a huge array of commercially available electric generators, batteries, and electronic components to work with. But assembling these components to get a desired result was, until very recently, no simple task. Designing even the simplest electronic projects required the assistance of an electrical engineer. In other words, even relatively simple electronics weren't user-friendly, and that was considered by many people to be a major obstacle to would-be scientists and engineers entering the electronics field. The learning curve was way too steep. But a small group of electrical engineers found a work-around that largely bypasses the problem. They invented a user-friendly product and gave it a friendly sounding name: Arduino (pronounced R-DWEEN-O).

Arduino has made building even sophisticated electronics accessible to amateurs, including very young amateurs. Now even young children are inventing and building their own electronic devices, starting with toys and games, but quickly moving to ever more sophisticated projects. The particularly creative kids have combined their Legos—those little plastic interlocking children's building blocks from Denmark—with their own Arduino circuitry to create a plethora of different kinds of robots.[24] But what exactly is Arduino?

Arduino is an *open-source electronics platform* that can be used to build small electronic devices. Open-source platforms are hardware and software programs that are available free of charge to the public. Users can modify them for their own use, in any way they like. In essence, the Arduino platform amounts to a small circuit board (called a *microcontroller*) paired with a piece of software (known as the *integrated development environment*) that runs on any laptop computer.[25] The software allows the user to write computer-coded instructions and to upload that code onto the microcontroller.

The birth of the Arduino platform (2005) is only slightly older than the founding of Backyard Brains (2009). The technology was an outgrowth of a doctoral thesis by Hernando Barragan. This Colombian graduate student was studying at the Interactive Design Institute in Ivrea, Italy, a small town near the city or Turin. The title of Barragan's thesis was "Arduino—the Revolution of Open Hardware." That's quite an audacious title for a graduate student to give his PhD thesis. But a revolution it did, in fact, spawn. Barragan and four of his engineer drinking companions had worked tirelessly to design an electronic wiring platform that was small, light, and inexpensive. What they ended up with was even better than inexpensive; it was largely free. It filled a long-standing need for a universal and cheap electronics platform that the group had previously contended to be an obstacle to progress in electronics, and the subsequent explosive popularity of Arduino supported their contention. It has also been a boon to STEM education, solving many obstacles to learning often encountered in traditional electronics teaching laboratories.[26]

As for the origin of its quirky name, Arduino is actually the name of the bar where Barragan and his colleagues did most of their brainstorming. The bar, in turn, is named after Arduin of Ivrea, a nobleman who served as a local military commander under the king of Italy from 1002 to 1014. Fittingly, the bar has been recognized for its contribution toward the technology born within its walls, with the recent invention of an Arduino-based electronic cocktail-dispensing machine named the Inebriator, which can dispense a drink called "Voodoo"—a mixture of coconut rum, coffee liqueur, and milk (but thankfully devoid of tetrodotoxin).[27] Who says engineers are nerds?

Backyard Brains has built its company's products around the Arduino platform. By layering the main microcontroller with specifically designed circuit boards, called *shields*, the company has been able to design a variety of different types of a basic *bioamplifier*—a standard electrical amplifier that has been adapted for electrophysiology work. Backyard Brains calls their little bioamplifier a *SpikerBox* because this small, box-shaped amplifier, about the size of a deck of cards, can be used to graphically display nerve fiber action potentials as abrupt "spikes" of voltage versus time on the screen of a laptop computer or tablet. The SpikerBox can also boost the relatively weak electrical signals coming from nerve fibers and muscles to levels that can be measured by standard electrical meters. The best part is that they have designed these bioamplifiers with teenagers in mind. The typical teenage student is given an Arduino board and a bag of electronic components, and in less time than it takes to watch the Super Bowl, she has built herself a SpikerBox sensitive enough to detect, amplify, and measure the relatively weak electrical signals coming from the nerve fibers in her own body.

The day finally comes and I'm ready. Cowdin's class is going to run the ulnar nerve experiment. I arrive shortly before the class starts and Cowdin briefs me on what I'm about to see. She tells me that during prior laboratory sessions the students had already used their Arduino-based SpikerBoxes to measure the action potentials passing through a cricket's leg. That required anesthetizing the cricket in ice-cold water, immobilizing it on a board, and sticking two electrodes, in the form of fine pins, into either end of its hind leg. By moving the electrodes around, the students were able to find where the nerves were located within the leg. They could see the action potentials generated by the nerve in the cricket's leg on their laptop computers each time they successfully located the neural circuit.

Moving beyond crickets, they went on to measure the action potentials in their own arms using the same approach, but without anesthesia and using adhesive stick-on skin electrodes rather than pins. The students saw that human action potentials weren't very different from cricket action potentials.

Today's experiment would move things to the next level. The action potential from one student's forearm would be detected the same as before with the SpikerBox, but the electrical signal also would be amplified to a higher voltage (recall that amplified mule) and rerouted to the forearm of another student. More specifically, the amplified signal would be sent to electrodes placed on the skin just above the ulnar nerve. The electrodes would transmit that amplified signal to the ulnar nerve, which, in turn, would be stimulated to initiate its own action potential, thus triggering the contraction of the muscles of the other student's forearm. Thus, the two students' forearms would move in synchrony, with one student controlling her own arm movement as well as that of the second student, who would be powerless to resist because the strong external electrical signal would overwhelm the weaker internal signals coming from her own brain.

The bell rings and the students file into the lab. Cowdin tells the students they have a visitor who is interested in observing their experiments. That would be me. I introduce myself and tell them not to mind me, I'll just be moving around the room asking questions about their work.

There are five groups working independently, with each group having three or four students. I move from group to group, observing and asking questions. I notice that, while some students are eager to be the controller, the job of controllee is somewhat less popular. The students know that stimulation of their ulnar nerve will mean feeling a slight shock and are reluctant to have that experience.[28] But never mind; each table has identified at least one willing volunteer and the controllers and controllees get themselves wired up. The remaining students are tasked with recording the data.

There are some initial mishaps and a few computer glitches to deal with, but soon everyone is ready to go. The first task is to adjust the electrical output of the amplified signal so that there is sufficient electricity to elicit a full muscle contraction in the controllee, but not high enough to give her an unnecessarily intense shock. That done, every group begins its initial test run.

The controller student must flex her muscles associated with the ulnar nerve. As we've said, those muscles are located on the underside of the forearm midway between the elbow and the wrist, and when fully contracted they flex the wrist, pulling the fist inward. (If you've ever arm wrestled, you'll recognize them as the muscles you use to overpower your opponent.)

The controllee student simply relaxes her arm. The ulnar nerve is primarily a motor nerve, so when the ulnar nerve receives the electrical stimulation from the controller, it initiates its own action potential that commands the same forearm muscles to contract in the controllee.

I notice the common reaction of the controllees to the involuntary contraction of their arms is uniformly wide-eyed surprise and laughter. They know what is coming, but it's still a bit "unnerving" to relinquish the control of your arm to someone else's wishes. They try to resist moving their arms but are unsuccessful; the electrical signal is simply too strong to be overridden by the nerve signals coming from their own brains.[29]

But this experiment is more than a stunt. This is not simply a parlor show used for entertainment, like the Flying Boy demonstrations of yesteryear. This is serious science. And that means there must be a hypothesis to test. Sure enough, Cowdin has asked each group to come up with a hypothesis that can be tested with their SpikerBox setups.

One group has reasoned the nerve signals from the controller's arm might be affected by the temperature of the forearm; that is, the temperature of the nerves within the arm. After all, wasn't the cricket's nervous system suppressed by a dip in ice water? That suggests cold slows nerve signals. They run a comparison between the controller's forearm prewarmed with an electric heating pad and precooled in a water bath. Cognizant that electricity and water present a safety hazard, they take special precaution to keep the heating pad and electronics away from the ice water. Their hypothesis is that a heated arm will enable stronger nerve signals. And they recognize the temperature comparison should be done on the same arm, not different people's arms, so as not to introduce between-individual variability into their data.

Another group thinks that nerve signals from a person's dominant arm should convey a stronger nerve signal than those from the nondominant arm. When I ask them why they are making that hypothesis, they tell me the dominant arm should be more physically robust because it is used more and, therefore, should be better able to produce strong signals. Sounds good to me. They decide to compare dominant and nondominant arm signals from all members of the group.

But the group that most interests me is the one that thinks it is irrelevant whether the controller's arm actually moves. They think that the mere act of trying to move the arm will result in a full signal being sent to the controllee. Their experiment is to either physically restrain or not restrain the controller's arm movement and see if she produces the same or a different response in the controllee's arm.

This particular group has three students—Isabella, Kate, and Emma. Kate is the controller and Emma is the controllee. In the first trial run, Kate's arm is free to move at will, and when she flexes her forearm,

Emma's forearm moves in synchrony, just as expected. But in the second run, Isabella forcibly pins Kate's arm to the table. Now when Kate tries to move her arm she cannot; nevertheless, Emma's arm still moves. The group has thus been able to uncouple the movements of the controller from the movements of the controllee. Emma reports her experience was the same in both cases: involuntary movement of her arm. Kate reports that in the second trial she could feel her muscles straining against the resistance of being pinned down, but lacked the strength to overcome that resistance, so her arm couldn't move. The group concludes that only the desire to move the arm is needed to send a nerve signal to the muscles of the forearm. Kate's muscles didn't actually need to complete the task with a full flex movement of the forearm for the signal to be transmitted to Emma's arm.

The bell rings again. Lab period is over. As the students clean up the lab before leaving, Cowdin tells them to consider the problems they encountered with their experimental designs and try to find ways they might improve their data collection methods. One group says that they could better quantify arm contractions if they used a scale to measure the force a contracting arm exerts. Cowdin commends them on their ingenuity and tells them to try using that approach when they replicate their findings in the next lab session.

As the girls get ready to leave, I ask Isabella, Kate, and Emma what they plan to major in when they go to college—all three are currently high school seniors. Isabella says she wants to major in astronomy, or possibly architecture. Kate says she's fascinated with biology but isn't clear yet where that interest may lead. Emma says she wants to do something in health, perhaps as a physician but maybe in some other capacity. I thank them and wish them all well. And I'm quite sure they all will do well. They are well prepared to study science in college and should excel at it, if they can just maintain their interest.

I think we can all acknowledge that Georgetown Visitation Preparatory School isn't your typical American high school. Rather, it's among the more elite college preparatory schools in the United States. Which begs the question: How much of the electrical and neuroscience training that Cowdin provides in her class, with the help of Backyard Brains' products, is transferable to the average high school? When I pose this question to Cowdin, I'm surprised at her answer: "All of it."

She explains that all students, regardless of their particular school environment, are inherently interested in both the human body and human behavior. Learning about the nervous system provides students with a window into how our bodies control the way we think, respond, and behave. Therefore, these types of neuroscience experiments grab the students' attention and keep them engaged in the learning process. Cowdin says she and other resourceful high school teachers have been going to hardware stores and raiding their own kitchens for years to scavenge needed materials for laboratory experiments. The big difference now is that the increased affordability and availability of technology-based equipment, like that produced by Backyard Brains, creates unlimited potential for stimulating interest in science experimentation at all levels and in all situations.

I agree with Cowdin. As mentioned above, because of the DIY approach to making the required bioamplifiers, the experimental costs are quite low.[30] And the low cost helps make it more widely available. I've been a Regeneron Science Talent Search reviewer for many years, and I've seen high schoolers do some amazing projects.[31] But I must admit that the students who submit the best projects often have access to expensive scientific equipment. Those without the needed equipment are at a considerable disadvantage because it limits their ability to ask and answer the most interesting scientific questions. Things like the DIY electronics kits help level the playing field in that regard, allowing the students' talent, rather than their access to expensive equipment, to be the major driver of the competition.

What about the expertise of the teaching staff? Do science teachers generally have the skills needed to take on Arduino construction and Backyard Brains' experiments? After all, how many high school science teachers have PhDs in neuroscience? Point taken. But again, this also appears to be a nonissue. Cowdin notes there are increasing numbers of web-based resources for teachers to find lesson plans for teaching neuroscience concepts and showing the electrical nature of the nervous system. Teachers don't have to go it alone. Cowdin feels confident other science teachers can handle it. She thinks that, if the schools provide teachers with enough time to develop their curricula and give them the relatively inexpensive equipment they need to run experiments, the teachers would embrace the challenge regardless of whether they have PhDs.

Having been a teacher for many years, Cowdin knows of what she speaks. She explains, "I have taught at low-income, inner-city schools as well as private schools, and feel that what I am doing here at Georgetown Visitation would be very applicable and beneficial to students across the board. This hands-on nature of this type of learning is what we need to do more of in schools." She's absolutely right.

It's easy to see how students can be both educated and entertained by Backyard Brains' human-to-human interface, but how can this technology be translated into everyday life? Maybe the best people to ask about that are the students themselves. Marzullo and Gage did just that. When they surveyed students about what value they thought such technology could provide, one of the answers they most frequently got was neurologically driven electronic prosthetics. Their idea was that, instead of having a human-to-human interface, one could design a human-to-machine interface, so that nerve signals to move an arm could be routed not to another person's arm but rather to a mechanical arm. Therefore, Backyard Brains came out with another teaching laboratory product (the HackerHand) to demonstrate to students how such a human-to-machine interface might work. It's a plastic model of an artificial hand that can be controlled by the nerves in a person's forearm.[32]

If biomedical engineers could actually invent such a real artificial hand, or even a whole artificial forearm, it might be a great option for people who have lost their forearm but still retain nerve function in their upper arm. The brain's signals to the missing forearm could be intercepted by electrodes placed on the upper arm, amplified, and routed to a mechanical prosthetic forearm to perform essentially the same functions that the person's natural forearm used to do. Very futuristic indeed . . . or is it?

Let's now take a quick look at the current state of neuroprosthetics and its practical application toward changing the lives of amputees at this very moment, as well as what we might expect to see soon, perhaps through the future hard work of one of today's high school students.

13

CROSSED CIRCUITS

NEUROPROSTHETIC LIMBS

Oh the nerves, the nerves; the mysteries of this machine called man!

—CHARLES DICKENS

Melissa Loomis lives in Canton, Ohio, and she loves dogs. In fact, she volunteers at her local dog shelter. So, when her own beloved dogs got into a tussle with a raccoon in the backyard one morning in 2015, she ran out to intervene, hoping to save the dogs from some nasty raccoon bites. When the ruckus was over, the dogs were fine and the raccoon was fine. Unfortunately, Loomis was not fine. The raccoon had bitten her right forearm—a nip to the wrist involving two small puncture wounds. But over the course of the next few weeks, those small wounds ended up producing a big enough infection that her arm needed to be amputated to save her life. The arm was amputated and her life saved but at a tremendous personal cost.[1]

Loomis has a lot of company in her suffering. In the United States alone, there are more than 2 million people missing at least one limb, and some projections anticipate that there will be 3.7 million amputees by 2050. Annually, approximately one million new limb amputations occur globally (or about one amputation every 30 seconds).[2]

Although organ transplants, and even face transplants, are saving lives and providing relief to many people, transplantation of limbs presents particular challenges because it involves connecting many different types of tissues, and the threat of the body's rejecting the transplanted limb is very high. The sad fact is, given the numbers of amputees and the obstacles to limb transplantations from cadavers, prosthetic devices are the only feasible remedy for most people suffering the loss of a limb.

Prosthetic limbs have been around for a long time. The *Capua leg*—a prosthetic leg found in a grave in Capua, Italy, dating to about 300 BC—is the earliest known artificial limb. It was made of bronze, which explains how it survived intact for over 2,000 years.[3] Prosthetics didn't progress much beyond Capuan technology until the nineteenth century. Then a revolution in the advancement of prosthetic limb engineering happened as a consequence of the American Civil War (1861–1865).

The astounding number of veterans returning home from the war without one or more of their limbs tremendously increased the demand for new and improved prosthetic limb products.[4] And the demand was further stimulated by the federal government's reimbursement of up to $75 (about $2,300 in 2020 dollars) toward a prosthetic limb for all Union veterans needing one. (As retribution for fighting against the federal government, Congress deemed Confederate veterans ineligible for the compensation.) This influx of money stimulated research and development in new and better prostheses. Between 1861 and 1873, there were 133 prosthetic limb patents issued by the US Patent Office.[5] Unfortunately, most of the prosthetic limb improvements were more cosmetic than functional. Since having an imperfect body carried a certain stigma at the time, many of the prosthetics were designed simply to disguise the infirmity rather than to replace limb function.[6] And none was electrically powered.

Today, war is still driving the demand for prosthetic limbs. Improvements in combat medical care, and the use of body armor that protects the torso but not the limbs, have resulted in more military personnel surviving severe battlefield injuries that involve limb amputations. While a third of all soldiers died during or shortly after their amputation surgery at the time of the Civil War, now an amputee has a much better chance of surviving, which further drives the need for prosthetics to serve this population of veterans who, in earlier times, likely would have died.

Today's amputees are more interested in regaining limb function than trying to hide their injuries. And the federal government is again spurring advancements in prosthetic research, mostly through the Defense Advanced Research Projects Agency (DARPA), which funds much of the prosthetics research done today. The goal of modern prosthetics is to emulate the movement of the natural limb as much as possible, as well as to control the limb naturally, even using an amputee's thoughts to drive the movements of the prosthesis in the same way thoughts once controlled his natural limb. Thoughts? How is that possible? Through *neuroprosthetics*, a branch of prosthetics that links the human nervous system with electronically controlled mechanical limbs.[7]

Imagine this. You are no longer using your ulnar nerve because your hand and forearm are gone, amputated due to accident or disease. Intellectually, your brain knows the arm is missing, but it still sends a signal to your ulnar nerve in response to your desire to move the fingers of your hand. The action potentials coming from your brain to your forearm stop at the end of the severed nerves because there is nothing to receive the signal, leaving the brain's message undeliverable. The consequence: wasted instructions from the brain.

But what if your brain's signals to your ulnar nerve were rerouted and repurposed? Let's say they were diverted to an electrical device that powered little mechanical "muscles" in a prosthetic hand, such that your brain's command to move your missing ring and pinky fingers—the two fingers controlled by the ulnar nerve—now results in the movement of the ring and pinky fingers of the prosthesis. That could be useful. But it is very difficult to pick something up using just your ring and pinky fingers. Try it for yourself. You'll see you're able to press the item against the palm of the hand, but not able to actually grip it. The three other fingers—pointer, middle, and especially the thumb—are also needed to allow a prosthetic hand to have a natural grasp. Getting control of these three fingers requires the input from another major nerve of the arm: the median nerve.

Just as we found the trunk of the vagus nerve running alongside the carotid artery of the neck, the trunk of the median nerve runs along the brachial artery—the major artery of the arm. You can feel your own brachial artery pulsating by pressing on the underside of your arm, up near your armpit. Press hard; it runs deep. At this location, the brachial artery and the median nerve are running together, side by side. Artery and nerve stay together as they move down the arm until they reach the area just below the elbow, at which point the artery and the nerve diverge. As the median nerve moves toward the fingers via the wrist, it goes through the *carpal tunnel,* a narrow passageway lined with bone and ligaments.[8] Compression of the median nerve at the carpal tunnel causes *carpal tunnel syndrome,* a common neurological disorder similar to bicyclist's palsy, except it involves numbness or pain in the hand's first three fingers rather than the last two.

As we can see, the ulnar and median nerves, working together, can close all the fingers of the hand. So why not repurpose both the ulnar and median nerves in order to get full mechanical control of the movement of all five fingers of the prosthetic hand? But even if we were to recruit both the ulnar and median nerves, the prosthesis would still have limitations because hands do a lot more than just close fingers. If you want to fully emulate hand movements, you'll need to recruit yet a third major nerve of the arm into the mix: the radial nerve. While the ulnar and median nerves control closing the fingers, the radial nerve is responsible for opening the fingers and extending the wrist. So it really takes all three nerves—ulnar, median, and radial—to make your forearm, hand, and fingers work naturally. Therefore, all three nerves are also needed to fully control the movements of a prosthetic arm.

Repurposing the three major nerves of the arm is no trivial task. But once all three have been recruited, the prosthesis will have nearly as many degrees of freedom of motion as a natural arm and hand.[9] Now that's a useful prosthesis. Unfortunately, for this scenario to be realized, a challenge must be overcome: one must be able to detect the signals coming from the nerves of the arm and then convert them into information

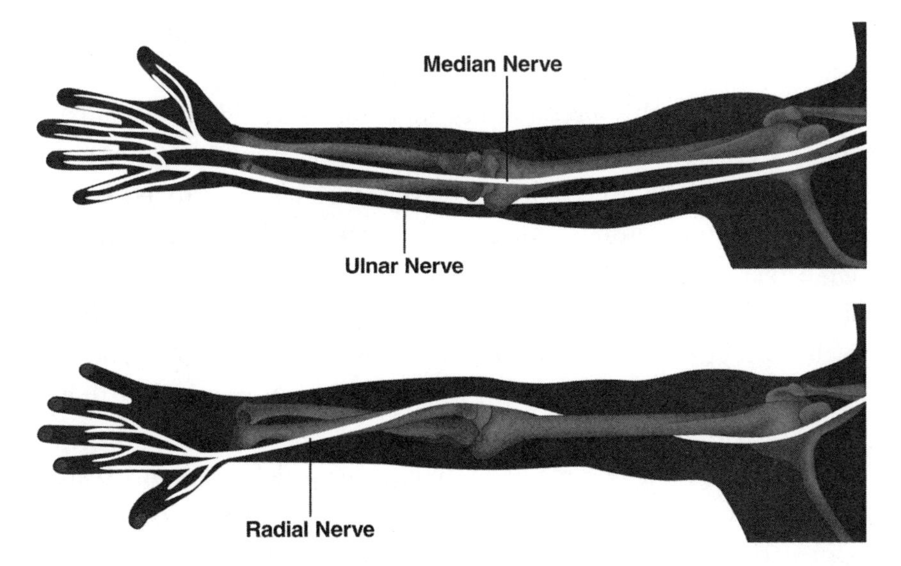

FIG. 13.1. Major nerves of the arm. Finger movements and touch sensations in the hand require the involvement of three major nerves of the arm: the ulnar, median, and radial nerves. For lower-arm amputees, these nerves are severed and no longer functional. But the severed nerves can produce a phenomenon known as *phantom pain,* in which the amputee experiences chronic pain that feels as if it originates from the missing part of the lower arm. Surgical redirection of the termini of these nerves to alternative target tissues can often both relieve phantom pain and even allow the nerves to gain novel functions, such as driving the movements of an electromechanical prosthesis and restoring some fingertip touch sensations.

that is *machine-readable*—a form that the mechanical devices in the prosthesis can understand.

When we first started talking about action potentials in chapter 10, we learned the electrical signals of the nervous system are propagated by bursts of positive ions crossing a neuron's cell membrane in a controlled manner. This differs in a fundamental way from most nonbiological situations where electrical signals are typically the result of electrons moving through wires. To use computer terminology, this means biological electricity and nonbiological electricity operate on two very different types of platforms. We might say they are "hardwired" differently. That presents a problem when you want to connect these two electrical platforms together.

At first glance, this discontinuity of platforms would suggest linking them for controlling a prosthesis would be a near insurmountable technological obstacle. After all, joining the two platforms together will need to be much more sophisticated than just splicing the ends of a nerve fiber and an electrical wire together. There needs to be some kind of "bridge" between the biological and nonbiological electrical systems. Such a bridge is often called a *brain-machine interface* (BMI), and it is designed to enable electrical information to pass back and forth between the two platforms.[10]

In computer engineering lingo, an *interface* is defined as any shared boundary across which two distinct compartments or systems exchange information. A BMI is a particular type of interface that takes nerve signals originating in the brain and translates them into electrical signals that can be relayed to a computer-controlled electronic device in order to instruct it to perform some desired action.[11] Thus, the main role of a BMI is to change ion currents into electron currents, a critical function since computerized devices can run only on currents produced by moving electrons. But the best BMIs work in the opposite direction as well—changing electron currents into ion currents—to allow signals to go both to and from the brain.

BMIs come in a variety of forms. The skin electrodes we previously encountered in the high school neuroscience laboratory are one of the simplest examples of a BMI. A skin electrode is a little metallic disk, about the size of a small coin, with an electrolytic jelly coating on one side. When attached to the skin with adhesive tape, it creates a little jelly sandwich, with skin taking the place of one slice of bread and the metal electrode substituting for the other slice. This jelly sandwich is able to convert ion flow into electron flow. It does this through specific electrochemical reactions, similar to the electrochemical reactions a voltaic battery uses to generate electric current.[12] The movement of positive ions through the jelly induces a movement of negative electrons in the opposite direction, and the electrode detects those moving electrons as a small electrical current. With the help of a bioamplifier, like a

SpikerBox, the voltage and amperage signals can be boosted to levels sufficient to drive electronic devices.

But there is still the requirement that the exchange must be machine-readable, which typically means the signal needs to be digitized. Fortunately, nerve signals are intrinsically binary by nature, and this allows them to be easily digitized. By "binary," I mean that an action potential is an *all-or-nothing signal* of constant magnitude—either a full action potential is sent or it isn't. The brain cannot send bigger or smaller action potentials; that's not how the brain controls the strength of its signals. Rather, if the brain wants to increase the magnitude of the signal it sends, it does so by sending more action potentials at a greater frequency. It's like knocking on the door of a room to get the attention of someone snoozing inside. If you want to enliven the sleeper's otherwise sluggish response, you could knock harder, knock faster, or both. That should stimulate him to get a move on. However, the brain can only tell a muscle to increase the strength of its contraction by knocking faster; it has no ability to knock harder.[13] From a muscle's perspective, faster knocking means the brain wants a stronger contraction, and it is biologically programmed to comply with the brain's wishes.

So, if the strength of the brain's message is proportional to the frequency of its action potentials, a machine needs to be able to count the number of action potentials arriving during a specific time interval. And that's exactly where electronic devices like the SpikerBox come into play. They detect "spikes"—the action potentials—picked up by the skin electrodes. They amplify those spikes and send an electrical signal to a computer each time a spike passes by the electrode. The computer assists by counting the spikes over time—a measure of the strength of the signal.[14]

Of course, BMIs can be much more sophisticated than skin electrodes and SpikerBoxes, but regardless of the type of BMI, the basic concept of converting ion currents into electron currents is the same.

As we also learned in the high school neuroscience laboratory, the easiest way to detect signals coming from the ulnar nerve is to intercept them as they arrive at their target muscle because the target muscle magnifies the signal as it spreads throughout the muscle. This signal magnification by muscle tissue increases the electrical signal to the point that it can be detected by just placing skin electrodes over the muscle.[15] No need to stick needle electrodes into the muscle. That would be too invasive, not to mention painful.

In the case of an amputee, however, the target muscles for the ulnar nerve often are gone—especially for someone with an amputation above the elbow.[16] Therefore, to use the same approach, one must give the ulnar nerve a new target muscle to act on, and that requires a surgical procedure called a *targeted muscle reinnervation* (TMR).[17] *Reinnervation* is the process of reintroducing a nerve connection for a part of the body that has lost it. The procedure was first developed by surgeons Todd Kuiken and Gregory Dumanian at Northwestern University in Chicago.[18]

In TMR surgery, severed motor nerves are surgically reconnected to remnants of upper arm muscles that have had their natural nerve connections damaged or removed, leaving them without any source of nervous system stimulation. Once these muscle fragments are reinnervated with an "alien" nerve, they twitch in response to brain signals intended for the nerve's original target muscle. These muscle fragments are acting, in effect, as surrogates for the intended target muscle. This creates a new muscle site over which skin electrodes can be attached to detect that nerve's signals arriving from the brain. Twitching of the reinnervated target muscle in response to brain signals, and the consequential movement of the prosthesis, are driven by the amputee's mental desire to move the missing limb. The amputee thinks about moving the fingers of her hand, the brains sends a message to the new alternate muscle via the redirected nerves, the muscle twitches, and skin electrodes detect the twitching of the muscle. A bioamplifier boosts the signal and sends it to a computerized device that translates the electrical signal into specific, machine-readable instructions. The device then quickly forwards those instructions to a motor inside the prosthesis, the motor reads the instructions, and then the prosthesis moves!

Since its debut in 2006, TMR surgery has been performed on hundreds of individuals with arm amputations and has greatly improved functional control of state-of-the-art prosthetic arms able to interpret those signals.[19] Such prostheses are called *advanced prosthetics*. If you're a *Star Wars* movie fan, you're already familiar with at least one advanced prosthetic. Luke Skywalker had one installed on himself after his father cut his forearm off with a light saber—an incident that severely strained their relationship. If you are not a *Star Wars* fan, let me explain.

An advanced prosthetic is a bionic arm with the dexterity, range of motion, and approximate weight of a natural human arm. One of the earliest advanced prosthetic prototypes was developed by DEKA Integrated Solutions Corporation with funding from DARPA. They named it LUKE—for life under kinetic evolution—in homage to Mr. Skywalker. This name surely must be the most incongruous combination of words ever grouped together just to create a cheesy acronym, but we're stuck with it. In any event, I'm sure glad that it was Luke and not Chewbacca who got the bionic arm. Regardless, the LUKE prototype made history and now several later generations of LUKE are in the field.[20]

Prior to her accident, Melissa Loomis had no familiarity with advanced prosthetics. She never knew anyone who had one, and she hadn't seen one in real life. But she had seen *Star Wars*. When I mention Luke Skywalker's bionic arm to her, she interjects, "Luke Skywalker, yup. That's about all I knew at the time." But she knew enough. When her physician floated the idea of her getting an advanced prosthetic, she was all in.

But before that could happen, TMR surgery would be required. Loomis was a good candidate for TMR, although her diabetes increased her surgical risk. Weighing her options, which were otherwise few, Loomis chose to have the surgery. She tells me, "I had had so many surgeries in connection with my arm's infection and amputation, I thought to myself, what's one more surgery?" She decided to move ahead with the TMR surgery, and that decision set Loomis on the path to making her own mark on the history of neuroprosthetics for reasons no one anticipated at the time.

All of this focus on the arm's severed motor neurons ignores another major consequence of losing an arm. It's not only that the amputee no longer has muscles for its motor neurons to direct. Almost as important is the fact that the sensory neurons—specifically, the sensory neurons of touch—also have been severed. What this means is that, even with TMR surgery and a bionic arm to replace the lost muscles, Loomis wouldn't be able to feel the things she grabs.

Loomis's surgeon was Ajay Seth, a private practice orthopedic surgeon in Canton. He was the surgeon who had amputated her arm, and he wanted to give Loomis the capability to fully utilize state-of-the-art advanced prosthetics technology. Upon investigating what "state of the art" really meant in 2015, he learned it might be possible for an amputee to have an advanced prosthetic that could actually feel as well as move, but it was currently uncharted territory. It would require a complex surgical procedure beyond standard TMR surgery. The added procedure was called *targeted sensory reinnervation* (TSR), and it had been attempted only a few times before. When Seth heard the details, he believed it possible for him to perform the procedure concurrent with Loomis's TMR surgery. That is, Seth thought he could give Loomis, in one very long operation, the capacity to use an advanced prosthetic that could both move and feel by doing a combined TMR/TSR procedure.

As they say, the devil is in the details, and the details of the TMR/TSR procedure are very important to its success. So I contact Seth, and he very graciously and patiently walks me through the details of exactly what he did. And when I hear what he actually did, I realize that his patience was the key ingredient to the procedure's success.

Seth tells me that he carefully teased apart the motor and sensory nerves in the stump of Loomis's amputated arm. He then redirected the motor nerves to patches of muscle (TMR) and the sensory nerves to patches of skin (TSR). From beginning to end, the process required 16 hours of operating room time. He explains the reason the surgery took so long is that the internal microanatomy of the three major nerves of the arm—ulnar, median, and radial—is so complex.

Within each nerve, the axons of the motor and sensory neurons are segregated into different *fascicles*, which are bundles of axons that twist around each other. Each fascicle is responsible for a different function. The situation is similar to that of a telephone line in a home. If you've ever had the experience of cutting through one, you'll know the outer plastic sheath of the line contains within it four fine wires, each of a different color—green, red, yellow, and black. They are color-coded so the telephone installer can tell which is which. Each of these fine wires is carrying its electrical current either toward or away from the telephone and controlling a different function. To work with the individual wires, the installer strips back the outer plastic sheath a few inches to expose the ends of the wires, and then separates the ends from one another so that they can be individually attached to their appropriate terminals in the phone.

It's a similar situation for the ulnar nerve fiber. If you strip back its sheath—a layer of connective tissue that surrounds and insulates the nerve—you reveal the individual fascicles that are the counterparts of the phone line's internal wires. But the ulnar nerve is far more complex than a phone line. Instead of just four wires, the ulnar nerve has 18 fascicles that need to be separated from one another. The median and radial nerves have comparable numbers of fascicles. That amounts to over 50 similar-looking fascicles that need to be isolated and redirected. And the fascicles are definitely not color-coded.

This was the challenge facing Seth. He had to cut the three different nerves to make clean ends, peel back their sheaths to reveal the fascicles inside, tease the fascicles apart, individually identify one from another, and then reroute them to their new homes.

It happens that only 1 of the 18 fascicles of the ulnar nerve is exclusively sensory; the rest contain motor neurons. Seth needed to find that particular sensory fascicle because he needed to route it to the skin. The remaining fascicles would be used to reinnervate fragments of muscle tissue from the biceps or triceps (the muscles on the front and back of the upper arm, respectively). The problem was how to identify which of the ulnar nerve's 18 fascicles was the sensory one. And to make things worse, the whole procedure needed to be repeated on the median and radial nerves, with their various fascicles.[21]

After many hours, Seth was able to successfully tease all the fascicles apart and separate them from one another. He then used a *somatosensory evoked potential* (SSEP) machine to identify the particular sensory fascicles he sought from the ulnar, median, and radial nerves. An SSEP machine has a tiny electrode probe that the surgeon can touch to the end of a fascicle to shock it. The shocking causes the neurons in the fascicle to fire off action potentials. A sensitive skin electrode placed on the scalp of the unconscious patient detects incoming action potentials from the fascicle, arriving at the brain less than a tenth of a second after the shock.[22] Only action potentials coming from fascicles containing sensory neurons are detected by the SSEP scalp electrode because motor neurons carry signals away from, not toward, the brain. So, if the brain gets a signal after a fascicle is electrically stimulated, it's a fascicle containing sensory neurons; if the brain doesn't receive a signal, it's a fascicle containing exclusively motor neurons.[23]

Seth shocked each fascicle with the SSEP electrode and identified the needed sensory fascicles. He routed the sensory fascicles to the skin on the underside of Loomis's remaining upper arm, and routed the rest of the fascicles to the available muscle tissue on the front and back of the arm. Once that was accomplished, he closed up Loomis's arm in the hope that her body would take over from there, with each transplanted nerve fascicle growing and thriving in its new body location.

There were a total of 41 steps in the complex surgery, and thus 41 opportunities for something to go terribly wrong. But, in testament to Seth's surgical skills, nothing did go wrong. From a surgical standpoint, the procedure was a complete success. What remained to be seen was whether it would be a functional success. Would Loomis be able to feel? Time would tell.

Feeling the prosthetic arm might not seem as important as moving the arm, but consider this: Why is it, when you pick up an egg or a glass with your hand, you don't squeeze too hard and break it? It's because your sense of touch tells you how much pressure you are exerting on the object, and your brain modifies your grip so as not to be too strong or too weak. The sensory nerves thus act like a feedback loop to fine-tune the motor nerves' commands. And this is where Loomis's case became special. She is the first amputee in the United States to successfully have

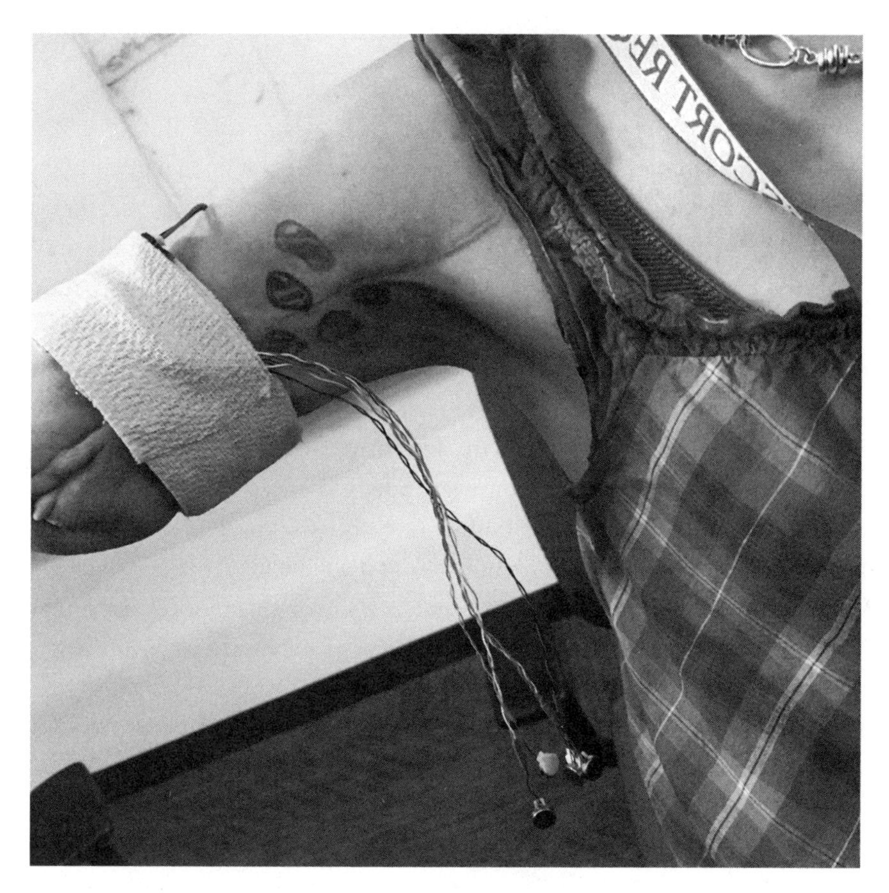

FIG. 13.2. An arm amputee regains a sense of touch. Amputee Melissa Loomis underwent targeted sensory reinnervation (TSR) surgery to reroute sensory fascicles from her ulnar, median, and radial nerves to five coin-size patches of skin on her upper underarm. Soon after the surgery, she began to experience pressure stimuli to each of those five skin areas as a sense of touch apparently coming from her five missing fingertips. Scientists at the Johns Hopkins Applied Physics Laboratory exploited Loomis's new sensory capability by equipping an advanced prosthetic arm with pressure sensors in its fingertips. By electrically connecting the prosthetic's fingertips to Loomis's underarm skin patches, Loomis was able to enjoy a simulated sense of touch when the prosthetic arm's fingertips touched an object. Pictured is Loomis's underarm with the areas of the five skin patches marked by her surgeon so that the scientists would know where to attach the electrical stimulators. (Photo courtesy of Ajay Seth)

had both targeted muscle reinnervation (TMR) and targeted sensory reinnervation (TSR) surgeries at the same time.

TSR is to sensory neurons as TMR is to motor neurons. But for TSR, it is a patch of skin on the limb stump, rather than muscle, that is reinnervated. Touching that patch of skin—an area about the size of a quarter—sends a touch sensation to the brain via the sensory nerves, and the amputee feels the touch as though it were coming from the natural target of the sensory nerve, such as the thumb. To get a better idea of what it's like, imagine probing the skin of your underarm with a pin along a straight line from your elbow to your armpit in 1-inch increments. You would feel the pinpricks moving progressively up your arm. But when the pin passed over a patch of skin that had undergone TSR involving the sensory nerves allotted to the thumb, that feeling of being pricked would jump from your underarm to your thumb, *even if you no longer had a thumb.*

TSR works because the brain is naturally programmed to recognize the signals arriving from that particular sensory nerve as originating from the thumb. It doesn't matter if the nerve is now responding to touch signals applied to skin on the stump of the arm; the brain thinks the sensation is coming from the finger of the missing hand. An interesting phenomenon, but can it be made useful to an amputee? It can.

The expectation was that it would take six to eight months for the nerves to heal and grow to the point where they would regain their function; that is, functional in the sense that they would transmit normal action potentials. But Loomis's progress turned out to be much faster than that. After just a couple of months, the fascicles had attached themselves to the tissue and were apparently thriving in their new locations. Over the next year, Loomis's sensory capabilities in her TSR skin patches continued to progressively improve, far exceeding Seth's most optimistic expectations. He tells me with astonishment, "She now has regrown [in her TSR skin patches] every nerve sensory unit she had in her fingers!"

An added benefit was that the phantom pain Loomis sometimes suffered was largely gone. *Phantom pain* is a feeling like an electrical shock or burning sensation that amputees often experience as though it were coming from a missing limb. The cause of such pain is unclear, but it seems that repurposing the severed nerves to a new muscle can greatly relieve it.

But there was a problem. All commercially available advanced prosthetics are currently able to work only with TMR; they do not yet have any TSR abilities. TSR-enabled prostheses with touch capabilities are still in the conceptual phase, even though the required neurosurgical capability has now arrived.

This leaves Loomis in limbo. Although she now has the biological capability to feel touch in her five individual fingers through the five patches of skin under her arm, there is no commercial prosthetic device that can send touch signals to those patches of skin. This is one of the reasons why TSR surgery has been slow to be deployed. And this situation presents a paradox. If there are no amputees who've had TSR surgery, there are no test subjects for prosthetic developers to work with on the problem of restoring touch. But if there are no touch-enabled prostheses available to amputees, why should people undergo TSR surgery? Thus, Loomis's combined TMR/TSR surgery makes her a remarkable amputee—unique enough that neuroprosthetics researchers were eager to work with her.

Loomis ended up going to the Johns Hopkins University's Applied Physics Laboratory (commonly known as the APL) in Laurel, Maryland, to work with scientists in its neuroprosthetics group. The APL is the largest university-affiliated research center in the United States—453 acres, 20 major buildings, and 7,200 employees—and its focus is on solving the nation's most complex engineering and analytical problems. Neuroprosthetics qualifies as such a problem, so a lot of DARPA-funded prosthetic research happens there.

I'm familiar with the APL, having received my PhD from the Johns Hopkins University in Baltimore. Although I never worked there myself, I have toured the APL's facilities a couple of times and once was the invited speaker at its Colloquium, a weekly scientific lecture series that has been continuously running since 1947. I still have some contacts at the APL, so I decide to call on them to see what's currently going on in their neuroprosthetics lab.

I'm introduced to Robert ("Bobby") Armiger. He's a biomechanical engineer who works on advanced prosthetics research and development. He was one of the people who worked with Loomis when she came to the lab. He tells me, "My research interest is in seeing how much intuitive and natural feedback control we can provide to advanced prosthetics users without going so far as to actually stick electrodes inside their brains." He believes TMR approaches will soon bring us very close to realizing the goal of completely natural movement control of limb prosthesis. But he says we haven't advanced quite as far in terms of an advanced prosthetic's ability to restore natural touch sensations. Nevertheless, when he became aware of Loomis's novel TMR/TSR surgery, he thought his group would be able to quickly adapt its current advanced prosthetic prototype to provide Loomis with a simulated feeling of finger touch. He told Seth to bring Loomis by the lab.

The idea was to put tiny pressure sensors on the tips of the five prosthetic fingers and to electrically connect those pressure sensors to little vibrating disks, about the size of a hearing aid battery, that would be applied to the five patches of TSR skin on the stump of Loomis's arm.[24] When the prosthetic finger touched something, the pressure sensor would cause the corresponding disk to vibrate and Loomis would feel it as though the tip of her missing finger was touching something vibrating. If you've ever had occasion to touch the housing of one of those small vibrating aquarium air pumps, that's very close to the sensation Loomis would feel. Picking up an object would feel to her as though she were picking up an aquarium pump. Armiger's research group proceeded to get their advanced prosthetic wired up accordingly in preparation for Loomis's visit.

For some time before, the neuroprosthetics research group had been working with amputee Johnny Matheny on nudging their advanced prosthetic prototype closer to perfection. Through practice, Matheny had become quite dexterous with the APL prototype. Matheny had undergone TMR surgery but not TSR. He also has had a highly novel surgery that Loomis has not: an arm *osteointegration,* a type of bone surgery performed on the end of an amputated arm. During this type of surgical procedure, a metal rod is inserted into the upper arm bone (the humerus) with a post protruding out through the skin. Matheny's external post allows him to attach the advanced prosthetic directly to his upper arm just by clipping the prosthetic onto the post. An osteointegration has three major advantages for the amputee:

(1) It creates a stable mechanical support for the advanced prosthetic to attach directly to the amputee's skeletal system, just as the natural arm had been anchored to the skeleton. (The alternative suction-based skin attachments are inherently less stable because of the skin's tendency to shift around.)
(2) It frees the surface of the skin to be used as a site for recording muscle signals and delivering touch feedback.
(3) It provides the user with better *proprioception*—the sense of body position and self-movement.[25]

Affable and gregarious, Matheny is a great ambassador for the field of neuroprosthetics in general. He sometimes accompanies the APL staff to scientific meetings and public events to demonstrate his proficiency with the advanced prosthetic. In fact, one feature of the current prototype is that the wires of the skin electrodes have been replaced with wireless Bluetooth technology—the same technology your laptop computer and cell phone use to send short-range wireless instructions to your printer.[26] Not only does Bluetooth simplify things by eliminating all the cumbersome wires, it also means Matheny can control the prosthesis from afar. As long as he's wearing the Bluetooth-enabled Myo band—a sensor-containing armband that looks like a leg garter for stockings—he can make the prosthetic arm move even when it is detached from his body and placed on the other side of the room. In fact, one of Matheny's favorite party tricks is to detach the limb and put it on a table. He then makes it crawl across the table on its own just by his thinking about his fingers clawing the table, to the amazement of everyone. Hardly a wallflower, Matheny's stunts make him quite the memorable party guest. We humans have sure come a very long way since the days when the Flying Boy amused us.

Unfortunately, because the prototype was designed to fit Matheny, who is a large man, it won't fit Loomis, who is a small woman. Therefore,

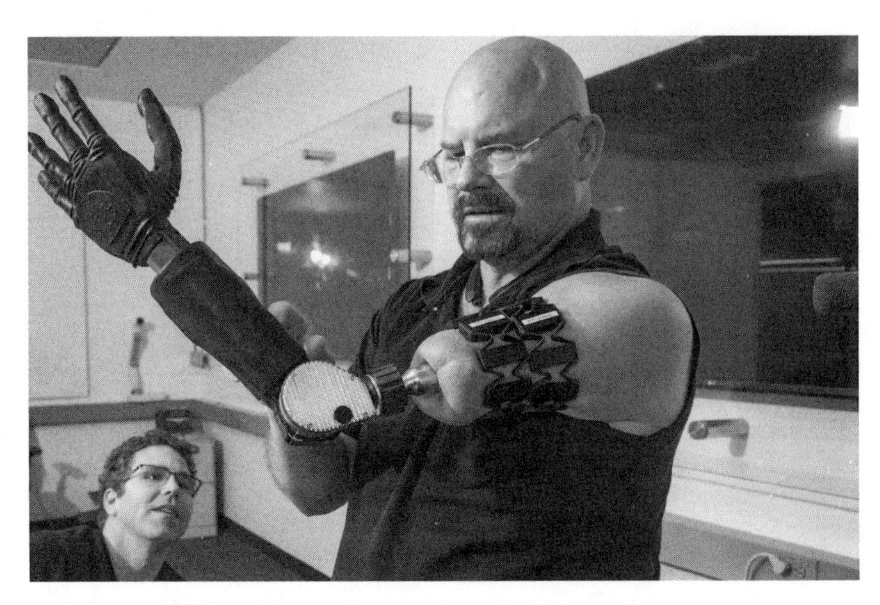

FIG. 13.3. Johnny Matheny demonstrates his advanced prosthetic arm. The neuroprosthetics research group at the Johns Hopkins Applied Physics Laboratory has been working for some time with arm amputee Johnny Matheny on their advanced prosthetic prototype. Matheny has had a highly novel surgery called *osteointegration*. This type of bone surgery is performed on the end of the amputated limb. A metal rod is inserted into the upper arm bone (the humerus) and a post protrudes out through the skin. The advanced prosthetic attaches to the post. In addition to providing the prosthetic arm with greater stability, osteointegration frees more skin on the stump of the arm to be used as a recording site to pick up the electrical signals that Matheny's brain is trying to send to his missing forearm. The brain's electrical signals are detected by sensors in an armband Matheny wears around the stump of his arm. The armband encodes those electrical signals and transmits them wirelessly via Bluetooth technology to a computer in the prosthetic arm. The arm's computer receives the signals and instructs the appropriate electromechanical devices in the advanced prosthetic arm to make the exact movements that correspond to the muscle movement commands coming from Matheny's brain. The prosthetic arm has more than 100 sensors that receive information that is crucial for movement—such as force, vibration, torque, and temperature—as well as 17 separate motors controlling 26 joints in the fingers, hand, and upper and lower arm. This design makes the prosthetic highly dexterous and capable of producing complex movements in a manner remarkably similar to an intact limb. (© 2021 The Johns Hopkins University Applied Physics Laboratory LLC. All Rights Reserved.)

the only way Loomis can experience the prototype is while it is detached from her body. But thankfully, using it detached wouldn't interfere with her being able to try it out.

It was June 2016 when Seth and Loomis made their trip to the APL. Loomis's father accompanied them to provide his daughter with some family support. The neuroprosthetic research group at the lab was just as interested in hearing about the details of Loomis's TSR surgery as Loomis and Seth were in having her try out the researchers' touch-enabled advanced prosthetic. After some discussion about what they would be doing, the research group brought out their advanced prosthetic prototype. As I've said, it wasn't possible for Loomis to actually wear the device, but the team hooked skin electrodes over the reinnervated musculature on the stump of her upper arm. The researchers then told Loomis to think about performing various movements with her missing arm, like bending her elbow, opening and closing her hand, and moving her wrist in specific ways. The electrodes were connected to a computer that recorded the patterns of electronic signals associated with each of Loomis's thought commands. The computer then converted those electric signals into machine-readable computer code that was fed to the advanced prosthetic's electronic motors. The code told the advanced prosthetic to perform a particular movement when it received the specific coded command. That done, it was time for the test.

The electrodes were rerouted to the advanced prosthetic, and Loomis then was asked to think about making the same movements. Remarkably, on her first try, Loomis was able to get the prosthetic arm to make the same specified movements just by thinking about them again. The hand and arm movements—the gross motor skills—were good enough, but were Loomis's thought commands enough to provide the dexterity to move the individual fingers of the prosthesis? The room was filled with anticipation. Loomis was asked to move just her index finger. She thought about the task and the prosthesis's index finger moved. She then was told to think about pinching something with her index and thumb and again the prosthesis made the appropriate pinching movement. Loomis had individual finger control! At this point, Loomis was able to control the arm just as prior amputees with TMR surgery, like

Matheny, had done before her. But was she also able to feel what she touched through all five of her prosthetic fingers?

The tension was palpable as the testing shifted from moving the prosthesis to feeling it. Everyone fell silent as the test began. The tiny vibrating motors were taped over the appropriate five patches of upper arm skin that were now home to what had once been the sensory nerves of her five fingers. When everything was ready, the scientists turned to Loomis's father and asked him to do the honors. He was happy to comply. They instructed him, "Just touch the thumb on the prosthesis—right here. When you do that, she should instantly feel it and tell us."[27] When her father touched the prosthesis's thumb, an infectious smile spread across Loomis's face and the whole room knew the TSR surgery had been a complete success. Loomis could actually feel through the prosthesis.

Loomis's father then moved one by one from finger to finger of the prostheses, touching the end of each. And each time Loomis felt his touch. All five fingers were working. Another milestone had been crossed in the goal of creating a fully functional prosthetic arm.

What Loomis felt when her father touched the fingertips of the prosthesis were the tiny vibrating motors taped to the reinnervated patches on her arm. To Loomis, the vibrations felt exactly like they were coming from her own missing fingertips, but they didn't feel like real natural touch. This kind of touch felt like vibrations, because they were actual vibrations. Although they would never be confused with natural touch, the vibrations do provide enough feeling so that an amputee can learn to modify the strength of the prosthetic's grasp accordingly. But the sense of touch Loomis most desires is one that encompasses all the nuances of natural touch, like the complex feelings of touch Loomis would get from petting her dogs.

That level of tactile sensation is still a long way off. Nevertheless, Armiger says they have two more touch projects in the works. One would replace the vibrating disks with little plungers that would press or tap on the TSR skin patch in a way more akin to our normal feeling of touch as pressure. Another uses temperature sensors in the prosthetic's fingertips to activate tiny heating films placed over the TSR skin patches. Armiger tells me the major challenge there is to match the very high-speed sensing abilities of the normal human nervous system. (It takes about an eighth of a second to sense the heat from touching a hot frying pan.) The reaction time of the prosthesis from temperature sensing to heating the skin patch is currently way too slow, providing a much delayed heat sensation.

I suggest to him that the delay might not present too big of a problem, since a prosthetic arm isn't as easily burned as a natural arm. But Armiger cautions me not to downplay the value of heat sensing. It might not seem as important as sensing pressure, but he explains that many amputees value it. They miss things like the cold-hand feeling you get from grasping an iced can of beer on a hot summer day. Or feeling the warmth of a loved one's hand. Feeling temperature is one of the many sensations that make tactile experiences pleasurable. He says amputees want to be made whole again, and that means regaining the full spectrum of tactile senses, including heat. He warns against discounting the value of any sensory experience.

In fact, touch gets no respect. People just don't understand that natural touch has all of the subtle nuances of our other four senses—and, arguably, to an even greater extent. The skin is innervated by multiple types of tactile neurons, each of which registers a different aspect of skin deformation. What people experience as touch is the integration of a complex pattern of activation of these different sensory nerve endings.[28] Touch isn't simply pressure or heat. Natural touch is so much more. That's what makes it particularly difficult to replicate in a prosthesis. The vibration Loomis feels coming from the prosthetic finger is just one limited dimension of the experience we call touch.

Perhaps getting to a true sense of touch will need a more sophisticated brain-machine interface. That's the approach to touch restoration being taken by a neuroprosthetics research group at the University of Utah, the home of the *Utah slanted electrode array* (or USEA).[29] The USEA is a small square of 100 microscopic needle electrodes arranged in a 10 × 10 matrix, resembling a tiny bed of nails. The "bed" that the needles protrude from is a tiny square (4 × 4 mm), covering an area about the size of Abraham Lincoln's face on a penny. Each needle can measure voltage independently. The needles become progressively shorter across the array so that they can measure voltage at different tissue depths, giving the surface of the array a slanted or wedge appearance, like the slope of a roof.[30]

The USEA can be stuck directly into a nerve, with its needle electrodes penetrating to varying depths. When a single needle detects a voltage change above a certain threshold, a spike is recorded, along with its time of occurrence and the needle's position in the array. The spike data can be analyzed to find "bursts"—the spikes originating simultaneously from a cluster of needles in close proximity to one another. Specific patterns of bursts are typically associated with specific touch sensations. If the incoming patterns of bursts can be linked with a particular aspect of the feeling of touch, it may be possible to emulate those burst patterns in an outgoing signal to artificially re-create the same feeling for a prosthesis wearer. The concept isn't much different than "teaching" a computer to recognize muscle twitching patterns as unique movement instructions from the brain and programming an advanced prosthetic to move accordingly when it gets that instruction pattern. It's just that, in this case, the USEA is doing this on a much smaller scale, and doing it for sensory neurons rather than motor neurons.

A very important feature of the electrodes on the USEA is that they can function in read or write modes. In other words, they can both measure the existing tissue voltage (read) at the tip of a needle and deliver a voltage or current (write) to the tip of the needle. What this means is that when the USEA is stuck into a nerve, it can enable a computer to read the pattern of voltage bursts resulting from a particular feeling, associate that specific pattern with that specific feeling, and later reproduce that same feeling by delivering the same voltage pattern to the nerve. In cases where the hand is missing, there are no sensory signals sent from the hand to the nerve because there is no hand, so no sensory signals can be recorded. But one can still record and interpret ("decode")

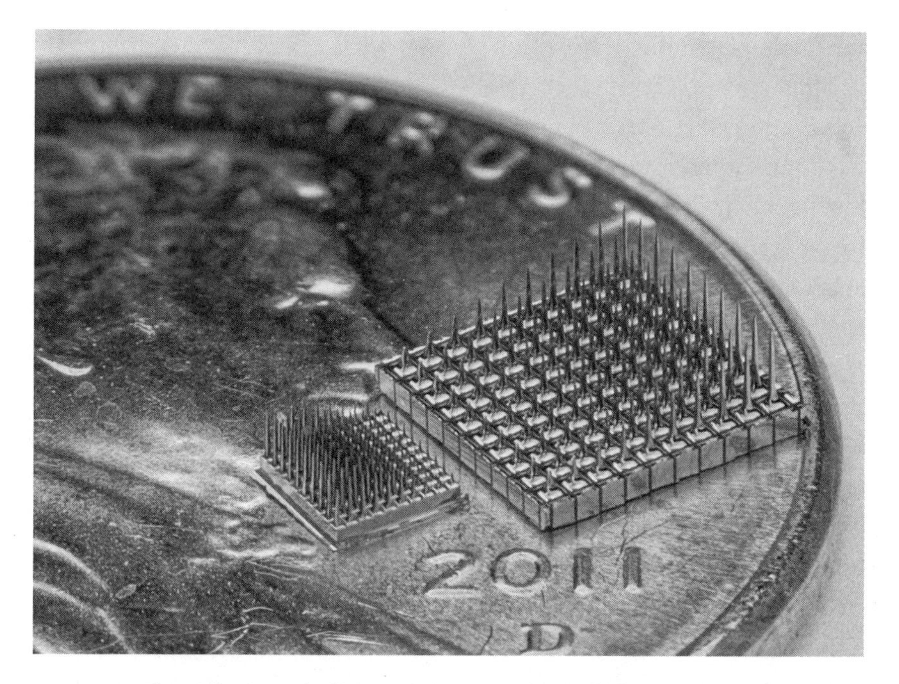

FIG. 13.4. Electrode arrays for brain-machine interfaces. The best electrodes for linking computers to the nervous system can be operated in read or write modes, thereby both receiving and sending electrically encoded information, respectively, to discrete areas of neural tissue. Miniaturized arrays with multiple electrodes are able to simultaneously process numerous channels of information coming to and from a small region of tissue. The Utah slant electrode array (USEA) represents one such type of array architecture in widespread use. It consists a 100 microneedles arranged in a 10 × 10 array. The needles progressively increase in length by fractions of millimeters from one row to the next, so that a range of tissue depths can be probed using a single array. Pictured are two USEAs of different sizes, placed on top of a penny for size comparison. (Photo courtesy of Richard Normann, University of Utah)

motor signals being sent from the brain through motor nerve fibers to the muscles and use those electrical signals from nerves, and from the muscles themselves, to control the advanced prosthetic. And, going the other way, one can still send electrical messages to the remaining sensory nerve fibers in the arm that came from the now-missing hand, activate those nerve fibers, and thus send a comparable set of digital pulses through those nerve fibers to convey sensory information to the brain. Yes, I know this theorizing is getting a bit complicated. So let's look at a concrete example of how this all might work for an arm amputee in real life.

Scientists at the University of Utah recently tested a USEA approach that may someday provide an alternative to TSR as a means of restoring touch to amputees.[31] Instead of surgically separating the various fascicles of a nerve and redirecting them to patches of skin, as Seth fastidiously did in the stump of Melissa Loomis's amputated arm, they instead stuck USEAs directly into the remnant arm nerves of an amputee. One was placed in the median nerve and one in the ulnar nerve, for a total of 200 needle electrodes. Because each needle penetrated the nerve fiber to a different depth, the 100 needles in each USEA provided good electrode coverage of all the approximately 18 fascicles within each nerve. The investigators then wrote to each electrode on the USEAs, one by one, and were able to evoke 119 different sensations (86 sensations from electrodes in the median nerve and 33 sensations from electrodes in the ulnar nerve). Each time they wrote to a USEA electrode, the amputee was asked to localize and categorize the feelings he experienced on a diagram of his missing hand. He was able to pinpoint all 119 to specific hand locations and categorize the feeling he experienced at each location as either vibration (37%), pressure (29%), pain (16%), tightening (12%), movement (3%), or tapping (3%). The result was a "feeling map" of the amputee's touch sensations.

Using the feeling map, the scientists linked the individual electrodes that evoked each sensation to the appropriate sensors (i.e., the congruent *receptive fields*) on an advanced prosthetic arm; that is, they matched each of the amputee's 119 sensations to the corresponding locations on the prosthesis. They then programmed the sensors to trigger a voltage burst into the amputee's appropriate nerve—median or ulnar. Tests showed that when someone touched the fingertip of the prosthesis or pressed on the palm of the prosthesis, the amputee experienced sensations that very closely mimicked the touch of a natural hand. And the amputee was able to use these feelings to improve his grip on muffins, whole eggs, egg shells, milk cartons, and soda cans, and, most importantly, to hold onto a wine glass. In fact, the sensory feedback from the prosthesis was able to significantly improve the overall performance of the advanced prosthetic for the amputee. As we've already discussed, "touch" is not a single sensation. Consequently, when the researchers made the digital pulse code that they sent to the nerve (and hence to the brain) more biologically realistic by mimicking the patterns sent by additional touch sensors present in a biological hand, the biomimetic stimulation patterns improved performance even more. I'll drink to that!

So where does all this leave Melissa Loomis? She tells me that when she originally asked for an advanced prosthetic with Luke Skywalker's full capabilities, she didn't really appreciate how truly futuristic her request was. "I just thought I'm going to get this fancy prosthesis and it will be totally awesome. . . . I didn't realize that it actually didn't exist. I didn't understand it wasn't something you could just go out and get." But that is the current reality for Loomis and other arm amputees. Loomis, at least, has been able to take the prosthesis of the future for a test drive, but touch-enabled advanced prosthetics aren't something amputees can currently take home and park in their garage. That option is still a few years away. And her peek into the future has somewhat frustrated Loomis. "I have this super ability, but I don't have the tools to make it work."

The APL's prototype advanced prosthetic, with its vibrating disk technology, isn't available for Loomis to use; it's still a research tool. And even if they could give it to her, it would need to be modified and shrunk down to fit her arm. It would also be too heavy for her to wear for any length of time because she doesn't have an osteointegration, like Matheny has, to better distribute the weight. So Loomis is in another "time will tell" situation. She must wait until commercial advanced prosthetics eventually become TSR touch-enabled, in order to make use of her remarkable capabilities. Likewise, the alternative Utah slanted electrode array technology is still in the research phase and is not soon to be commercially available for human use. Even if it were, using it would require Loomis to undergo more surgery to install USEAs in the nerves of her upper arm, since the Utah approach is completely different and doesn't make use of skin TSR to generate touch sensations.

But Loomis maintains her optimistic outlook, an outlook that already has served her well through this whole ordeal, starting on that first day when she encountered the raccoon. She tells me the TMR/TSR surgery has nearly eliminated her phantom pain, which was considerable before the surgery. So she would have had the surgery even if that was the only benefit. And although her current commercial advanced prosthetic is heavy and uncomfortable, has no touch capabilities, and is often in the shop for repairs, she does find it useful at times. More importantly, she thinks the future is bright for touch-enabled advanced prosthetics.[32] She says she's happy to work with researchers to push the technology forward, and is currently working as a test subject with a number of different prosthesis research teams interested in TSR. She is also hopeful that future generations of amputees will have ready access to affordable commercial advanced prosthetics with touch capabilities.[33] So Loomis is satisfied with her current situation and has no regrets. She doesn't even harbor any resentment toward raccoons, as long as they stay away from her dogs.

Touch is just one of the human senses, and amputees are a somewhat special situation. What about the sensory losses people suffer apart from loss of limbs? Are there electrical options to help them regain sensory input? Indeed, there are. We'll talk about them next.

14

SOUNDS OF SILENCE

SENSORY NEUROLOGICAL IMPLANTS

> Deafness is a much worse misfortune [than blindness]. For it means the loss of the most vital stimulus—the sound of the voice that brings language, sets thoughts astir and keeps us in the intellectual company of man.
>
> —HELEN KELLER

> She [Helen Keller] is fellow to Caesar, Alexander, Napoleon, Homer, Shakespeare, and the rest of the immortals.
>
> —MARK TWAIN[1]

As infants, both my son and daughter had chronic ear infections. With my son, Matthew, we were able to manage the infections with antibiotics until he grew out of the problem. However, for his younger sister, Anna, the antibiotics weren't working well enough. Our pediatrician was testing her hearing monthly so that we could keep tabs on how the infections were affecting her sound perception. The pediatrician was pushing for her to get ear tubes—tiny plastic cylinders that are surgically inserted into the eardrum to allow fluid to drain from the inner ear—but I was pushing back. The ear tube surgery required general anesthesia, and there was no guarantee the tubes would be effective in reducing the infections. Also, Anna didn't seem to have any ear pain; it was just that she had reduced hearing. I didn't like the idea of such a small child undergoing the risk of anesthesia and surgery for a problem that didn't seem to be significantly bothering her. Besides, I thought, how much does a one-year-old need to hear anyway? She doesn't even talk yet.

When Anna's hearing dropped to 50% of normal, our pediatrician became impatient with me and somewhat insistent that Anna have the ear tube surgery, implying to do otherwise amounted to child abuse. I told him my rationale that a one-year-old didn't need to hear much and the problem would resolve itself anyway, just as it had with her brother. He countered my argument, saying a child must have normal hearing in order for normal speech and language skills to develop. I expressed skepticism and he expressed anger. In the sternest voice he could muster, short of yelling, he snapped at me: "If you had a 50% hearing loss, you would find the situation intolerable!" That shook me into action. Anna had the surgery. The pediatrician turned out to be correct. After the surgery, Anna's ear infections did subside, her hearing did increase into the normal range, and I was proved an idiot.

So, if 50% hearing loss is such a significant detriment to an infant, imagine the difficulties for a child with no ability to hear. Around 3 of every 1,000 children in the United States are born with some level of hearing loss, and approximately 1 of those 3 hearing-impaired children is completely deaf.[2] There are multiple causes for being born deaf, including infectious agents, like cytomegalovirus, rubella, herpes, and toxoplasmosis. Some medicines taken during pregnancy can also increase the risk of having a deaf child, like the antiepileptic drug Dilantin and the acne medication Accutane. There are also inheritable forms of deafness, where both parents carry a mutated hearing gene. Although the parents typically aren't deaf themselves, approximately 25% of their offspring are born deaf.[3]

For children who are totally deaf, there are few good options. Ear tubes won't work, and even hearing aids often can't help. That's because hearing aids simply boost the volume of sound. If there is no ability to hear sound, increasing the volume is not going to do anything. But for many, there is a remedy.

For the past 50 years, deaf children and adults have been treated with *cochlear implants*.[4] The word *cochlea* comes from the Latin word for a spiraling snail shell, and it perfectly describes the appearance of this tiny organ of the inner ear. The workings of the human ear are complex, but the sequence of events needed for proper hearing is not. Sound waves traveling through the air hit the eardrum, causing it to vibrate. The vibrating eardrum sets tiny bones (the *ossicles*) into motion, which has the effect of amplifying the vibrations. These amplified vibrations then enter the outer opening of the cochlea and move down the spiraling tube, like a snail retreating into its shell. The internal walls of the cochlea are lined with little sensory, hair-like cells, giving it the internal appearance of a progressively narrowing pipe lined with long-pile carpet. When the hair cells start to vibrate, they send a pattern of action potentials to the brain, which the brain then translates into the sensation we know as hearing. In many cases, deafness is caused by damage to these sensory hair cells in the cochlea. No hair cells, no action potentials, no hearing. Just that simple.

But the ability to hear normally also requires the ability to differentiate *pitch*, the sensation of the frequency of the incoming sound wave. Pitch is to sound as color is to light. Just as colors represent different frequencies of light, pitch represents different frequencies of sound. And just as color is a visual sensation you perceive, pitch is a hearing sensation you perceive.

Understanding spoken language is all about deciphering words encoded as interruptions and changes in pitch. This is where the location of the specific hair cells transmitting the action potentials becomes very important. The frequency of a sound wave determines how deeply the vibrations get into the cochlea. This is due to the physics of the vibrations and the way they get absorbed by the walls of the cochlea and not due to any sound frequency discrimination on the part of the hair cells. High-frequency sound waves produce high-frequency vibrations, which are absorbed quickly into the walls of the cochlea and thus don't make it very far down the tube, while low-frequency vibrations are not as easily absorbed and thus penetrate much deeper. Therefore, the depth the vibrations can penetrate into the cochlea is inversely related to the sound frequency. In short, the hair cells respond to any vibrations they encounter; it's just that the frequencies of vibrations they encounter are determined by how deep inside the cochlea they reside.

As it happens, the brain actually "knows" which populations of hair cells in the cochlea are sending the action potentials it's receiving. That knowledge allows the brain to deduce the frequency of the sound waves that must have stimulated the incoming action potentials. If the signals are coming from the hair cells near the opening of the cochlea, the brain deduces the frequency of the sound wave must be high, so the brain translates those nerve signals into a sensation we perceive as a high-pitched sound. If the hairs sending the action potential reside deeper in the cochlea, the brain registers it as a low-frequency sound wave and produces a hearing sensation of low pitch.

Now that I've explained how the detection of pitch works, perhaps I should step back and define the concept of sound itself before going any further. Physicists often define sound as an *acoustical wave*—a vibration with a frequency that can be heard—traveling through a gas, liquid, or solid. But physiologists sometimes define sound from a biological perspective: the reception of acoustic waves by the ear and the brain's perception of them. What then is the answer to that centuries-old philosophical question, "If a tree falls in the forest and there is no one present to hear it, does it make a sound?"[5] The answer depends upon whom you ask. The physicist would say yes because the falling tree produces acoustical waves, but the physiologist would say no because you need an ear and a brain to convert acoustical waves into the experience of sound. I'm glad that important question is finally answered. Don't you just love philosophy?

We learned in chapters 4 and 5 that both Volta and Duchenne were able to elicit a sensation of sound by sticking electrodes inside their ears and shocking themselves. Duchenne was even able to generate sound sensations in deaf people by shocking their inner ears. But the sounds the electricity produced were just noise. If you want to produce a true natural hearing experience, one that reflects all the nuances of hearing sounds with various pitches, you must selectively stimulate specific regions of the cochlea. Indiscriminately stimulating the whole cochlea, or just part of it, only generates noise.

Modern cochlear implants do not make noise; they provide a genuine sound experience for deaf patients. This is how they do it. The business end of the cochlear implant is a slender, flexible, tapered probe, similar in appearance to a tiny blade of grass. The probe is surgically placed at the relatively broad opening of the cochlea and then carefully inserted so that it works itself deep inside, all the way to the very end of the spiraling chamber. At precisely spaced intervals along the probe are different electrodes. By selectively applying electrical stimulation to different electrodes, different spots along the cochlea can be artificially stimulated. The specific location stimulated determines how the brain interprets the signal it receives from the electrodes. The brain records signals originating from near the cochlea's opening as high-pitched, and those coming from deep inside as low-pitched. Likewise, those coming from the middle of the cochlea are sensed as intermediate-pitched.[6] But to emulate a full hearing experience, you need more than three levels of pitch. In fact, the more electrodes that can be placed along the length of the stimulating probe, the better the hearing experience will be. That's because having more probes means higher resolution of the incoming sound waves and, consequently, better frequency discrimination and a richer hearing experience for the patient. Each electrode on the cochlear implant's probe represents a range of frequencies, known as a *channel*. Increasing the number of channels narrows the range of frequencies each channel needs to cover. In other words, more channels give better resolution of pitch and thus a sound experience closer to natural hearing.

In the early days, implants had only 6 channels, which resulted in human speech sounding very robotic to people with cochlear implants; accordingly, their own speech became robotic as their brains tried to copy the speech patterns they heard. In contrast, some of the latest high-resolution cochlear implants now have as many as 120 channels, which provides a hearing experience very close to the natural experience.[7] As a result, the speech of patients with modern cochlear implants has dramatically improved.

But the implant is only part of the story. There is other electrical hardware to consider. A cochlear implant's probe is connected by a fine wire running from the cochlea to a little electronic device about the size of a poker chip that is implanted under the skin near the base of the skull, just behind the ear with the implant. (If a person has implants in both ears, there are two subcutaneous devices on either side of the skull.) This device sends the signals to the appropriate electrodes in the cochlear implant. By definition, this is a "write-only" brain-machine interface. The cochlear implant is not receiving any information from the ear or the brain; it is just transmitting information in the form of an electrical stimulus to the brain by way of the cochlea.

From where then, you may ask, is the subcutaneous device getting its auditory information? The answer is that there is an external microphone sitting over the device, held in place on the skin by magnetic attraction. The microphone picks up sound from the air and wirelessly sends that information to the subcutaneous device. Thus, there are no wires penetrating the person's skin, which greatly reduces the potential for infection. The subcutaneous device decodes the electrical signals from the microphone and recodes the information into signals that go to the appropriate electrode channels. The brain takes over from there, and the deaf person experiences sound, sometimes for the first time.[8]

Cochlear implants were one of the earliest types of implanted neuroprostheses to be developed. They were first introduced in 1957 but initially encountered problems. They were easily dislodged, and there were other technical obstacles related to their implantation. Research funding also was hard to find. Some auditory surgeons became discouraged by these circumstances and, consequently, development of cochlear implants nearly came to a halt. But in 1958, the first internal heart pacemaker was implanted in a patient.[9] (External pacemakers had been in use since the 1930s.) These heart stimulation devices need only a one-channel electrode. They work by placing in the heart an electrode that is programmed to emit small electric shocks at a frequency of about once per second to stimulate the heart muscle to contract at a normal rate. The concept is much simpler than a cochlear implant in that a simple shock from a single electrode, placed at the correct location within the heart, will spread across the heart muscle and produce a normal synchronized contraction; there is no need to cover the heart with electrodes. Also, the heart is not as small, delicate, or surgically inaccessible as the cochlea. Progress with implanted heart pacemakers, therefore, moved much faster than for cochlear implants.

The success of implanted heart pacemakers largely can be attributed to the miniaturization of electronics, made possible by the invention of the transistor in 1948. Prior to the introduction of transistors and integrated circuits, the cumbersome pacemaker controllers—about the size of a book—were too large to implant and were, therefore, worn externally. As batteries and circuit boards became ever smaller, it eventually was possible to move the entire pacemaker into the body itself, such that the patient didn't need to wear any external hardware. The modern hardware of an implanted heart pacemaker resembles that of a cochlear implant. The stimulating electrodes are placed within a chamber of the heart, and the pacemaker controller, a matchbook-size electronic device, is placed under the skin just below the collarbone, rather than at the base of the skull as is the case for cochlear implants. Both devices seek to electrically stimulate their target organs into better performing their natural functions—either getting the heart muscle to beat at a normal rate or getting the cochlea to send the appropriate nerve signals to the brain.

Heart pacemakers are heavily used to this day to treat people suffering from an irregular heartbeat. Although it is hard to get reliable figures, some estimates for the number of Americans living with pacemakers are as high as three million, or a little less than 1% of the population of the United States. About 70% of pacemaker patients are over the age of 65, so the proportion of seniors with pacemaker implants is considerably higher.[10]

Heart pacemakers also served as a model for the recent invention of a type of "breathing pacemaker" for people who suffer from sleep apnea, a disorder characterized by a stoppage of normal breathing during sleep. This disease is often caused by the relaxation of the muscles of the tongue and other soft palate tissues of the throat during sleep. The flaccid tissues collapse into the airway and obstruct normal breathing. This new chest-implanted device monitors breathing. When breathing stops, it sends an electrical impulse to an electrode in the throat that shocks the relaxed tissues into contracting, thus reopening the airway. The device is called Inspire, a play on words of the term's double meaning—to breath in or to stimulate someone to do something. (I, for one, find this clever moniker a welcome change from the tiresome and witless acronyms typically used to name new electronic devices.) Inspire was approved by the US Food and Drug Administration (FDA) in 2014 for people with obstructive sleep apnea.[11]

Internally implanted heart pacemakers were a success from the start. The first recipient lived until the age of 88, although he had upgraded pacemaker models 26 times during his lifetime. In fact, it was the great success of implanted heart pacemakers that renewed the audiologists' desire to resolve the issues associated with cochlear implants. And resolve them they did.[12]

Today, if you consider the combined numbers of heart, cochlea, and throat electrical implants, the total number of Americans with electronic implants is substantial. And it is likely there will be greater numbers in the future, as more and more organs become targets for treatment with electricity. But I'm getting ahead of myself. This chapter is supposed to be about sensory neuroprosthetics. We should get back to that.

Yes, cochlear implants have been around a long time. They have been steadily and incrementally improving for over half a century to the point where they can provide a high-quality auditory experience for people who otherwise would be totally deaf. But there isn't much that's fundamentally new about how they work. What then is new with them? The age of the patients.

Cochlear implants are being used on younger and younger children, some as young as 3 months old.[13] Although the FDA guidelines for cochlear implantation still include an age threshold of 12 months, more and more congenitally deaf children are actually receiving implants at less than 12 months. This is being driven by two factors: (1) the risk of the surgery turns out be relatively low, even for infants; and (2) the children who get their implants early have significantly better language skills by the time they get to preschool.[14] Yes, science has confirmed what my children's pediatrician told me over 20 years ago: even infants need to hear well or their future language skills will suffer.

But cochlear implants aren't a panacea for treating deafness. They only help people suffering from a very specific hearing defect: nonfunctioning cochlear hair cells. The implant patients need an otherwise intact cochlea in order to work. Although defective hair cells are the most common problem underlying deafness, this isn't the only type of hearing problem. What if the structure of the cochlea is damaged or there is some other situation that makes a cochlear implant impossible? Up until now, there have been no good options for such patients. Recently, however, scientists have found a way to bypass the intermediary—the cochlea—entirely, and send sound signals directly to the brain. In this strategy, the electrical stimulation doesn't target the cochlea; it targets the auditory nerve as it enters the brain stem.

The auditory nerve is the eighth (or VIIIth) cranial nerve. Unlike the vagus nerve, this cranial nerve stays local, never leaving the head. The auditory nerve is attached directly to the cochlea, and it's the nerve that passes all the action potentials generated by the cochlea to the brain. Just as the ulnar nerve is left jobless when someone has his arm amputated, the auditory nerve has little to do when the cochlea is unable to emit nerve signals.[15]

The auditory nerve contains only sensory neurons, so it can only transmit signals from the ears to the brain, not in the opposite direction.[16] Because the auditory nerve is devoted to hearing, if you electrically shock the nerve, you will hear a sound. Nevertheless, there originally was skepticism that directly shocking the auditory nerve would be able to provide a deaf person with a useful perception of spoken words. As it turned out, a direct shock was found to produce something quite close to speech perception if the auditory nerve was electrically stimulated at the spot where it enters the brain stem—at the cochlear nucleus.

In a generic sense, the word *nucleus* means the center or core of something—the area where all the action takes place. A cell has a nucleus; it's the place where its genes reside. An atom has a nucleus; it's the place where its protons and neutrons reside. Well, brain stems have nuclei too. They are the places where networks of neurons performing a similar function reside.

The auditory nerve interacts with the brain at the cochlear nucleus of the brain stem. So, if you want to substitute an electrical signal for a missing nerve signal from the cochlea because the cochlea is damaged or missing, the cochlear nucleus is a good place to deliver the electricity. You would be directly stimulating the same brain stem neurons that the nerves from the cochlea stimulate in response to sound detected by the ear.

Fortunately, since much of the cochlear nucleus is located near the surface of the brain stem, the electrodes needn't actually penetrate inside it. Rather, they can be positioned on the surface of the brain stem just over the cochlear nucleus to stimulate it into hearing, just as you would stick skin electrodes over the ulnar nerve of the arm to stimulate it into moving the arm. So, although such surgery is invasive, it doesn't require actually opening the skull and cutting into the brain itself. You might say the surgery amounts to taking electrical stimulation right to the doorstep of the brain without actually stepping over the threshold and penetrating deep inside. For this reason, the surgery is considered relatively safe.

This type of surgery is called *auditory brainstem implantation* (ABI). It was first performed in 1979 using a pair of electrodes placed on the cochlear nucleus about 1.5 millimeters (roughly 1/16th of an inch) apart. This rather crude device allowed patients to be aware of sounds in their environment but did not produce an authentic hearing experience. Today, the implantation device most commonly used looks like a tiny canoe paddle with 21 evenly spaced electrodes on one side. The set of 21 channels is a great improvement over its predecessor, but the hearing experience is still not nearly as good as with cochlear implants. That's why an ABI is considered only in cases where deafness is total and cochlear implants aren't feasible. Many of the patients fitting these criteria are, again, very young children. So, as is the case with cochlear implants, ABI is now being performed on younger and younger children. Some estimates place 2.1% of deaf children in the United States as potential candidates for ABI.[17] Most of these children are suffering from neurofibromatosis type-2, a rare genetic neurological disorder that often results in damage to the auditory nerve. Although the auditory experience continues to get better with improvements in ABI technology, patients with successful implants (about 80%) still cannot understand the spoken word with the implant alone, but they can use what they are hearing to improve their lip-reading capabilities. And 93% are better able to understand sentences. Again, ABIs, like cochlear implants, are not a panacea for providing hearing to the deaf, but they do offer additional options and a promising future as the electronic technology continues to improve.

So much for deafness; what about blindness? As you might imagine, the challenges are even more daunting for artificially restoring sight than they are for restoring hearing. There are three primary reasons for this. First, the required spatial resolution for useful sight is much greater than the pitch resolution required for useful hearing. Second, the required brain-machine interface must be much more sophisticated. Third, the required implantation surgery is much more invasive. Nevertheless, significant progress in providing an artificial sight experience through electrical nerve stimulation has been made, and there are even people walking around now with such neuroprosthetic vision devices. The details of the different vision implants currently under investigation are varied, technically complicated, and beyond the scope of this book.[18] But there are many parallels with what has already been achieved for hearing loss. Let's now briefly discuss artificial vision by focusing on how it is similar to and different from artificial hearing.

The mechanisms of human vision are much more complicated than the mechanisms of hearing, but just as for hearing, the major components of the vision process are fairly simple. They are similar to the parts of a camera. If you have any familiarity with how a camera works, you'll know it has a lens in the front that admits the light, just like a human eye has. The lens focuses the light on a piece of photographic film (or an array of electronic light detectors in the case of an electronic camera) in the back of the camera housing to produce a two-dimensional image. In the human eye, the "film" component is called the *retina*. Much of human blindness can be attributed to problems with either the eye's lens or its retina.

The major vulnerability for the lens of the eye is that it can become cloudy—a condition known as a *cataract*—which prevents the normal passage of light to the retina. This clouding can occur due to age, some types of drug use (including anabolic steroids), and genetic factors. Clouding is, in its essence, an optical problem. And optical problems can usually be effectively treated with an optical solution—an artificial lens implant. The cloudy natural lens is surgically removed and a clear plastic lens is inserted as a replacement. The passage of light to the retina is thereby restored, as is the person's vision. In the United States, over 3.8 million lens replacement surgeries are performed every year, most all of them highly successful at restoring sight.

I developed cataracts a few years back and had my natural lenses replaced with artificial ones in both of my eyes, so I can count myself among those who are living with implanted lens prostheses. And I can personally attest to the fact that this treatment is highly effective. Virtually overnight, I went from significantly impaired sight to 20/20 vision. It was a quick and permanent fix of a serious problem. But fortunately, my eye disease was restricted to my lenses; there was nothing wrong with my retinas.

Retinal problems are much more challenging to deal with. Some retinal problems can be mechanical in nature, such as a retinal tear or detachment—a situation where the retina has somehow torn away from the back of the eye. This type of mechanical problem can sometimes be remedied with a mechanical solution—the retina is surgically reattached to its original location. Unfortunately, there are many other retinal problems where the retina is structurally intact but not functioning properly on a neurological level. In such cases, the retina is not sending the correct signals to the brain, so the brain cannot create an image for the mind. This is inherently a neurological problem and requires a neurological solution: a neuroprosthetic eye implant.

The bad news for the field of neuroprosthetic eye implants is the small size and delicacy of the eye. It is much harder to work with an eye prosthesis than an arm prosthesis or even a cochlear implant. On the other hand, the eye is surprisingly tolerant of having foreign materials inserted into it. Although body rejection was a persistent problem with the development of heart and ear implants, our history with cataract surgery has taught us the eye's interior can accommodate a number of foreign materials. With the notable exception of metals like iron and copper, which are highly toxic to the eye, other materials, such as acrylic, polymethyl methacrylate (a clear plastic commonly know as PMMA), and silicone, can remain in the eye indefinitely without causing any problems.[19] (My lens implants are made of PMMA.) This unique property of the eye means scientists have a somewhat broad selection of materials available to them for the design of neuroprosthetic eye implants.

We learned in chapter 5 that, in addition to electrically inducing sounds within his ears, Volta was also able to stimulate visual sensations, including colors, by applying electrodes to the surface of his eyes. He interpreted this observation as evidence for the involvement of electricity in the sense of sight. So the potential to artificially stimulate vision with electricity has been appreciated for about two centuries, starting with Volta—the man who discovered the electrical battery.

But modern visual neuroprosthetics research has a somewhat shorter history. Its origin is often credited to the work of Otfrid Foerster, a German neurosurgeon working in the 1920s. Rather than electrically stimulating the eye, Foerster electrically stimulated the occipital pole of the human brain. The occipital pole is a region of the occipital lobe of the brain, and it's located way in the back of the brain, about as far away from the eyes as you can get.[20] The optic nerve (cranial nerve II)—the sensory nerve fiber that emerges from the rear of the eyeball—wends its way toward the occipital pole.

Due to this anatomical feature, the occipital pole had long been suspected to be important to vision, and Foerster's findings seemed to confirm this. When electricity was applied to the occipital pole, patients reported seeing lights in the form of stars, clouds, and pinwheels. Neuroscientists call such a visual sensation a *phosphene*—a phenomenon of seeing light when no light is actually entering the eye. Foerster's work suggested that perhaps, one day, it might be possible to provide a visual experience to blind people by electrically stimulating the occipital poles of their brains. This premise subsequently became the scientific rationale for developing a prosthetic brain implant that would allow the blind to see. Various groups have taken somewhat different approaches to the problem, but all rely on electrical stimulations of neurons somewhere along the nerve pathway from the retina in the eye to the occipital pole of the brain. Scientists have been working toward that goal for nearly a century now and have recently met with some notable successes.[21]

In all the different approaches, the camera-like function of the natural eyeball itself isn't used. Instead, the electrical output of a real external camera, worn on the head, is routed to a brain-machine interface that encodes the information into the appropriate electrical output signals needed to stimulate neurons at some point along the visual nerve pathway. Four different populations of neurons have been used as the target: (1) the retina, (2) the optic nerve, (3) the surface of the occipital pole (known as the *visual cortex*), and (4) the lateral geniculate nucleus (LGN). The LGN is in the middle of the visual pathway. It acts as a relay center, forwarding incoming signals from the optic nerve to the occipital pole.

Regardless of where they enter the pathway, the goal of the electrical input is to stimulate the neurons of the visual cortex so that the brain can convert the signals into a two-dimensional image for the mind. The visual cortex is, in effect, the "mind's eye."

Of the four possible target populations of neurons, stimulation of the retinal neurons has shown the most progress toward producing artificial sight. A retinal stimulation system—the Argus II by Second Sight Medical Products—was one of the first visual neuroprostheses to be approved by both the European Union (2011) and the US FDA (2013) for use in patients suffering from retinitis pigmentosa, a group of eye diseases that involve the loss of retinal cells. Since then, other retinal systems likewise have been granted regulatory approval.[22]

The technology shows promise, but to date the patients are still only able to see light and shadows. This is of some assistance to them in living their daily lives, but it's nothing like a natural visual experience. One of the problems related to this retinal stimulation approach is the inability to confine the electrical stimulation to a small enough area of the retina. Retinal neurons theoretically can be stimulated into firing an action potential by even a single interaction with one photon of incoming light. Thus, the theoretical limit of the natural resolution of a retina is at the level of the width of a single neuron. But it isn't yet possible to confine electrical stimulation to such a small area of the retina. In practice, the electrical current spreads across an appreciable area of the retina, encompassing many neurons and greatly reducing resolution compared to natural vision.

But a major advantage of using a retinal target for electrical stimulations is that the required electrical implants only need to go into the eye and not the brain. Eye surgery, though challenging in itself, is not brain surgery. There are a whole host of issues related to highly invasive brain surgeries that aren't posed by relatively straightforward eye surgeries. Besides, there may be little to lose if the eye surgery is unsuccessful, since the eye already is not functioning. But a mishap in the brain can have dire consequences beyond just the eyes.

One area of visual neuroprosthetics that has been making significant progress lately is the improvement of the brain-machine interface. We learned in the last chapter that, for the sense of touch, University of Utah researchers were able to construct an electrical "vocabulary" of hand sensations for a particular amputee by pretesting the amputee for his touch sensation experiences. The benefit of this approach is that the researchers don't need to understand the basic principles the nervous system uses to encode its touch signals; they just have to create a battery of random patterns of electrical stimulation and observe how the amputee's brain translates those individual patterns into specific touch sensations. Then they can produce an individualized electrical "dictionary" of touch sensations that can be programmed into the encoder of the amputee's brain-machine interface. The encoder can then translate the prosthesis's sensor outputs into the appropriate electrode stimulation patterns that the amputee's brain will recognize as the corresponding touch sensations.

In principle, you should be able to do something similar with the optic nerve of the eye. That is, you could implant a multielectrode array into the optic nerve, deliver a random pattern of stimulations, and ask the person to describe the visual sensations she perceives. But if you did that, the best you could probably hope for would be that the blind person would say, "I see a blue spot to the left," or "I see a diagonal black line." She would not say, "I see the face of a cat." Visual imagery is just too complex for such an empirical approach to work. But suppose, rather than random electrical signals, you cheat a little. Suppose it was possible to narrow down the electrical pattern options by getting some hints from a normal eye.

As it happens, there is a research group doing just that using mouse eyes.[23] Under the premise that all mammalian eyes work more or less the same, scientists at Cornell University are inserting electrode microarrays into the optic nerves of mice, and then "reading" the nerve signals that are moving through the optic nerve when the mouse is looking at various pictures. They record the nerve signal patterns coming from the mouse's optic nerve and use computer algorithms to identify which specific image features are associated with which specific signal features. The goal is to produce an encoder for a brain-machine interface that would be able to take images from a digital camera, decode them, and then electrically "write" those signals on the optic nerve so that the brain recognizes them as the correct incoming image.[24] Again, the advantage of this approach is that you needn't understand anything about the way the retina electrically encodes images; you just have to be able to duplicate the encoded signal as a patterned electrical output signal you can send to the brain. It's like saying, "Ich muss schnell ein WC finden!" You don't need to understand the literal meaning of the sentence; you just need to know that when you say it to a German, he will immediately point in the direction of the nearest bathroom. And every time you say it, you will get the same pointing response, regardless of which particular German you choose to say it to. So it's a useful sentence to know when you are traveling in Germany, even if you never learn what it means literally. In the same way, we don't really need to understand the electrically encoded language of the nervous system to be able to use its language to our advantage; we just need to memorize the key expressions. This approach has yet to move to humans, but the proof of principle has been established with the mouse's eye, which bodes well for future human studies.[25]

Another approach is to skip the optic nerve entirely and take the electrical signal directly to the visual cortex of the brain. Second Sight—the same company that produced the retinal stimulation system Argus II—has launched a new device called Orion, which bypasses the eyes and optic nerve and applies electrical stimulation directly to the brain's visual cortex. The device uses a 60-channel electrode implant, similar to the 21-electrode implant that is placed on the cochlear nucleus of the brain stem during ABI surgery. Also similar to ABI, the implanted electrodes deliver a pattern of electrical stimulation that the brain learns to interpret. Just as stimulation of the cochlear nucleus produces a sound experience, stimulation of the visual cortex produces a visual experience in the form of spatially resolved light sensations. But it's an image with quite limited resolution.

The 60 channels represent a theoretical maximum resolution equivalent to 60 picture elements, or *pixels*—the small areas of illumination on a screen display from which a digital image is composed. For comparison, a checkerboard has 64 squares (8 × 8). Envisage creating images on a checkerboard by moving the checkers around into different board patterns and you'll have some idea of the image resolution that currently can be achieved. For comparison, Apple's Retina Display on its computer screens—so named because it has a resolution rivaling the human retina—has over 3 million pixels (2048 × 1536). Consequently, blind people with an Orion system see nothing close to what a normally sighted person can see, *but they can see.* And even this level of sight can be very valuable to them in their daily lives.

The Orion system is currently in clinical trials with six blind patients.[26] All of these patients once had sight but lost it later in life due to accident or disease. This is an important feature of the trial, because it means all six patients know what it's like to see and what common objects look like. Therefore, they are able to describe to the researchers what visual sensations they are having: "I see something that looks like a sunrise." As of this writing (May 2020), the trial is in its twenty-third month of a planned study duration of 60 months. It is the first-ever trial of a brain implant to create bionic vision. It's a landmark study and the patients are true pioneers. I decide I need to talk to one.

Richard McDonald was the fifth patient to be admitted into the trial, and he's had his Orion implant for nearly two years now. He and his wife, Charina, have come east from their home in Los Angeles on a business trip and, at my invitation, they agree to meet me for dinner at a seafood restaurant in Baltimore. They have been together for 17 years. McDonald had suffered with glaucoma—an eye disease where fluid pressure builds up with the eye—since birth. He had been gradually losing his sight for his whole life. He finally became totally blind 13 years ago. McDonald tells me that he is thankful he had four years with his wife before becoming blind because it means he knows what she looks like, and he carries her image around with him in his mind. Charina has been at his side the whole time, so she's experienced his losing his sight as well as partially regaining it. They both are thrilled and thankful that McDonald was chosen for the study and how well the implant is working for him.

McDonald is wearing the Orion headset while we dine. The camera has a tiny lens, similar to the size of a lens on a cell phone. It sits on the bridge of a pair of wraparound sunglasses that look similar to any pair of plain dark sunglasses. He has a black "sweatband" around his head to hold the electronics in place, and a controller pack, about book size, that unremarkably dangles at his side like a shoulder purse. As we talk, he looks around, turning his head to "see" various things on the table and the faces of speakers, including our waiter's. His body and head movements seem completely appropriate to what he is doing. I don't think the waiter distributing the menus to each of us even realizes that this man wearing strange headgear is blind.

After some discussion of the menu, we order. Soon our dinner arrives. McDonald can see the dinner plate in front of him, but Charina tells him the positions of the various side dishes: "String beans are at nine o'clock." He looks directly at my face when he speaks, as though he

FIG. 14.1. Richard McDonald demonstrates his artificial vision device. McDonald is a blind patient who is participating in a clinical trial of a device called Orion, which bypasses the eyes and optic nerve and applies electrical stimulation directly to the brain's visual cortex via surgically implanted brain electrodes. The electrodes deliver a pattern of electrical stimulation that the brain learns to interpret as vision. Orion produces a visual experience in the form of spatially separated light sensations. It's an image with quite limited resolution, but even having low-resolution sight can greatly enhance the daily lives of people who are otherwise totally blind. The device consists of a tiny camera lens that sits on the bridge of a pair of wraparound sunglasses. (The shaded lenses are for cosmetic purposes only; they serve no function.) A headband holds the electronics in place. A controller pack, about the size of a small book, can be worn dangling from the shoulder. Next-generation devices promise to provide increased image resolution by employing brain implants with higher-density electrode arrays and incorporating better vision-producing algorithms into the controller's software. (Photo courtesy of Richard McDonald)

knows exactly where my eyes are. I ask him what seeing with his brain implant is like and he describes it in detail.

"Imagine," he says, "viewing one of those old-style radar screens you see on submarines in vintage World War II movies. The experience is very similar to that. I don't see color, just monochrome. And the light I see is pulsating." He goes on to explain that, because the device is set to deliver electrical impulses to his brain at a 6 Hz frequency (6 pulses per second), the images he sees appear to be flickering at 6 Hz; almost too rapid to be perceived as pulsating but not quite. He says when the frequency is increased to 10 Hz, the flickering subsides, but the device doesn't routinely deliver 10 Hz, just 6. He's looking forward to being able to get at least 10 in the future as the flickering is a bit of a distraction.

Despite the flickering, McDonald is pleased with how his artificial vision has improved his everyday life. The most important thing, he tells me, is being aware that something is in front of him, be it a person, car, trash can, etc. It gives him confidence as he is walking around that he won't bump into things. But he still needs his cane to navigate the specific features of the ground, like stair steps.

His brain is constantly learning to interpret the shadowy figures he sees into images of actual objects. For example, he's learning to distinguish a triangle shape by deciphering the light pattern he sees. He tells me, "I don't see a triangle like you see a triangle; rather, I learn that a certain visual pattern I am perceiving represents a triangle. So my mind creates an image of a triangle when I see that light pattern." McDonald says seeing a triangle in this way can be good enough for a blind person like him, but he says the scientists he works with think there is much greater potential. Their goal is for McDonald to see a triangle like everyone else sees a triangle, and they believe the key might be in improvements to the software, or what McDonald calls "the app."

Although we can't go into detail here, the app is the software that processes the camera image and converts it into an electrical stimulation pattern for delivery to the brain. The idea is to make the app do what McDonald's brain is trying to do on its own; that is, it tries to recognize a particular pattern of stimulation as a triangle. What the app can potentially achieve is to recognize the camera image as a triangle and send an electrical signal to the implant that the brain already associates as being an eye image of a triangle. This is similar to what researchers are doing with decoding the signaling in the optic nerve from a mouse's eye that we discussed earlier.

Yes, it's true that McDonald only has 60 electrodes in his implant, which does limit image resolution. But signaling patterns also convey image information. Therefore, it is very likely improvements can be made to the app that would enhance visual perception well beyond what doubling, tripling, or even quadrupling the number of electrodes in the implant could achieve on its own. Which is good news for McDonald. He says he's not sure whether he'd want to start from scratch with another implant and the additional brain surgery it would require. At age 53, he thinks next-generation implants will arrive too late to be useful to him. He's going to leave the next generation of brain implants to the next generation of blind people.

McDonald is, however, looking forward to whatever vision enhancements the app improvements can achieve for his current implant. He is very optimistic. But even if future app versions provide him with no better vision perception than he has now, he says he's still satisfied. Before the surgery, McDonald had no vision and now he has some vision. And he says, "I've had the privilege to be one of the first blind people to experience artificial vision from an electrical brain implant." McDonald likes new experiences.[27]

Tonight's dinner actually provided another new experience for McDonald. He says he had never eaten a crab cake before, but finds he likes them a lot. This is the only claim made during tonight's dinner conversation of which I am skeptical. As a resident of Maryland—the home of the Chesapeake Bay—it's hard for me to believe anyone hasn't eaten a crab cake.

So far, we've covered neuroprostheses for touch, hearing, and sight. That leaves us with only two senses left—taste and smell. Conceptually, there should be little difference in how these senses potentially could be tackled by neuroprosthetics: identify the nerve pathways that must be targeted between the tongue or nose and the brain, develop artificial sensors that can detect the chemicals that produce taste or smell sensations, design an appropriate brain-machine interface, and electronically hook everything together. Unfortunately, there are some unique obstacles to deal with.

The first problem is that the tongue and the nose share sensory pathways; that's why, when you cannot smell because of a cold, neither can you taste food very well. The tongue has receptors for the five basic flavors—salty, sweet, bitter, sour, and umami. But our sense of smell assists our taste buds and adds complexity to our palates. Recent research has shown the nose can discriminate between one trillion different odors.[28] Were it not for our noses, our tongues would be quite satisfied with cheap wine. So functional prostheses for the tongue and nose would likewise need to have highly sophisticated chemical receptors to convey natural taste and smell experiences. And despite Armiger's earlier admonishment to me that no sense should ever be discounted, the invasive surgical procedures required to regain the senses of taste and smell are quite daunting.[29] For all these reasons, there has been little work on neuroprosthetics for taste and smell, and there isn't likely to be much anytime soon. These types of sensory prostheses will probably have to wait until the technology for touch, hearing, and vision neuroprostheses have been better perfected, and then that technology will be commandeered to deal with taste and smell deficiencies.

But the tongue hasn't been completely left out of the neuroprosthetic picture; some have sought to repurpose it to provide blind people with "sight" by *sensory substitution* through their tongues. This approach exploits the *neuroplasticity*—the ability to reshape neural functions—of the brain. Let's pause here and talk a little about sensory substitution and neuroplasticity.

If you're like me, you may already have had some experience with sensory substitution. When I was a child, I would sometimes play a game with my friends in the summer while lying on the beach. One of us would draw a large letter, or stick figure of some object, on the bare back of another. The objective of the game was for the person feeling the drawing to guess what had been drawn. This task requires the brain to convert a spatial touch sensation into its visual equivalent. With practice, I became quite good at it. Little did I know at the time that "seeing" with your back would become an important area of scientific research.

The pioneer in the scientific field of sensory substitution was Paul Bach-y-Rita, an American neuroscientist who essentially invented the field in 1969 and remained its most prolific researcher until his death in 2006.[30] Bach-y-Rita took the back-writing game to the next level and applied a scientific approach to studying it. He started by modifying an old dental chair. I know, the last time I mentioned modifying a dental chair we ended up with a prisoner execution device. But this time the dental chair modifications were done for a nobler purpose: to allow the blind to see.

Bach-y-Rita installed a grid of 400 small vibrating motors in the seat back of the dental chair. These vibrating motors were similar to the tiny vibrating motors applied to the skin patches on Melissa Loomis's underarm to give her a sense of touch. Bach-y-Rita's system consisted of a video camera that sent its camera images to a computer, which translated the visual image into a vibration pattern. A blind person sitting in the chair would use the video camera to scan an object placed in front of her, and the camera's black-and-white image would be converted into an image "drawn" on her back with vibrations, as though the image were painted on her back using vibrating ink. The amount of vibration would be proportional to the blackness of that part of the image, with white areas completely devoid of vibrations. With training, the blind test subject became skilled at identifying objects written on her back as vibration images. Soon a vest was constructed with vibrators sewn into the back of the vest. Wearing the vest allowed the blind person mobility, such that she didn't need to be confined to the dental chair to experience the vibration images.

But a problem with this approach was the limited resolution of the touch senses of the skin of the back. Bach-y-Rita's studies showed that *two-point discrimination*—the minimum distance the touch stimulators need to be apart for the person to perceive them as two separate touches—is about 50 millimeters (roughly 2 inches) on the back. In contrast, the two-point discrimination of the human tongue is about 1.1 millimeters (less than 1/16th of an inch)—significantly better resolution. So Back-y-Rita transferred his sensory substitution technique from the skin on the back to the surface of the tongue. He also changed the form of stimulation from vibrations to weak electrical shocks.

You may recall, from chapter 5, how Volta had used his tongue to detect weak electrical signals from his voltaic pile and, from chapter 2, how it's possible to feel the electricity coming from a 9-volt battery by bridging the positive and negative terminals of the battery with your tongue. Thus, the tongue is indeed very sensitive when it comes to identifying electrical current. So it made sense to use electricity to stimulate the tongue.

Bach-y-Rita showed the tongue was much superior to the back when it came to forming images in the blind person's mind. He also showed that blind people were much better at identifying the images than sighted people who were merely blindfolded. He further showed by monitoring brain activity during the picture identification testing that, when a blind person was identifying images using her tongue, areas of her visual cortex were activated.

These brain studies suggested the visual cortex is the area of the brain that forms images for the mind to see, and not just for seeing with the eyes. Although the eyes are its preferred information source, the visual cortex can substitute touch sensations to form images when visual inputs are missing due to blindness. The findings underscore Bach-y-Rita's long-held conviction that the brain processes image information the same way, regardless of which organ is sending it information. He had often claimed, "You don't see with your eyes; you see *with your brain*."

These findings were consistent with earlier studies that showed the visual cortex to be active when blind people were reading braille. It seems like the visual cortex is adept at substituting senses when forming images. It can use spatially distributed touch signals when visual signals aren't available.

Bach-y-Rita's tongue vision device has been commercialized and is currently sold as a product called BrainPort.[31] The device consists of a grid of 400 tiny electrodes spread over one side of a square-inch piece of thin plastic. The device is placed, electrode side down, on the surface of the tongue. In addition to spatial information, the device also can deliver the equivalent of a "grayscale" electric image, in that the amount of current fed to each of the 400 electrodes is proportional to the pixel's darkness. The pixel darkness information is derived from a small camera held on the forehead with a headband. Digital images from the camera are processed by a brain-machine interface into a 400-pixel grayscale image of electrical stimulation to the tongue. Blind wearers of the device describe their image experience as that of "pictures being painted with tiny bubbles."[32]

This phenomenon is an example of *neuroplasticity,* the adaptability of the brain and its uncanny ability to use one part of the body's neurosensory system when another isn't available to it. Substituting senses is just one aspect of neuroplasticity. The brain also can swap out brain function areas when one part of the brain is not working due to a stroke or some other type of brain injury. By the utilization of both normal sensory perception and sensory substitution, and leveraging the brain's innate neuroplasticity, it's possible for the brain to achieve some truly amazing things. Perhaps it's time we take a closer look at the electrical life of the brain.

15

INNER SANCTUM

THE BRAIN

All roads lead to Rome.

—MEDIEVAL PROVERB

No matter what route you take—electrical or neurological—all studies of electricity's interactions with the human body ultimately lead to the brain. It is the destination to which I have been both alluding and eluding throughout this book. "Alluding" in that I have often hinted about the brain's central electrical role, and yet "eluding" in the sense that I've intentionally avoided any direct discussion of how the brain works. Until now. With this chapter, we have finally arrived at the topic of electricity and the brain.

Why have I been so reticent to deal with the brain? Despite its direct role in all things electrical in the human body and its overall importance to sustaining life, it would be very difficult to tackle what's going on in the brain on the electrical level without first understanding how its component electrical units—the neurons—actually work. But we have reached a point in this book where we now have a working knowledge of how individual neurons electrically transmit signals and how they are organized within nerve fibers that reach out from the brain, like tentacles, to all corners of the body. With that knowledge, we are ready to tackle the electrical life of the brain itself. But don't get your hopes up, because we really don't know that much. The brain remains largely an electrical black box. We send electrical signals in and we get electrical signals out, but what it all exactly means is open to a lot of interpretation and some intense controversy. Nevertheless, now that we've finally arrived at the brain, we need to start somewhere. I think it best to begin with a topic relatively devoid of controversy: the brain's power consumption.

The brain is very "green." The adult human brain runs continuously, whether awake or sleeping, on only about 12 watts of power. For comparison, a typical desktop computer draws around 175 watts and a laptop somewhere around 60 watts. And the brain's power source is renewable; it's the solar energy stored in food. In short, the human brain is the greenest computer on earth.

Oops! Did I just equate the brain with a computer? I think I did. And in so doing, I've stepped directly into the first of the controversies—because some neuroscientists loathe the model of the brain as a biological computer. They claim it is nothing of the sort. Well, I'm in the thick of it now. I might as well finish what I was saying before giving equal time to my detractors.

The secret to the brain's greenness is its ultrahigh computational efficiency; that is, it can generate a tremendous amount of computational output for the little power it draws. Studies have shown the brain has higher computational power efficiency than electronic computers by orders of magnitude.[1] This has led to efforts to make computers work more like the brain. Not only would mimicry of the brain's circuitry make computers more intrinsically efficient, it would also help with the development of more efficient brain-machine interfaces. As we've already discussed, the nervous system and electronic systems are hardwired differently, hence the need for brain-machine interfaces in neuroprosthetics. But it is also true that the greater the difference between any two platforms, the more power it takes to link them. Since we can't (and wouldn't want to) make the human brain work more like a computer, it seems the only prudent option is to make computers and other electronic devices work more like the brain. If computers' circuitry were to become more brain-like, the power requirements for brain-machine interfaces would go way down, which would translate into smaller batteries and longer times between charging.[2] Great news, by the way, for neuroprosthetics.

But apart from the practical application to neuroprosthetics, attempts to develop electronics that emulate the circuitry of the nervous system in terms of its efficiency may lead to a better understanding of how the nervous system works at its most basic level. In fact, some biologists argue that evolution among higher animals has been largely driven by natural selection for neurological efficiency.[3] So, by emulating the nervous systems of higher organisms in our electronic designs, we might be exploiting design strategies that have already withstood millions of years of natural vetting.

Let's take just one example to illustrate the point, using some parameters we previously discussed in chapters 10 and 11.[4] As we know, the giant squid neuron is 1,000 times larger than a human neuron. The larger size means the squid neuron's cell membrane has more surface area. And because the cell membrane houses the voltage-gated channels, more membrane area roughly translates into more voltage-gated channels for the neuron. We've also learned it's the individual voltage-gated channels, working in coordinated synchrony, that propagate action potentials—the fundamental unit of all nervous system signals. Thus, in electronic design terms, we can consider each neuron to be a distinct *electrical compartment*. Since the squid neuron's electrical compartment contains a multitude of voltage-gated channels, it has a lot of *information capacity*, which is a good thing when transmitting a lot of information. But the metabolic energy consumption of cells increases dramatically with their size; that is, larger cells consume disproportionately larger amounts of energy. So the increased information capacity enjoyed by having larger neurons comes at a price, and the price will be paid with a great deal more metabolic energy.

Given the energy costs associated with increasing the size of neurons, the question simply comes down to this: How much energy can the body afford to allocate to increasing the information capacity of its neurons? It's similar to the internet choice you face for your home. You may want the 300 megabytes per second (Mbps) speed to satisfy all your gaming, computer, and television needs, but your wallet says you must lower your internet speed in order to afford food for your dog. So you decide to get by with just 50 Mbps, and the dog gets to eat.

This is likely the reason why the animals with the most advanced nervous systems—the mammals—have relatively small neurons. Mammalian species with larger brains (e.g., whales and elephants) have achieved their greater brain size by adding more neurons, not by making their neurons larger. There seems to be a trade-off between information capacity and metabolic energy requirements, and the mammals have worked out a balance by standardizing on an optimized neuron size that diverges only slightly among mammalian species. So, when we talk about the best size for an electrical compartment in an electronic circuit design, we might want to heed the lesson from evolution: when it comes to neurons, size matters . . . but bigger is not always better.

This illustration shows that some significant inferences can be made about how the brain works by combining the things we have learned through earlier electrical and neurological research. And all that knowledge was gathered largely through a *reductionist approach*; that is, we try to learn about the larger picture by focusing our research on the smaller components that make it up. But that isn't the only approach to studying electricity and the brain. And the proponents of the alternate approach are most incensed by the brain-as-computer model I put forth just now.

My detractors no doubt would claim there are severe limitations on what we can learn about the brain solely through reductionism. They say true insight into the electrical nature of the brain can be gained only by studying the brain's electrical activity as a whole. These antireductionists contend reductionism ultimately fails to gain true insight, because such an approach cannot see the forest for the trees. They, therefore, advocate for complementary large-scale approaches, saying such approaches are critical to understanding the workings of the brain.

Some neuroscientists are actually hostile toward the long-dominant metaphor of the brain as a computer. These neuroscientists say the metaphor has outlived its usefulness and is now holding us back. Holding us back because the brain-as-computer model ignores what they call *emergent properties*—the properties that emerge as a system functions and that cannot be predicted just from studying its components. They contend the things we most want to know about brain function, such as the mechanism of consciousness and the nature of sleep, are emergent properties, and thus inaccessible to us as long as we keep trying to find an understanding of the brain in terms of corresponding computer components. This group of neuroscientists generally believes insight into the brain will be obtained through studies of behavior, not computers.

This criticism of the brain-as-computer model has been around for a long time. As early as 1951, neuroscientist Karl Lashley decried the use of any machine-based metaphor for the brain. Said Lashley:

> Descartes was impressed by the hydraulic figures in the royal gardens, and developed a hydraulic theory of the action of the brain. We have since had telephone theories, electrical field theories, and now theories based on computing machines. . . . We are more likely to find out how the brain works by studying the brain itself, and the phenomenon of behavior, than by indulging in far-fetched physical analogies.[5]

This is a common sentiment among the modern-day haters of the computer metaphor of the brain. In particular, they believe the heavy focus on studying the brain's interaction with the senses (as we have been guilty of in chapters 13 and 14) ignores the true marvel of the brain: its control of behavior. It is the processing and translation of sensory information into appropriate behaviors that, they believe, is the key to understanding how the brain actually works. Unfortunately, we know little about how the brain controls the body's behaviors and, they argue, we are never going to get there by studying the details of things like eye-to-brain visual circuitry. According to them, we will never be able to figure out why, when the eyes see flames, the nose smells smoke, and the ears hear an alarm, the legs then get the body out of the building as fast as possible. When we understand that, we will understand how the brain actually works.

The problem, however, is that a scientist's brain works best when it's following a metaphor. We have seen the power of metaphorical approaches many times in this book. We learned early on how the metaphor of electricity as a single invisible fluid, as expounded by Franklin, served the scientific community well for many years because electricity actually behaves in many respects like water. Of course, electricity isn't a fluid. We now know electricity is the movement of electrons, and we even know electrons flow in a direction opposite to the direction of Franklin's alleged flow of fluid. But even when we finally did discover the electron in 1897, it really didn't affect our practical use of electricity in any significant way. In fact, we are still using some of the old fluid terminology and analogies from the eighteenth century in modern electronics. And we still pretend the "current" of electrons moves in the direction Franklin first postulated for his fluid—opposite to the reality of the directional flow of electrons. Remarkably, the old fluid metaphor actually works better for us than does the modern knowledge of the electron, because we still don't really understand what an electron actually is. The electron remains a physical entity that challenges the human intellect. But we all have a pretty good idea of what water is all about.

Personally, I don't think Franklin himself really believed electricity was literally a fluid, and Faraday certainly didn't believe it. It's just that the fluid metaphor worked well for them, and they didn't have an alternative metaphor that worked any better to guide them in their research. And that, in my view, is the problem with abandoning the metaphor of the brain as computer. We all know it's not a computer. But just acknowledging that it isn't a computer doesn't give us any better insight into what the brain is. And with regard to those "emergent properties," proponents of the brain-as-computer metaphor would counter that the term is just a catchall expression for brain activities we don't yet understand . . . but soon might, if we stick with the brain-as-computer metaphor and a reductionist approach.[6]

At this point, you as a reader may be saying to yourself, "I was fine with all the neuroscience and the electrical physics, but things now seem to be getting a little too metaphysical for my taste." I hear that. But sometimes metaphysics can make us reflect on the actual meaning of the things we do in science and, in that way, improve it. Recall that tree-falling-in-the-forest question the philosophers posed so many years ago; it helped us to better define the scientific concept of sound. So, too, these questions about the fundamental nature of the brain will help guide future brain research. At least we hope so.

Fortunately, my purpose in this chapter is not to explain how the brain works or to settle the debate about the brain-as-computer versus the brain's emergent properties. The goal here is much more narrow and focused. I will remain agnostic on those broader issues and just try to describe the brain in terms of its electrical activity. We'll look at the electricity the brain sends out and the electricity we send in, and simply regard the brain itself as a black box. I hope, through this admittedly limited approach, we will at least be able to deduce something about how the black box works even if we cannot fully explain it.

First, let's deal with the electrical signals coming out: brain waves.

What are brain waves? What indeed. An interesting question that brings us to our second major controversy. But let me start by telling you what they are not: brain waves are not electromagnetic waves, such as radio waves or microwaves. As a radiation scientist myself, I find the name *brain waves* unfortunate because it causes a lot of confusion. Brain waves are "waves" only in the sense that they are detected in the form of oscillating, wavelike patterns of varying frequencies. The fact that the frequencies of brain waves are measured in hertz units, just like electromagnetic waves, further exacerbates the confusion.

As to what they are, although the answer has evolved greatly over the last century, we aren't quite certain even today. Some neuroscientists say they represent another dimension of nervous system signaling—a dimension yet to be explored—and believe brain waves are the key to understanding how the brain functions. Others say brain waves are just the background noise of the brain performing its normal electrical functions at the level of the neuron: transmitting and receiving signals via action potentials. For now, let's sidestep their alleged function and just define brain waves electrically, as best we can, so that we're all on the same page regarding what they amount to in electrical terms.

Brain waves are best defined operationally; they are the periodic fluctuations in voltage that can be detected using skin electrodes placed on specific areas of the scalp. They are produced by the electrical discharge of different populations of neurons near the surface of the brain, with each localized population firing in synchrony with its own characteristic wavelike pattern of voltage oscillation.

There are brain wave patterns associated with normal brain activity for both awake or asleep states, and aberrant patterns associated with abnormalities such as brain diseases or sleep ailments. Brain waves are, in expert hands, very useful in the diagnosis of some brain disorders, particularly epilepsy and sleep problems. This much everyone agrees on. After that, the going gets rough.

Even before the giant squid axon made the scene, people were studying brain waves. But little progress was made regarding how they are produced or what they might do. We still don't know what they do, if anything, but we now have a fairly detailed knowledge of how they are produced.

We now know, through reductionist neuroscience studies, that individual neurons segregate positive and negative charge across their cell membranes. We also know, largely from Faraday's work, that segregation of charge produces an electric field. So each neuron has around it a tiny electric field. When a neuron produces an action potential, the charge segregation flips its orientation across the membrane and, accordingly, so does its electric field. If neurons in the brain fired off action potentials randomly, all the electric fields would average out, and the brain itself would have no net overall electric field. But that's not how the brain works. Within the brain, local populations of neurons congregated within its various functional nuclei fire off action potentials together, in synchrony. This coordinated action of millions of neurons produces a net electric field large enough that it can be detected outside of the brain. Different regions of the brain increase and decrease their electrical activity because of what a person is doing or thinking, and their local electric fields fluctuate as well.

It is possible to detect these fluctuations in electric fields inside the brain by placing skin electrodes on the scalp and measuring small changes in the surface voltages in that particular area of the scalp. These voltage changes are being driven, in turn, by the local electric field changes in the area of the brain underlying the electrode. The surface electrodes are able to detect only the electric field changes on the outermost surface of the brain—the cortex. The skin electrodes can provide little information about the state of electric fields deep within the brain. It's like the story of the man looking for his car keys on the ground underneath a streetlight because that's the only place where he can see. Measuring scalp voltages with skin electrodes allows us to see the electric fields of the brain directly under the skin electrode, and nothing further. And the signals they detect are very weak, such that any electrical activity in the environment can corrupt the measurements by producing *noise*—electrical artifacts unrelated to the brain activity being measured.[7]

Another problem is the poor cellular resolution of brain waves. It is a truism that the smaller the electrode and the closer it is to the brain tissue, the higher its spatial resolution. For example, if a microscopic electrode is actually embedded within cortical brain tissue, it has the theoretical ability to record a single action potential from an individual neuron (resolution of about 0.2 millimeters). Slightly larger embedded probes can measure the collective activity of a cluster of neighboring neurons (resolution of about 1 millimeter). Somewhat larger probes placed on the surface of the cortex can measure the electrical activity of a small portion of a single fold of the cortex (resolution of about 5 millimeters, or 0.5 centimeters). But each scalp skin electrode, positioned outside of the skull, covers an area of underlying cortex equivalent to the width of about three brain folds (slightly more than 1 inch, or 3 centimeters). This is a large area; it's larger, in fact, than many of the gross anatomical features of the brain.[8] It is hard to imagine that detailed electrical signaling information, such as instructional signals sent from the brain to close the ring finger of the left hand, can be deduced from the crude electrical measurements gleaned from a scalp electrode. This is why brain waves aren't typically used to control neuroprosthetic limbs; the signal resolution is just too poor to drive fine motor skills.[9]

Despite the limitations of a scalp electrode for measuring the brain's electrical activity, the hope is that even this small snapshot of brain activity, when combined with other scalp electrodes blanketing the entire surface of the skull, will provide a pattern of information useful for deducing what the brain is doing at the moment.

In terms of the brain's anatomy, the scalp electrodes primarily are measuring the electrical activity of the brain's cortex—the outermost layer, which has a thickness of just 2 to 4.5 millimeters (about the thickness of a pumpkin seed). The good news is that the cortex of the brain is the place where most of our sensory processing takes place. For example, as we've seen in the last chapter, the visual cortex at the back of the brain processes vision. The bad news is that the cortex is the place where most of our sensory processing takes place. Why is it also bad news? Because it means brain waves are very much affected by what the body is sensing and doing at the time of the measurements. For example, brain waves with eyes open are different than brain waves with eyes closed. Opening one's mouth also affects the brain waves because of the jaw muscle's electrical action. Consequently, it is fair to ask whether brain waves are a window into the state of the brain or the state of the body. This question has persisted since brain waves were first reported in 1929.

The German physician Hans Berger was the first to investigate human brain waves, but his fellow scientists showed little interest in his work.[10] Berger's research occurred at a time when other neuroscientists were focused on a reductionist approach, studying action potentials of individual neurons.

It also didn't help that Berger had a strong interest in paranormal psychology and a belief in telepathy between individuals. He saw brain waves as a potential mechanism for transferring thoughts between individuals over great distances. None of these fringe ideas endeared him to the mainstream scientific community of the day. (And the fact that he was a Nazi sympathizer didn't help his historical image either.) Some people began to disparagingly call brain waves "Berger waves," not to honor Berger but rather to imply that they were of interest only to him.

The road to respectability for brain waves has been a long one, and some would say brain waves still don't get any respect. Their many critics say brain waves have only limited potential to tell us anything significant about how the brain works. They liken brain waves to steam.

Pardon my use of yet another dreaded machine metaphor, but if the brain were a steam locomotive, brain waves would be the released steam. It's very hard to deduce how a steam engine works by measuring fluctuations in its steam output rate.

The proponents of brain waves often claim the very fact that brain waves oscillate suggests they have some kind of functional significance. After all, don't radio waves oscillate as they transmit coded information? Perhaps brain waves are communicating coded information from one area of the brain to another. But others strongly disagree. These scientists note that all systems trying to maintain a *steady state*—a condition of equilibrium—tend to oscillate. It doesn't matter whether it's the population of mice in the Canadian tundra, which oscillates around its optimum sustainable size level, or the oscillations of a hummingbird's wings as it tries to hover in one place, or the oscillating air temperature in your house as the furnace tries to maintain the setting on your thermostat. The occurrence of oscillations is often just an indication that some type of balancing act is going on.

In the case of brain waves, the primary cause for oscillations is likely the need for neurons to maintain a balanced baseline state of excitability. If a neuron's basal activity state is too high, firing action potentials at an accelerated rate, its ability to go higher in response to an increased stimulus is limited. In contrast, if its basal activity is too low, it has limited ability to decrease its signaling when the stimulus diminishes. Thus, by maintaining a position somewhere in the middle on the activity scale, the neuron's ability to quickly move either up or down is maximized.[11] Since whole groups of cortical neurons are fine-tuning their basal activity state in synchrony, we see the phenomenon as oscillating scalp voltages, or what we call "brain waves."

But it is also true that the oscillations of brain waves occur at different frequencies. In fact, brain waves are categorized and named by their frequencies: delta (0.2–3 Hz), theta (3–8 Hz), alpha (8–12 Hz), beta (12–27 Hz), and gamma (27–100 Hz). Some people have tried to define a distinct functional role for each of the frequency categories. They claim to be able to diagnose psychological abnormalities by comparing a person's brain wave activity for specific frequency categories with that of a "normal" individual. Then they use neurofeedback therapies to try to bring the person's "abnormal" brain wave oscillations back within the normal range. Mainstream medicine considers these brain wave correction therapies nonsense, and few health insurance companies pay for them. But there is a kernel of truth in the claim that abnormal brain wave patterns can be indicative of some brain function abnormalities, particularly epilepsy.

Epilepsy is a disease characterized by seizures, which can be likened to an electrical storm in the brain. While a person is having a seizure, their brain waves are highly erratic. But even epileptic patients who aren't in the midst of a seizure tend to have a characteristic brain wave pattern.[12] For that reason, measuring brain waves, using a diagnostic medical procedure called an electroencephalogram (EEG), is useful in the diagnosis and management of epilepsy patients.[13] Electrodes are placed all over the individual's scalp, and brain waves are recorded. It is analogous to an electrocardiogram (EKG), where multiple skin electrodes are placed over the chest of a patient to try to deduce abnormal electrical activity of the beating heart within.[14] The EEGs of epileptics show high-amplitude rhythmic brain waves, produced when large populations of neurons across the brain are firing off action potentials at nearly the same time. Although some degree of brain wave synchrony is normal, excessive synchrony, known as *hypersynchrony*, is a hallmark of epilepsy; thus, EEGs are very useful in diagnosing and monitoring the state of the disease.[15]

The war among neuroscientists about the exact function, or nonfunction, of brain waves continues to this day. In fact, whole books have been written about it.[16] We are not going to be able to resolve this controversy here, so perhaps we should end our discussion of electrical activity coming out of our heads and turn our attention to the electrical activity we put into our heads. And yes, we will need to deal with yet another controversy.

The public knows electroconvulsive therapy (ECT) by the outdated term *shock therapy*, but this term does not mean what most people think. The term *shock*, when used in a medical context, refers to a dangerous condition that occurs when the body suffers from an insufficient flow of blood, typically caused by an extreme drop in blood pressure. Similarly, although the shock trauma unit in a hospital often treats people who have been accidentally electrocuted, the "shock" in the unit's name refers to medical shock, not electrical shock.

Shock therapy started in the 1920s. The strategy of the therapy was to induce a state of controlled medical shock using drugs, often high doses of insulin that would produce medical shock by causing very low blood sugar levels. Physicians thought that putting a person in a prolonged state of shock, which involved their being unconscious for days or weeks, somehow gave the brain a rest from its torments and had beneficial effects for different forms of psychoses. It didn't. So the therapy had largely been abandoned by the time ECT came along.

In contrast, ECT does not produce medical shock; it produces seizures. And the induced seizure (frequently called a *convulsion*) has bona fide beneficial effects for at least one psychiatric disorder: depression. While the origins of medical shock therapy are a little murky, with several physicians claiming to have invented it, the first use of ECT can be pinpointed to April 11, 1938, in Rome. And the events leading to its initial use are quite clear.[17]

Epilepsy, as we have already discussed, is a brain disease that involves periodic seizures. Schizophrenia is a severe, long-term psychological disorder affecting a person's ability to think clearly and behave normally. These two very different disorders are fairly common, but physicians in the 1930s believed it was extremely rare for both diseases to occur in the same individual. Theories abounded as to why this should be the case. Some thought epilepsy and schizophrenia might represent opposite extremes—the pathologically overactive and underactive states—of some unidentified normal brain function. We now know none of this is true. In fact, the two diseases occur together much more frequently than previously believed. Nevertheless, physicians at the time thought, because of the two diseases' alleged mutual exclusivity, that there was some physiological connection between the two. Some physicians theorized the diseases didn't occur together because epilepsy—more specifically, the seizures that accompanied the disease—might actually protect against schizophrenia. That led to the idea that it might be possible to treat schizophrenia by inducing epileptic seizures.

At first, physicians tried inducing seizures in schizophrenic patients with drugs. But drug dosing was hard to manage and patients suffered multiple complications.[18] There also had been some accidental deaths. Because of these issues with drug-induced seizures, there was a need for a better and safer alternative.

When Ugo Cerletti, a psychiatrist practicing in Rome, learned of this, he immediately thought of using electricity to induce seizures. This was because he had been studying epilepsy in the laboratory for some time, using a dog model. Cerletti would artificially induce epilepsy-like seizures in dogs using electricity, and then study the effects on the dogs' brains. Remarkably, his first attempts at inducing epilepsy didn't involve electrifying the brain itself. In fact, Cerletti was trying to avoid electrifying the brain entirely because he thought running electricity through the brain might damage the very tissue he was trying to study. Instead, Cerletti attempted to induce seizures by electrifying the dogs' torsos. He placed one electrode in the dog's mouth and the other in the rectum, which kept the dog's brain out of the direct route of the electrical current.

If these laboratory experiments with electricity in dogs sound a lot like the gruesome ones Edison and Brown performed 50 years earlier, they should, because they are essentially the same. As such, Cerletti rediscovered what his predecessors had learned long ago: the heart is the critical target for electrocution. By running the current from mouth to rectum, it went directly through the dog's heart, thereby killing it.

But then Cerletti learned the local slaughterhouse was stunning pigs with electricity to their heads before dispatching them. As we have learned earlier, electrical stunning is commonly used to this day to humanely induce instantaneous unconsciousness in animals before slaughtering them. Cerletti paid a visit to the slaughterhouse and learned that the pigs, which were being shocked by applying a pair of electrodes to their temples, virtually never died from the stunning itself. Rather, if not immediately slaughtered, the pigs exhibited symptoms indistinguishable from *grand mal epilepsy*, a severe form of epilepsy involving both muscle spasms and prolonged unconsciousness. But they later recovered, suffering no apparent long-term effects. With this insight, Cerletti had a solution to his problem of a high dog mortality rate. He adapted the slaughterhouse's pig electrodes for use on dogs, and applied electrodes to his dogs' temples. Henceforth, he had excellent results in inducing the seizures needed for his epilepsy research without killing any dogs.

Repeating Edison's and Brown's mistaken emphasis on voltage, Cerletti further tried to determine the optimum voltage needed to induce the seizure without killing a dog, completely ignoring the fact that amperage, not voltage, drives lethality. He ran a series of voltage dose response experiments with dogs and determined that, with his particular apparatus and procedure, 120 volts would induce a seizure in a dog within a few seconds, while it took the application of 400 volts for over 60 seconds to kill one. Cerletti judged this to be a large *margin of safety*—the ratio of the dose that will kill to the dose needed to treat—and he was thus able to continue his experiments without losing any more dogs to electrocution.

Consequently, when the clinical need arose for a safe and effective alternative to drugs for the treatment of schizophrenia in humans with seizures, Cerletti had a "shovel-ready" solution: electricity was both highly effective and very safe. In his mind, the technique had already been safety tested in animal trials (i.e., his prior five years of experiments with dogs). And since there was no animal model for schizophrenia, the only logical next step was to try it on patients. And that's exactly what Cerletti did. He modified the dog electrodes for use on humans, and he was ready to treat his first patient.

Cerletti had his opportunity when the Rome police brought a severely schizophrenic man, who had been causing a disturbance at the local railway station, to Cerletti's psychiatric clinic rather than to jail. On April 11, 1938, this man became the first patient ever to be treated with ECT. It took several attempts with escalating voltage to determine his threshold for seizures, but once it was found, he was treated about 10 more times over the next few weeks, after which he made a remarkable recovery. Cerletti started treating more patients, getting similar results.

Soon Cerletti expanded his ECT treatments to include patients who suffered from clinical depression. He found depression was even more responsive to ECT than schizophrenia. This meant ECT had a much larger potential patient population, because depression is about seven times more prevalent than schizophrenia. Just as Franklin's reputation with Leyden jars had once caused a stream of paralyzed people to arrive at his door seeking electrical shock stimulation of their malfunctioning muscles, Cerletti soon found his clinic deluged with depressed and schizophrenic people seeking ECT for their malfunctioning brains, and he treated as many as he could.

The effectiveness of ECT for treating major depressive disorders has held up over the years.[19] Today, physicians worldwide see ECT as a valuable clinical option for patients suffering from drug-treatment-resistant depression.[20] In short, ECT is still used after all these years because it's highly effective, painless, safe, and works quickly—patients typically report some benefit after just the first or second treatment. And it's been shown that ECT combined with antidepressive drugs (often called *antidepressants*) produces results superior to drug treatment alone.[21]

No one is sure exactly how ECT works. The most touted hypothesis regarding its mechanism is the *anticonvulsive theory*. This theory is based on the observation that most patients (nearly 95%) become increasingly resistant to electricity-induced seizures during the course of their therapy, requiring the electrical dose to be steadily increased with each subsequent treatment.[22] The thought is that the mechanism causing seizure resistance is somehow linked to the treatment effect. But, of course, this "hypothesis" just kicks the can down the road because we don't know the mechanism of the acquired seizure resistance.

Another theory is the *enhanced neuroplasticity theory*, which posits that ECT works because it enhances the neuroplastic abilities of the brain. In the last chapter, I described how the brain's innate neuroplasticity allows it to swap out spatial touch signals from the back or the tongue to form images in the mind when vision signals from the eyes are not available. The thought here is that an enhancement in neuroplasticity caused by ECT may enable the brain to construct a work-around solution to the brain changes brought about by depression (e.g., atrophy of dendrites and decreases in the volume of the hippocampus). But the theory remains unproven due to a lack of suitable experimental techniques to measure changes in the brain's neuroplasticity potential.[23]

Measurable biochemical changes in the brain following ECT abound. But it's hard to know which of those changes might be relevant to the treatment effect. However, recent studies in mice are providing some clues. Researchers at the Department of Psychiatry at the Johns Hopkins School of Medicine have studied changes in neurons of the mouse *hippocampus*, a seahorse-shaped brain structure believed to be involved with learning and memory, after giving ECT to mice. The researchers were able to confirm what others had previously observed: ECT stimulates the production of new neurons in the hippocampus.

But the Johns Hopkins group was able to extend that observation further by using genetically engineered mice. They found a protein called Narp was required for those new neurons to produce normal dendrites.[24] Mice that lacked Narp had both fewer dendrites sprouting from their new neuronal cells after ECT and less improvement in their depression levels.[25] "What all of this tells us is that Narp seems to regulate communication with other neurons by forming new synapses, or connections, and this may be the way that, in part, it enacts its antidepressive effects following ECT," says Irving Michael Reti, the senior researcher on the study.[26] What's more, Narp doesn't seem to modify the effectiveness of antidepressant drugs, such as ketamine, suggesting ECT and drugs relieve depression by different mechanisms. So perhaps Narp will eventually lead us to a molecular explanation of ECT. Until then, we lack a robust theory to explain how ECT actually works.

As for the modern patient's experience, ECT as practiced today is very different than it was in Cerletti's clinic. ECT is now done under general anesthesia—the patient is fully asleep—with use of muscle relaxants to minimize the muscle contractions during the induced convulsion. Patients feel no pain during the procedure and are completely unaware that the treatment is happening. And the short-term memory loss that was a side effect of the treatment in the early years has been reduced by the use of brief pulses of electricity rather than continuous current. Another welcome change is that the electrical dose for ECT is now appropriately and universally measured in terms of coulombs (i.e., electrical charge) delivered to the brain rather than by the voltage (i.e., electrical pressure) between the ECT electrodes. Voltage is an electrical parameter relevant only to a specific ECT device and is thus of limited value for comparing electrical doses between patients treated in different clinics.[27]

Despite its successes in alleviating depression, ECT remains one of the most controversial of psychiatric treatments. Why? Many things have shaped the attitude of the public toward ECT, including an inherent fear of electrocution. But the most serious blow to its image occurred in the film *One Flew over the Cuckoo's Nest* (1975), in which ECT is depicted as a means of torture and control of patients.

Another factor that contributes to ECT's negative image is that the exact mechanism of its therapeutic effect remains unknown. Physicians practicing science-based medicine aren't very comfortable with treatments where the mechanism remains mysterious. Nevertheless, the depression patients who have undergone ECT, and their family members, express high levels of satisfaction with ECT and its effectiveness.[28] Slowly but surely, ECT is rehabilitating its image and will likely see increased use in the future, unless a more effective treatment comes along.

And how will we know when a more effective treatment comes along? Can't it be argued the history of electrical therapies suggests that all these various treatments are merely fads geared to address the prevalent disease theory of the day? Aren't they soon supplanted by the next treatment fad based on the latest disease theory? Is the erroneous theory of epilepsy protecting against schizophrenia any better than the erroneous theory of disease being a localized nutritional problem, as many of the early electrotherapy physicians believed? Are modern treatments spawned by misguided theories any better than older treatments inspired by misguided theories? All excellent questions worth considering here.

The main difference between electrotherapy as practiced at the turn of the twentieth century and electricity treatments increasingly being practiced today is that we are now living in the age of the *randomized clinical trial,* a powerful scientific method for assessing the value of medical treatments. The basic design of a randomized clinical trial includes three fundamental principles:

(1) Two or more therapeutic treatments should be simultaneously compared to each other, one of which is typically the standard treatment and another a sham treatment.
(2) The patients must be randomly assigned to the treatment groups.
(3) There must be an objective measure of the therapeutic effect, with a statistical analysis of the possibility of error in the measurements.[29]

Of these three criteria, the randomization of patients is the most important for getting at the truth.

Prior to the birth of the randomized clinical trial, most knowledge about the effectiveness of medical treatments was based on the professional opinions of practitioners and testimonials from patients, each of which is highly unreliable. As we've seen, patient testimonials were the only thing the Pulvermacher belt had going for it. There were no randomized clinical trials to validate the preposterous health claims of the belt's manufacturer. When the Pure Food and Drug Act made the scene in 1906, the pressure was on to substantiate all treatment claims, but it wasn't really clear by what means claims could be scientifically demonstrated. The birth of the randomized clinical trial provided the means.

The modern randomized clinical trial is an outgrowth of earlier epidemiology studies of the distribution of diseases within populations. (The term *epidemiology* literally means the study of epidemics.) In the nineteenth century, the field of epidemiology adopted the actuarial methods used to determine death rates for different types of workers. Later, these same methods were used to identify the likelihood of succumbing to specific diseases and then, conversely, to identify the likely causes of those diseases. Most famously, London physician John Snow used such statistical techniques to show the cholera epidemics that regularly struck London were specifically due to drinking water from wells contaminated with sewage. But it was a long time before these statistical approaches, which were already widely employed for public health protection and prevention purposes during the late nineteenth century, started to be employed in clinical situations to compare the fates of patients who had received different treatments.

It really wasn't until the book *The Design of Experiments* by the acclaimed British statistician Sir Ronald J. Fisher was published in 1935 that medical practitioners became enlightened regarding the power of tightly controlled clinical experimentation to answer their questions about the effectiveness of various treatments for specific diseases. Fisher strongly advocated randomization as a fundamental and powerful statistical technique. Shortly afterward, the randomized clinical trial came into widespread use. The major motivation was to determine the effectiveness of various drugs and newly discovered antibiotics in the treatment of infectious diseases like tuberculosis, a chronic and often lethal bacterial lung ailment that once plagued the world. Since then, randomized clinical trials have been applied to every type of medical treatment, from drugs to surgery.

So, why is a randomized clinical trial so powerful? Because randomization of patients reduces or even eliminates *selection bias*—the tendency to steer patients to certain treatments depending upon the severity of their disease. And it can guard against *observer biases*—the predilections physicians and patients may have toward certain treatments. Randomization, in effect, makes professional opinions and patient testimonials obsolete.[30] The truth is found in the numbers. This is why I can so confidently say that ECT is effective against depression—a string of randomized clinical trials have repeatedly and reproducibly shown that is so. In fact, the body of evidence currently is so strong that further ECT clinical trials for depression with sham-treated controls could be considered unethical, because ECT for depression has been clearly established as among the more effective and safe treatments in medicine.

This is not yet the case with schizophrenia. Ironically, although ECT has been shown to be a benefit for treating depression, its effectiveness for treating schizophrenia—the disease that originally spurred ECT's development—is not nearly as clear. Almost from the start, psychiatrists were suspicious of Cerletti's claims regarding the effectiveness of

ECT for schizophrenia, since few other psychiatric clinics could match his remarkable patient outcomes. It is likely that Cerletti was himself guilty of a little observer bias in favor of the treatment of schizophrenics, since the theory that explains ECT's effectiveness was specific to schizophrenia. In contrast, there was no rationale for how ECT might work for depression.[31] Recent randomized clinical trials have shown that the addition of ECT to standard care for treatment-resistant schizophrenia patients hasn't shown a significant advantage over adding sham ECT. Both ECT and sham ECT produced a measurable benefit of about twofold, suggesting that the improvement is mostly due to a placebo effect and has little to do with actually receiving the electricity. There was also no clear benefit for treating with ECT alone in this group of schizophrenia patients.[32] So modern randomized clinical trials point to a conclusion counter to the original schizophrenia-driven treatment hypothesis; they show that patients with clinical depression gain the most from ECT, while schizophrenics benefit much less, if at all.

Yes, randomized clinical trials are very powerful tools to compare one treatment with another (or with no treatment at all). Armed with the randomized clinical trial, we are now able to consider a variety of electrical brain therapies and objectively assess their effectiveness. Let's now look at another one.

For a long time, some physicians feared that, because ECT caused epileptic-like seizures, it might actually cause epilepsy. Recent studies indicate it does not.[33] Other physicians theorized that it might actually be possible to suppress the brain seizures of epilepsy with electricity, if the electricity was applied to the vagus nerve instead of the head.

The idea of vagus nerve stimulation as a treatment for epilepsy resulted from an accidental finding in a 1952 study of brain waves in cats.[34] Researchers discovered that electrically stimulating the vagus nerve of a cat desynchronized its brain waves. Since, as I mentioned earlier, hypersynchronization of brain waves is the hallmark of epilepsy, it was logical to conclude that stimulation of the vagus nerve of epileptics might lessen their seizures by reducing or eliminating their hypersynchronized brain waves. The treatment was first introduced in 1988, and its safety and effectiveness was confirmed in randomized clinical trials during the 1990s. It has seen steadily increased use since then.

The vagus nerve stimulating device works similarly to the cochlear implants and the heart pacemaker implants we discussed in the last chapter, and it shares a good deal of their technology. Electrodes are surgically attached to the vagus nerve on the left side of the neck, at about the level of the Adam's apple. These electrodes are connected by wires to a subcutaneous controller that sits just under the collarbone, similar to the location of a heart pacemaker. The device sends out an intermittent electrical stimulus to the vagus nerve. There is some discomfort to the patient during the stimulation, so its duration and frequency is kept to the minimum level needed to suppress seizures. Typically, stimulation once every five minutes is enough. As with ECT for depression, the mechanism of vagus nerve stimulation in relieving seizures remains unknown, but its effectiveness in controlling seizures is well established.[35]

Using the vagus nerve as a portal to control electrical activity of the brain, as done to treat epilepsy, is a relatively noninvasive procedure compared to actually putting electrodes inside the brain. But for some brain diseases, such as Parkinson's, putting electrodes deep inside the brain is essential to the treatment. Let's now look at how electricity is used to treat Parkinson's disease.

Parkinson's disease (PD) is a neurodegenerative illness that primarily affects the neurons that produce the neurotransmitter dopamine. Such neurons are clustered within the *substantia nigra*—Latin for "black substance" and so called because this brain structure has a remarkably dark appearance when examined at autopsy. The dark color is caused by melanin, the same pigment that darkens human skin. The melanin in skin helps to protect against damaging ultraviolet light from the sun. Melanin in the substantia nigra, however, has another function. It is part of the biochemical pathway that makes the neurotransmitter dopamine.

Neurotransmitters chemically stimulate neurons into firing off action potentials. But for a neuron to be stimulated, it must have a receptor that can bind to that specific type of neurotransmitter. It's similar to a lock-and-key situation, where the receptor is the lock and the neurotransmitter is the key. As with locks and keys, when they don't match each other, they don't work. If the neuron doesn't have a dopamine receptor, it is unaffected by dopamine. But the neurons that do have dopamine receptors are dependent upon the dopamine to function normally. In Parkinson's disease, the neurons producing the dopamine are slowly dying off for some reason, so the neurons in need of dopamine aren't getting any. These dopamine-deprived neurons are largely involved in control of muscle movements, so the dopamine shortage causes the muscles, particularly those of the limbs, to move erratically such that the patient suffers from tremors.

You may remember that we briefly touched on the function of neurotransmitters when we discussed the possible biochemical mechanism of electrotaxis (the phenomenon of fish swimming toward a positive electrode) in chapter 8. In fact, we specifically discussed the recent evidence that electrotaxis might somehow be mediated through the same dopamine pathway that's involved in Parkinson's disease: the D2-type dopamine receptor pathway. That's because chemical inhibitors of the D2-type dopamine receptor both induce Parkinson's disease-like symptoms in mice and abolish electrotaxis in fish.

Parkinson's disease is commonly treated with levodopa (L-Dopa), a synthetic chemical that enhances natural dopamine production. It can be taken orally and it satisfies the dependency of neurons that have dopamine receptors. It is effective in temporarily suppressing the tremors associated with Parkinson's disease. But L-Dopa has a host of side effects, including significant psychiatric problems, and patients can become resistant to the drug. Clearly, better treatments are needed.

Although the use of deep brain stimulation (DBS) to treat Parkinson's disease began in the early 1990s, its history can be traced to the early 1930s when neuroscientists Wilder Penfield and Edwin Boldrey used electrical stimulation to identify the different functional areas of the brain's cerebral cortex. Their approach, called *cortical stimulation mapping*, was to stimulate the various locations of the surface of the cerebral cortex with electricity and see what bodily responses it would elicit. In that way, they could map different bodily functions to specific points on the cerebral cortex.[36] They did this for both the sensory and motor functions of the nervous system. Working methodically over the whole brain surface, they were able not only to map the locations on the cortex that were associated with every body part but also to measure the total cortical area devoted to each body part. They found that, as far as the brain was concerned, all body parts did not warrant the same amount of its attention. As might be expected, for sensory functions, the eyes, nose, mouth, and hands consumed a lot of cortical tissue, while the rest of the body parts used relatively little.

Some have sought to dramatically visualize the vastly disproportionate attention the brain's cortex devotes to the sensory needs of different body parts by depicting, in three dimensions, how the male human body would look if the sizes of its body parts were proportional to the brain area dedicated to their service. Depicting the male body in this way results in a distorted little humanoid figurine affectionately known as the *homunculus*.[37] He is the antithesis of the *Vitruvian Man*—Leonardo da Vinci's sketch depicting the exact proportions of the perfect male body. In contrast to the *Vitruvian Man*, the homunculus is grossly distorted, looking like a troll with an enormous head and huge hands.

Extending Penfield and Boldrey's approach of mapping brain function using electrical stimulation from the cerebral cortex to other areas of

FIG. 15.1. The sensory homunculus. The disproportionate demands of different body parts for sensory brain function and the magnitude of that disproportion is depicted in the form of a three-dimensional figurine of the male human body, where the size of each body part reflects its relative brain tissue needs. The result is a misshapen little humanoid, often humorously referred to as the homunculus. Historically, *homunculus* was the term used for the minuscule deformed being that allegedly resided in a man's sperm. This nonsensical sixteenth-century attempt to explain how human sexual reproduction works first appeared in the writings of the physician Paracelsus (1493–1541), but the bizarre concept has captured the imaginations of people for centuries, so the homunculus is mentioned in a wide variety of historical medical writings and literary works in different contexts. Nevertheless, modern scientific references to the homunculus typically relate to the sensory brain function figurine pictured here. (Mpj29/WikiMedia Commons)

the brain wasn't feasible in the 1930s. As I've mentioned, the cerebral cortex represents just the outer surface of the brain, so the electrical stimulation approach was relatively noninvasive. But electrically mapping the function of inner brain tissue would require sticking electrodes deep inside the brain, and that would injure the overlying brain tissues.

However, in the 1940s, neurosurgeons developed stereotactic devices that allowed them access to the inner sanctum of the brain with electricity without causing significant brain damage.

Stereotactic neurosurgery is a minimally invasive type of brain surgery that uses a three-dimensional coordinate system to locate and gain access to small anatomical structures deep within the brain. It is often used to perform delicate surgical procedures, such as the biopsy of a small lesion or the precise injection of a drug. Damage to overlying brain tissues is minimal because of the thinness of the surgical probes and because it's often possible to select a route to the innermost structures that avoids penetration through the most sensitive overlying tissues.

The development of stereotactic devices allowed the electrical stimulation technique for mapping brain function to move beyond the brain's surface to deep inside. In this way, neuroscientists were able to electrically map the functions of internal brain structures in the same way that previously had been done for the outer cerebral cortex. For example, the role of the substantia nigra in controlling muscle movement was identified, and the center of the brain's reward and pleasure systems was localized to the *nucleus accumbens,* a small brain structure nearby.

The electrical stimulation of internal brain structures for treating Parkinson's disease was actually an outgrowth of *ablation surgery*. Ablation is the removal or destruction of tissue. Often it is used to destroy tumors, but it can also be used to destroy dysfunctional normal tissue when necessary. Ablation surgery was often used to treat Parkinson's disease. It was found that a small structure called the *globus pallidus* had direct neuronal outlets to the substantia nigra, and that specific destruction of one side of the globus pallidus relieved the tremors of Parkinson's disease on the opposite side of the body. But it couldn't be used to relieve tremors on both sides of the body, because ablating both sides of the globus pallidus also resulted in impairment of both cognition and speech. And, of course, the most serious disadvantage of ablation surgery is its permanency. It cannot be undone.

At first, electrical stimulation was used during brain ablation surgery just to confirm that the tissue targeted for ablation was the correct tissue. That is, electricity was applied to the brain tissue to be removed in order to briefly stimulate its function and thus confirm that the surgical target was actually the one controlling the desired functional target. But often, when doing ablation surgery for Parkinson's disease, the prior electrical stimulation itself momentarily stopped the patient's tremors.[38] This ultimately led to the idea that the electricity alone should be the treatment, and thus DBS for Parkinson's disease was born.[39] Whereas it had always been just an educated guess as to how much tissue to ablate, electrical stimulation could be easily adjusted and optimized. And electrical stimulation had the advantage over ablation in that the amount of electricity (i.e., the dose) could be modified at will. Most importantly, electrical stimulation was not permanent; the electricity could be shut off and the electrodes could be surgically removed from the brain.

From a patient's perspective, DBS involves the implantation of a pair of electrodes deep inside the brain using stereotactic surgical techniques. But DBS also involves implantation of a subcutaneous neurostimulator device similar to other implanted electrical therapy controller devices. The surgeon implants a neurostimulator in the chest and then runs a pair of subcutaneous wires up the neck and under the scalp. The wires are attached to two electrodes that penetrate the skull and terminate within the brain structure to be electrically stimulated. An external wireless programmer is used to adjust the settings of the internal neurostimulator.

What is the mechanism of the therapeutic effect of DBS in Parkinson's disease? Once again, no one knows. It is another example of an electrical treatment accidentally discovered, and empirically perfected, without a solid understanding of how it works. But it does work. Randomized clinical trials have proved it so.[40]

DBS stimulation has been applied to many other brain disorders and, again, there have been controversies. In particular, attempts to use DBS on the *nucleus acumens*—the pleasure center—in order to relieve anxiety disorders, such as obsessive-compulsive disorder, have resulted in people overstimulating themselves to get opioid-like highs or to generate erotic feelings.[41] Some have even become physically "addicted" to the electrical pleasures of DBS. The electrical tingling sensation produced by the Pulvermacher belt might have been pleasurable, but it wasn't addictive. At least DBS is not something you can do at home. A physician needs to be involved, and physicians are aware of the dangers of too much happiness. People need some basal level of anxiety, since the desire to alleviate anxiety drives self-preservation behaviors. People who are too happy don't worry about even the life-threatening things they should worry about, so they tend not to live very long.

I am able to cover here only these few examples of brain electrical therapies—ECT for depression, vagal nerve stimulation for epilepsy, and DBS for Parkinson's disease. I chose to feature these particular electrical brain therapies because

(1) The diseases are relatively common.
(2) The electrical therapies used to treat them have been around for a relatively long time and are currently in widespread clinical use.
(3) Their effectiveness has been established by randomized clinical trials.
(4) Each therapy demonstrates a different means to electrify the brain. Electricity is administered either through the skull (ECT), via a peripheral nerve (vagus nerve), or with an electrode inserted deep into the brain (DBS)—three very different and increasingly invasive approaches.

But I don't want to leave you with the impression that these are the only brain diseases that are amenable to treatment with electricity. DBS, in particular, is being investigated for diseases such as dystonia (a diseased characterized by uncontrollable twisting muscles), Tourette's syndrome (a disease of uncontrollable tics), and even pain relief, including the phantom limb pain that amputees suffer. But most of these treatments are still in the research phase and have not yet entered routine clinical practice.[42]

Although the three electrical therapies discussed here differ in many ways, they all have something in common: they are fairly simplistic in terms of the number of electrodes they use. ECT, vagal nerve stimulation, and DBS all use just a few electrodes placed at specific locations. Compared to neuroprosthetics, they seem crude. As we've seen, the brain stem implant used to treat deafness has 20 electrodes, and the visual cortex implants used to treat blindness currently have up to 60. It doesn't seem right that the number of electrodes used for brain therapies should be so scant, particularly when a cochlear implant in one ear can have as many as 120 electrodes. The brain is a large and complex organ responsible for many complex bodily functions; it does a lot more than an ear. If we want to get serious about treating brain-based diseases with electricity, we are going to need to step up our game. We're going to need to significantly increase the number of electrodes to have any hope of meaningfully impacting the plethora of brain disorders out there.

Fortunately, we already have a person on a quest to supersize the number of electrodes that can be placed in the brain. And this guy has an impressive record of accomplishment in scaling up technology. In fact, many consider him a hero in the area of futuristic technology. So there is good reason to be optimistic that he will get the job done. Unfortunately, this hero also has a knack for attracting controversy and stoking conflict. But, as we've seen, when it comes to the brain, there is always going to be controversy. And as for conflict, the best stories are always driven by conflict and involve a hero on a quest. So I'd say we have here all the makings of a great story. Let's end this book on electricity and the human body with that story.

16

FUTURE SHOCK

ARTIFICIAL INTELLIGENCE

It is difficult to make predictions, especially about the future.

—DANISH PROVERB[1]

It's July 16, 2019. Elon Musk is hosting a major event, presented live at the Morrison Planetarium at the California Academy of Sciences and watched worldwide over the internet. It is a galactic venue and a very big day. Musk is publicly announcing the launch of his newest company: Neuralink.[2] Maybe you've heard of it. Even if you haven't, you've probably heard of Elon Musk. Musk is the celebrity engineer, entrepreneur, futurist, and founder of SpaceX, the reusable rocket company, as well as Tesla, the battery-powered automobile company.[3] You remember Tesla and its Model S car. We spoke of the Model S earlier when we discussed the evolution of the battery from Volta's original pile to the modern 18650 lithium version. Each Model S automobile is powered by 7,104 of these lithium batteries.

Musk is a provocative person. He seems to attract controversy because of his eccentric business behavior and unusual lifestyle, but I don't care anything about that. One thing, however, is certain: no one ever has accused him of being stupid. So when Musk talks technology, I listen, because he knows of what he speaks. And he doesn't waste his time talking about unimportant things. Today, he's revealing Neuralink's next-generation brain-machine interface (BMI)—something that Musk predicts will transform civilization. With our civilization on the line, how can I not help but give him my full attention?

Musk proclaims, "It is important for us to address brain-related diseases. . . . Whether it's an accident, congenital or brain-related disorder, or a spinal disorder, . . . we can solve that with a chip. And this is something that most people don't understand yet." I know I don't; tell me more. But when he goes on, I'm disappointed. "All this will occur . . ."—he stops midsentence, as though contemplating, and then continues, "Actually, I think, quite slowly. I want to emphasize that. It's not going to be all of a sudden Neuralink will have this neural lace and start taking over people's brains. It will take a long time." Then he adds with a chuckle, "And you'll see it coming." Not taking over my brain anytime soon? Well, that's a relief. But what is this "neural lace" thing anyway?

Although I'm disappointed that civilization's transformation will be very slow, I am happy that I'll be able to keep control of my own brain for a bit longer. And I'm relieved that, regardless of when in the foreseeable future you happen to be reading this, what I am about to tell you now isn't likely to be too far out of date.

In addition to treating brain diseases, another of Musk's goals for Neuralink is to "link" the human brain with computers. (Suddenly, I realized the significance of the company name.) In its spare time, Neuralink will work on this secondary goal. But Musk warns that connecting Neuralink's brain-implanted BMI with computers also will enable the linking of human consciousness with artificial intelligence, and that would be an existential threat to humanity. So we must always be vigilant against the possibility of artificial intelligence taking control of our minds and bodies. It would amount to a computer entering our minds and enslaving our bodies in order to have us do its bidding. And we wouldn't be able to outwit the computer because artificial intelligence would, of course, be smarter than human intelligence.

The existential threat to humankind posed by artificial intelligence is something Musk has been preaching about for a long time, so his mentioning it doesn't surprise me. What does surprise me, however, is his enthusiasm that Neuralink's BMI might enable artificial intelligence to do the very thing he has been so worried about. Isn't that playing with fire? Perhaps in anticipation of the question, he preemptively states that we must "get ahead of" the artificial intelligence threat by "achieving a sort of symbiosis with artificial intelligence," and Neuralink's technology somehow provides the means to do that. "Symbiosis"? Does that mean artificial intelligence would become something like a benevolent parasite? Musk is losing me.

I'm the first to admit I don't have a mind like Musk's, so right about now my head is spinning. I'm trying to absorb neural lace, curing brain diseases with a chip, artificial intelligence, and having my brain taken over by a computer in a symbiotic way. And though I'm amazed at all the great things Neuralink is going to do for humanity with regard to brain diseases, I can't help becoming just a bit skeptical. I'm beginning to feel like I'm listening to a sales pitch for a Pulvermacher belt.

Let's pause here and try to unpack everything in Musk's big bag of promises and prophecies. By teasing apart his claims about Neuralink into smaller pieces—taking a reductionist approach, so to speak—perhaps we can begin to comprehend the larger picture Musk wants us to see. Maybe we'll even start to understand the prospects and problems of BMI technology as he does.

First of all, what's all this about a "neural lace"? I'm somewhat familiar with BMIs, but I haven't come across this term before. After a little online searching, I find the reason I've never heard of it. I've been too much focused on the science (silly me). "Neural lace" is a term from science fiction. Since childhood, Musk has been a science fiction enthusiast, and he got the term from the books of Ian Banks, his favorite science fiction author.[4] Banks describes a neural lace as a BMI that is implanted into the brain of a young person. As the person grows, so does the BMI, spreading and penetrating into and around his brain, providing a means for outside artificial intelligence to control his mind and body. Sounds wonderful, doesn't it? At this point, I'm beginning to wonder: If this company is all about doing good, why use terms that refer to the more nefarious side of the technology?

And what about "curing brain diseases with a chip"? What does that actually mean? It turns out that Neuralink's BMI is what could best be described as a high-density brain-implantable electrode array, with all the electrodes anchored to computer chips. We've already seen an implantable electrode array when we talked about restoring tactile senses to amputees. The Utah slanted electrode array was applied to the ulnar and median nerves of the arm to give amputees with prosthetic arms a sense of touch. You'll recall that it consists of 100 microelectrodes in a square 10 × 10 array. But beyond just nerves, Utah arrays (either the slanted or level versions) also can be applied to the brain's cortex. In fact, the Utah array is currently the state of the art for implanted cortical brain BMIs. It is the technology that Neuralink's BMI seeks to replace or, rather, supersize.

Neuralink hopes to outdo the Utah array by 100-fold, providing 10,000 electrodes, rather than the Utah array's mere 100. Neuralink's electrodes, which the company manufactures in the form of tiny "threads," each have a diameter of 24 microns—about one-tenth the diameter of a human hair—and are barely visible to the eye. Thus, these threads have dimensions comparable to the individual axons from which they will receive, and to which they will deliver, electrical signals. As I mentioned earlier, implanted electrodes this small would have theoretical resolutions so high they could potentially interact with a single neuron. The threads are designed to be "stitched" into the brain's cortical tissue using a physician-guided surgical robot that looks like a sewing machine from hell. Neuralink apparently has all the required technology in place. To prove it, Musk shows us a short movie (with a Hollywood sound track, of course) that depicts a lot of young engineering types hard at work. But the film has no narration, so it's hard to tell from the visuals who's doing what and why.

Regarding the "artificial intelligence" that Musk thinks will present an existential threat to humanity, he isn't talking about ordinary artificial intelligence (AI)—the kind that's programmed to perform a single task, like read the handwritten dollar amount on the checks you deposited in automatic teller machines, or play against you in a game of chess, or drive your car to work for you. Although impressive, that type of AI is very limited in scope. The AI that beats you in chess is not capable of driving your car. And the AI driving your car can't even play chess, let alone take over your brain. No, the AI Musk frets about is more properly called *artificial general intelligence* (AGI), and it doesn't exist yet. A computer with AGI would understand the world as well, or better, than any human. And it would have a superhuman capacity to learn a vast array of tasks, well beyond playing chess and driving cars.

If you've ever seen the classic science fiction movie *2001: A Space Odyssey* (1968), you'll likely recall the computer affectionately called "Hal" by the crew of the spacecraft. Hal seemed to be a nice "guy"—he had a soothing male voice—and was always responsive to the crew's wishes, helping out with spaceship chores and such. But Hal's AGI programming not only made him smart; it also had some kind of programming glitch that gave Hal a very humanlike paranoid personality disorder. Hal soon began to see humans as his own existential threat. I don't want to spoil things for you if you have yet to see the film, but let's just say it didn't turn out so well for the human part of the crew. Hal's AGI is the nightmare that Musk worries about when he speaks of the danger humans face from artificial intelligence. He doesn't mean that his AI-equipped Tesla cars are going to stage a mutiny against humanity. (But wouldn't that make a great sci-fi flick?)

Filming in the 1960s, Stanley Kubrick, the producer of the movie, envisioned that computers programmed with AGI would have arrived by the year 2001. They didn't. It's now the 2020s, and they're still not here. Some AI experts think their arrival by 3001 would be wildly optimistic, and others say AGI is simply impossible. I don't know which experts are right. Regardless, I'm more worried about being hit by an AI-driven car with a programming bug than by the possibility of AGI taking over my brain. But maybe that's because I don't have much to lose. In any event, let's forget about AGI for now and focus a bit more on run-of-the-mill AI, because there is something about ordinary AI that's pertinent to our story.

Ordinary AI has very much arrived. Although AI for driving your car still may not be ready for prime time, if you want a computer program to search the internet for pictures of yourself (or someone else), AI is up to the task. And the way much of AI works is relevant to that controversial brain-as-computer metaphor I spoke of in the last chapter.

The goal of most AI geeks is to produce a program that emulates the brain's profound ability for insightful learning, an endeavor known as *machine learning* programming. A subfield of machine learning programming, called *deep learning,* uses algorithms inspired by the structure and function of the human brain. Such computer algorithms are referred to as *artificial neural networks.*

The objective of an artificial neural network is to enable a computer to learn from its own experience, similar to the way humans do. Artificial neural network programs are designed to mimic the way that neuroscientists believe the human brain teaches itself. Brain learning is thought to start at the level of the individual neuron. A neuron receives its electrical inputs from numerous input sites on its many dendrites, the branch-like projections that bristle from its cell body. (A single neuron can have as many as 100,000 input sites spread over its 10,000 dendritic branches.) The neuron merges all the incoming electrical stimuli from all its inputs and, if the combination reaches a certain threshold, the neuron fires off an action potential (spike) that heads to the next downstream neuron via its axon.

But not all of the neuron's input connections are of the same "strength." Some are much stronger than others. The stronger connections contribute more electrical signal than the weaker ones and thus have a bigger influence on the neuron's total membrane voltage. Most neuroscientists believe that human learning relies on the brain's ability to modify these connections—strengthening some and weakening others—thereby giving positive feedback to the connections resulting in the desired outcome and negative feedback to those associated with an undesired outcome. This is the human brain learning strategy that programmers of artificial neural networks attempt to imitate in their algorithmic designs.

An artificial neural network, at its most elementary level, does mathematically what neurons do electrically. The basic computation is to add together weighted inputs and then to output either a 0 (no) or 1 (yes), depending upon whether the sum of the inputs is less than or greater than, respectively, a certain preprogrammed threshold value. The important ingredient of an artificial neural network is a feedback loop based on an objective outcome: producing the correct answer. The initial weightings among the connections are often assigned somewhat randomly by the programmer. But as the program runs for longer times, it begins to recognize the input connections most often associated with correct answers, and it increases the weight of those input connections in doing its mathematical calculations. At the same time, connections rarely associated with correct answers have their weights decreased. It's as though the program is teasing out the relevant signals from the background noise. With each iteration of this process, the weightings move closer to an optimal combination of input weights; that is, a specific combination of weights that gives the highest likelihood of producing the correct answer. Consequently, the percentage of correct answers keeps getting higher the longer the program keeps at the task. As they say, "Practice makes perfect." In effect, the computer program is teaching itself to better do the job being asked of it. In other words, *it is learning*.

I've just described the workings of an artificial neural network in terms of its most basic unit, a unit first proposed by psychologist Frank Rosenblatt in the 1950s and dubbed the *perceptron*. A perceptron is the computer program's equivalent of a neuron. But a complete artificial neural network has many of these perceptrons organized in layers, similar to the way neurons are layered in the brain. The layering greatly enhances the power of the network to arrive at the correct answers. And it even allows the network to adjust its own thresholds for decisions, in a process very similar to what we believe the human brain is doing when it learns a new task.[5] (It's because the best artificial neural networks are often many layers deep that the programming field is called *deep* learning.)

You know that old adage "Give a man a fish and he'll eat for a day, but teach a man to fish and he'll eat forever." It's the same way with computers programmed with artificial neural networks; you must train them to learn on their own. The typical way to train a neural network program is to take a very large data set and divide it in half to produce two smaller data sets—a training set and a test set. You first provide the computer with the training set with which to learn. Later, you use the test set to measure how well it's learned.

For example, if the data set is a collection of animal pictures, give the program half the data set and tell it to identify the pictures that show a cat. All the program needs to do is consider each picture, one at a time, and answer the simple yes/no question: Is this a cat image or not? The computer assesses the various image parameters of the picture—the presence or absence of a pair of triangles (ears), multiple horizontal fibers (whiskers), etc.—and makes its decision. After each answer, you immediately tell the computer whether it is right or wrong (positive or negative feedback), and it tries again on the next picture. When the computer is finished with all the pictures in the training set, you give it the test set, ask it to identify the cat pictures without telling it the answers, and then grade its performance in successfully picking out cats once it's finished the job. If the computer has learned its lesson well, it should be able to identify correctly the vast majority of the pictures of cats in the test set.

When first developed in the 1960s, artificial neural networks didn't perform very well; that is, they got lousy grades on their test sets. As a result, people lost faith in artificial neural networks as a useful approach for achieving AI, and artificial neural networks were largely abandoned for several decades. But back in those days, it was hard to find large data sets for training and testing. The problem basically turned out to be that the original artificial neural network programs were unable to learn very well because they were being given skimpy data sets to learn from. But the advent of the internet made it possible to obtain massive data sets. (If you doubt that, search Google for images of animals. I did and found nearly 10 billion animal pictures, 3 billion of which were cats.)

With the arrival of "big data," artificial neural networks have redeemed themselves, reemerging as one of the most utilized and promising approaches for achieving humanlike learning through AI. They are now widely used to do many practical tasks. The next time an automatic teller machine (ATM) correctly reads the handwritten dollar amounts scribbled on the personal check you are depositing, remember the program taught itself to do that. And it did so using an artificial neural network—software that imitates the architecture of your brain.

The point I'm trying to make with all this AI talk is that it's not just that emulation of the brain's electrical compartments and architecture can produce better computer hardware that runs more efficiently, as we discussed in the last chapter. It's also that software emulations of the way the brain's neurons interact with each other result in computers being able to learn from their own experiences. And they do so in ways that look very much like what we humans call intelligence, artificial though it may be. So designing both hardware and software that mimics the way the brain works produces big payoffs in terms of both energy efficiency and computational capacity. It enables the computer to do many more things on much less energy. Apparently, the human brain knows what it's doing.

Does the fact that artificial neural networks perform as intended mean that the brain really learns the way we surmise? In other words, is it fair to say that our current understanding of how human learning works must be correct, simply because software written based on that understanding allows computers to learn for themselves? Not necessarily. Perhaps it is just fortuitous programming that happens to give us correct answers by a means completely different from how our brains would do the job. But I will say this: the fact that artificial neural networks learn as intended does suggest it is unlikely that our ideas about how the brain learns are very far from the truth.

But I've digressed. I really want to know more about how Neuralink hopes to deliver in the near term with regard to those brain diseases. Let's now get back to the launch event and find out.

Our celebrity futurist has now finished his spiel. He surrenders the stage to Max Hodak, the president of Neuralink. Hodak seems more grounded in this century, and appears to be well aware that there are skeptics in his audience. In fact, he confides to us straightaway that when Musk first approached him, he too was doubtful that available technology was ready to do such futuristic things in the human brain. But, Hodak sheepishly explains, "Elon has this ability to pierce through imagined constraints and show you a lot more is possible than you really think. . . . You'd better be very careful telling him something is not possible, unless it's limited by a law of physics, otherwise you're going to look really stupid." Not wanting to look stupid, I perk up to attention. Then Hodak, having completed his public pandering to the boss, grasps the remote control slide changer and gets on with his technical presentation.

Hodak appears more interested in convincing us than wowing us, and he's using terms that I can recognize—the terminology of science rather than science fiction. He briefly reviews the current landscape of BMI technology, and I'm happy to see he mentions nothing we haven't already covered in this book. So he and I are definitely on the same page. Hodak makes the case for the need for BMIs with higher electrode capacities, which is the same case I made to you in chapter 14, so I'm still following. And then he starts to tell us, in detail, the things Neuralink already has achieved toward realizing the goal of making a BMI with 10,000 brain-implanted read/write electrodes, and I am all ears.

Hodak explains that Neuralink's technology revolves around a computer chip dubbed the N1 (presumably for Neuralink model 1). Each chip serves as an anchor for 3,070 electrodes distributed across 96 threads (32 electrodes are entwined within a single thread).[6] And everything is extremely small. The intent is to implant several of these chips into a single individual's brain.

Hodak goes on to explain exactly how these chips will be placed in the brain. He claims the medical procedure will be no more bothersome to the patients than standard LASIK eye surgery, a fairly common outpatient procedure for correcting vision. An area of the patient's head will be shaved to expose the scalp. A small cookie cutter–like device will punch out a little round hole in the scalp, similar in size to the punched holes in loose-leaf binder paper, thus exposing the skull. Then a similar-diameter hole will be bored through the skull. This will expose the underlying *meninges*, the membranes that cover the brain's surface, and a similar hole will be made there to expose the cortex of the brain.

With a circular patch of brain surface now observable through the hole, a neurosurgeon will insert the threads into this visible area of cortex tissue, with the assistance of a dexterous robotic "sewing machine" and high-powered magnification to guide him. With the help of the robot, it's possible for the surgeon to insert up to six threads per minute into the cortical tissue (i.e., 192 electrodes per minute). If you think putting standard sewing thread through the eye of a needle is a talent, this threading job is comparable to sticking a nearly invisible thread through the pupil of Abraham Lincoln's eye on a penny. It's the epitome of precision threading.

The other ends of the threads are attached to an array chip. Miniature custom electronics are used to stream electrical data through all the electrodes simultaneously. Custom spike detection software processes the electrical output data to measure spikes generated by individual neurons neighboring the implanted electrodes.

One nifty feature is that the chip and its associated electronics are hermetically sealed within a little "button" that serves the dual purpose of protecting the electronics inside and acting as a plug to seal up the hole in the skull. A wire is run from the button, under the scalp, to a subcutaneous controller and battery pack that sits behind the ear. The controller transmits its data via Bluetooth to a cell phone or other electronic device outside the body, so the patient wears no external wires or hardware. All the components are implanted within the body. The controller behind the ear can service multiple buttons. Neuralink's initial objective is to put four buttons on one side of the brain—three in motor cortex areas and one in a sensory cortex area.

As a proof of concept, Neuralink has already implanted its threads into the brain of a rat and then recorded spike activity as it ran around in its cage. (Musk also made an oblique allusion to an ongoing experiment with a monkey that he said was yielding some promising results.) Company scientists hope to be able to statistically analyze the spike patterns from these experiments and others, and deduce some important stuff about how the brain processes data. As Hodak puts it, "Everything that we care about is found in the statistics of spikes." Who could argue with that?

After Hodak, a string of Neuralink scientists, each an expert in one part of the required technology, fills us in on the details of the grand plan that Musk and Hodak have just sketched for us. I take it all in. And then, as abruptly as it started, the show comes to a close. Musk ends the launch event by asking for people in the audience with expertise in any

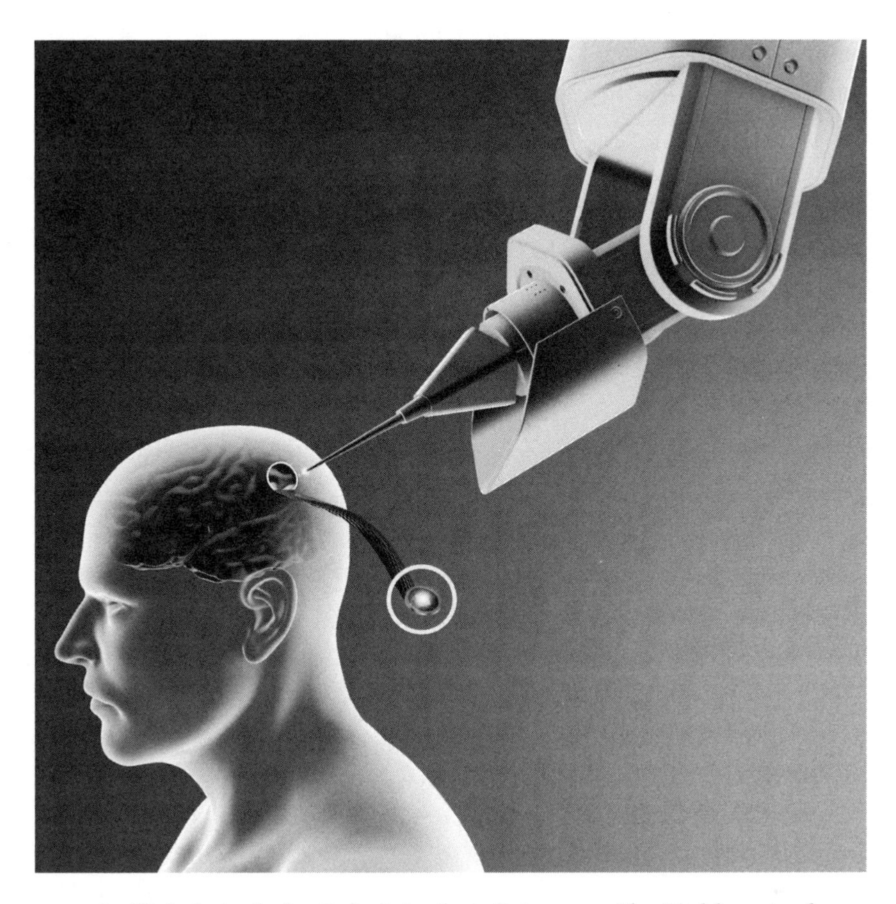

FIG. 16.1. High-electrode-density brain implants. Entrepreneur Elon Musk hopes to advance the state of the art for electrical brain implants by greatly increasing their electrode density. His company, Neuralink, has developed a giant sewing machine that a neurosurgeon can use to stitch electrical wires ("threads") into the surface (cortex) of the brain via a small opening drilled through the skull. Once the stitching is complete, the hole in the skull is plugged with a circular computer chip to which the ends of the many threads are attached. Using this approach, Neuralink hopes to increase the electrode density of brain implants by manyfold, and thereby greatly improve medical devices where performance is limited by the number of electrodes implanted in the brain, such as artificial vision devices. But there are also fears that this technology could potentially be used to control a person's thoughts and behaviors, and might even provide a window for computers programmed with artificial intelligence to take over our brains and bodies. This sewing machine approach for brain implantation may seem futuristic, but Musk claims a human clinical trial is imminent. (Naeblys/Alamy Stock Photo)

of the technology discussed today to contact Neuralink about a possible job. The company is hiring.

I am impressed by Neuralink's story, but what do I know? I decide to behave like a lowly perceptron. I'm going to accept inputs from independent sources, weigh each of them, and then output a binary decision about Neuralink's prospects: Will Neuralink's technology transform civilization or not?

I decide that this single input from the Neuralink launch event might be more propaganda than science, since the scientific presentations seemed to have been orchestrated by the company's marketing department. I assign the information coming from the Neuralink launch presentation a low weight. But I need some high-weight inputs if I'm to arrive at a valid decision. I find some.

Stuart Russell is a professor of computer science at the University of California at Berkeley, an Andrew Carnegie fellow, and an expert on AI. He believes the technical obstacles that Neuralink faces are significant but not insurmountable. Rather, Russell thinks the real problem with regard to treating brain diseases is that "we understand almost nothing about the neural implementation of higher levels of cognition in the brain, so we don't know where to connect the device and what processing it should do."[7] But, despite this obstacle, he sees reason for optimism. Russell goes on:

> It used to be thought, for example, that we would have to understand the code that the brain uses to control arm muscles before we could connect the brain to a robot arm successfully. . . . It turns out instead that the brain does most of the work for us. It quickly learns how to make the robot arm do what the owner wants. . . . It's entirely possible that we may hit upon ways to provide the brain with additional memory, with communication channels to computers, and perhaps even with computer channels to other brains—all without ever really understanding how any of it works.[8]

I find Russell's point persuasive. You can't wait for full understanding to arrive before moving forward. Didn't humans go on to exploit the potential of electricity without understanding how it worked, even producing an electric power industry without even knowing of the existence of the electron?

In fact, we've seen multiple examples in this book of how electrically probing around in the darkness led to some impressive insight into neurological science. The University of Utah researchers were able to deduce which of their 100 individual electrodes were neurologically connected to specific regions of the amputee's lost hand even without the knowledge of the particular ulnar nerve fascicle the electrode was touching. They simply electrically stimulated the nerve through each electrode, one at a time, and inferred its former sensory connection from how the patient described the feeling. That allowed them to draw a usable feeling map of the hand. And the Utah researchers' approach wasn't much different than the tactic Penfield and Boldrey used to functionally map the brain's cortex. You'll recall that, back in the 1930s, they simply electrically stimulated different areas of the cortex, monitored the patient's responses, and depicted their findings as the homunculus—the 3-D human figurine depicting the relative proportions of the brain's cortex devoted to various areas of the male body. Going back even further, Galvani and Volta did something similar centuries ago by shocking the nerves in frog legs at various locations. Suppose they had decided not to work with frog legs because they didn't fully understand the mechanism of frog leg movement. Where would we be now?

No, to require complete neurological understanding of the brain before employing electrical stimulation methods for treatment is not reasonable. We're already successfully treating Parkinson's disease and depression with electricity, even though we have virtually no understanding of what these treatments are doing in the brain. In fact, we know from centuries of scientific experience the value of blindly shocking neurons to observe what happens; it can lead to very important neurological insights that aren't otherwise achievable.

Regarding Musk's claims about Neuralink's goal to reach a "symbiosis"—a truce, if you will—with our existential enemy AGI, Russell is more dismissive: "If humans need brain surgery merely to survive the threat posed by their own technology, perhaps we've made a mistake somewhere along the line."[9] I absorb what Russell's saying, weigh it highly, but seek another independent source of input.

Andrew Hires, an assistant professor of neurobiology at the University of Southern California, saw the Neuralink presentation. He thinks Neuralink has improved on the existing technologies in three ways. First, the floppy electrode wires that Neuralink has developed are very useful. Hires says, "Stiff wires in the brain cause a lot of damage because the brain can move around." Second, he thinks the idea of using a "sewing machine" to insert the tiny electrodes is a good idea. He notes, "These are very tiny things . . . it's very hard to have a steady enough hand to do these things manually." Third, Neuralink is using a superpowered chip located very close to the brain itself to translate the brain's activity. Hires explains, "There's a problem with getting electrical signals out of the brain, and that is that they are very small. And the farther they travel down a thin wire the more they are going to get distorted by noise, because there is always some electrical noise going on in the world around us. You want to be able to digitalize the signal as close to the source as possible." Neuralink is doing exactly that by implanting its BMI hardware right at the surface of the brain.

It's not that other groups haven't been working on all three of the technological areas that Hires describes. Rather, it's that Neuralink is combining these three state-of-the-art technologies into producing a single brain-implantable device. And that would amount to a significant achievement. But Hires doesn't think much about Musk's remarks regarding integrating brains with AGI. Hires scoffs, "To get to the level on integrating with AI [i.e., AGI], this is where he [Musk] sort of is going off into aspirational fantasyland."[10]

Hires makes some valid points. What most impressed me about Neuralink's launch presentations was the absence of anything fundamentally new. Spikes *can* be "read" from individual neurons if the electrodes are small enough, and you *can* "write" spikes into nerve tissue using the same type of electrodes. Even the high schoolers at Georgetown Visitation were able to do that with their SpikerBoxes and skin electrodes. In fact, the students were able to amplify their spikes using bioamplifiers they made from do-it-yourself Arduino kits. And digitalization of spike signals *can* be done even with downloadable apps that run on an iPad. And you *can* restore hearing to the deaf and sight to the blind using electricity, as we've seen with brain stem implants and visual cortex implants, respectively. And you *can* improve the performance of these various electrical treatment devices by increasing the number of electrodes. All of this is established science. There is no need to validate any of Musk's claims about the possibilities afforded by treating brain diseases with electrical stimulation. They already have been validated by over a century of scientific research, as we have seen throughout this book.

Rather, the challenge of today is not to establish the possibilities of all these things but to convert the possibilities into a reality. Neuralink's fundamental strategy to achieve the reality is through miniaturization. Reducing the scale of the hardware will allow for a significant increase in the number of read/write electrodes—what Musk refers to as "increasing our bandwidth"—and that should significantly push the science forward. And Neuralink has presented preliminary data showing that the company can do such things on a microscopic scale. So, is Neuralink's BMI a product of Musk's genius, or is it just the next logical step in the progression of our understanding of how electricity interacts with the nervous system and the practical exploitation of that knowledge to treat disease? Hasn't that been the goal since Franklin first shocked paralytics with his Leyden jar?

At this point, I'm starting to agree with Hodak's opening remark: telling Musk such things are impossible does make you look stupid because every step of the process he envisions for Neuralink has a solid scientific foundation. The only thing that's really new is bringing all these different technologies together on a really grand, yet tiny, scale. And Musk's track record for merging technologies to produce real products is also well established. So all the essential ingredients for success seem to be in place.

This, in fact, was the underlying theme of the series of scientific presentations at the Neuralink launch. Every company scientist who spoke after Hodak gave homage to their discipline's scientific predecessors and tried to show how we are actually halfway to our destination already. The subliminal message was, if you think all this stuff is impossible, it's just because you don't fully understand where we are now. In reality, there is no fundamental law of physics being violated in what Neuralink proposes to do. Rather, the laws of physics are being exploited to influence the biology. It's just another iteration of the same centuries-old dance between electrical and neurological science that we've watched throughout this life story of electricity. As this dance marathon has continued over the decades, one or the other of the two scientific fields has taken the lead, but neither dance partner has ever left the dance floor. Perhaps now it's just Neuralink's turn to call the tune.

Considering all of this, I reason that what Hires says about the value of Neuralink's combining of various technologies into a single implantable device makes a lot of sense. So I weigh Hires's input highly. The experts seem to be confirming my own gut feelings about the Neuralink launch. The futuristic stuff is bunk, but the technology could have significant implications for the here and now. But there's one issue I still need to resolve: the safety of Neuralink's proposed implants. I need to get some input from an expert on brain implant safety. I find one.

Autumn Bullard is a biomedical engineer at the University of Michigan. She just completed a review of the scientific literature for every reported device-related complication with a deep brain stimulation (DBS) or Utah array brain implant through September 2018. She was able to find 240 papers that gave estimates of the risks associated with DBS implants and 76 such papers for Utah array implants. She and her colleagues read them all and recently reported in the journal *Neuromodulation* exactly what they found.[11] I meet with Bullard to pick her brain about the safety issues that need to be considered for any type of brain implant and, more specifically, for Neuralink's.

The first thing I learn from Bullard is that our knowledge about DBS risks is extensive because there are so many patients treated over so many years. The papers she collected accounted for a total of 34,089 different DBS patients. Bleeding (hemorrhage) is always a major concern with surgery, and many of the reports (133 articles covering 19,389 patients) included data on hemorrhages. The overall hemorrhage rate was 2.86%, and the DBS-hardware-related hemorrhage rate was 2.49%. But most of these hemorrhages were minor and medically manageable; only 6 of the 483 DBS-hardware-related hemorrhages resulted in death.

Infection is also a major concern. The infection rate for DBS implantation surgery was 3.79%, but, remarkably, almost half (44.2%) were associated with implantation of the controller under the skin, not implantation of the BMI in the brain. This rate is similar to the controller infection rate for heart pacemakers and other electronic devices with subcutaneous controllers.

Bullard says there is much less information on risk for Utah array brain implants since there were only 48 reported patients. The low number of patients is because Utah array technology has arrived more recently, and all the patients with brain implants come from a limited number of research studies of short duration. Nevertheless, it is noteworthy that no significant complications have been reported for Utah arrays implanted in the brain. But Bullard also notes that none would have been expected in such a small group if the complication rates are as low as for DBS. Which begs the question: If there are major risks associated with the Utah array brain implant, could they even be detected with so few patients?

But Bullard has already anticipated that question. She tells me she did a power calculation, which is a statistical technique that provides an estimate of how large a complication rate is needed to be able to detect it with very few patients. Her power calculation showed that, if rates of major complications for the Utah array were more than five times higher than that for DBS, it is likely they would have seen those complications even with just 48 patients. This is reassuring. Given that the complication rates for DBS are so low, even fivefold higher rates for the Utah array is a medically manageable number. Many common surgical procedures have complication rates higher than that.[12]

I ask Bullard what she thinks about the safety of Neuralink's proposed implants. She says she doesn't see why Neuralink's brain implants should have any higher hemorrhage or infection rates than the Utah array's. She remarks, "From what I know about Neuralink's technology, the surgical opening in the skull is smaller than that required to implant a Utah array. I believe the small size of the individual electrodes in the Neuralink system would allow their placement to avoid major blood vessels. This suggests that Neuralink's hemorrhage and infection rates might be even lower than for Utah array implants."

With regard to infections associated with the subcutaneous controller, Bullard expects Neuralink's rates will be similar to infection rates for other subcutaneous controllers. But even there, she says, "Mitigation procedures for infections due to the implantation of the subcutaneous controller, including antibiotics and or revision [corrective] surgeries, frequently allow the intended electrical therapy to continue." So even infections are not necessarily a showstopper.

Bullard's insight is valuable and relevant to assessing the prospects for Neuralink's implant technology. I thank Bullard for speaking with me. As scary as having a BMI implant might sound, the surgical risks, at least, seem relatively low and manageable. And Neuralink's plan to robotically implant its BMI into the human brain is, evidently, not likely to be particularly dangerous . . . or crazy.

All expert inputs collected, it's time for me to make my decision: Is Neuralink's plan solid science or mere science fiction? A tough call. And even if it is science, will it succeed? Another tough call.

I believe Neuralink definitely has all its ducks in a row. They have accumulated all the needed expertise and produced the necessary hardware. If anyone can do this, the team of scientists assembled at Neuralink can. But there is also a strong possibility that Neuralink will fail, blocked by some fundamental flaw in our understanding of the mechanisms of neurobiology. Yet again, even in failure, we may learn completely new things about the nervous system and the brain, and wouldn't that be wonderful? In the meantime, we likely will end up with more flexible electrodes, smaller bioamplifiers, better surgical robots, more powerful spike analytic tools, etc. So there may be a significant silver lining in terms of advancing all the relevant electrical brain technologies, even if the final goal isn't immediately realized. Besides, there is no shame in taking on a grand challenge and failing. Failure certainly has happened before—many times before—in the process we call *science*.

Regarding Musk's claim that all brain diseases can be cured with a chip, I'm not putting much stock in that. But even if Neuralink's chip technology doesn't purge the planet of all brain illnesses, it may allow significant improvements for those brain diseases where electrical stimulation of neurons can have a beneficial effect. And this may be particularly true for those brain diseases where today's electrical therapies would benefit from higher electrode densities, such as for restoring vision to the blind by stimulating the visual cortex. And as we learn more about the electrical pathologies of other brain diseases, perhaps increasingly more of them will reveal potential electrical stimulation targets for their treatment. Since many of these brain-related diseases currently have no effective treatments, even some level of success would be very welcome indeed.

But if there is an Achilles' heel to Musk's plan for Neuralink, it's that he's promising too much too soon. Inconsistent with his opening remarks about all of this taking a very long time, Musk says that Neuralink has aspirations of treating the first human by the end of 2020! It is now early 2021, and this hasn't been achieved, and I'll bet it still won't have been achieved by the time you're reading this.[13] He may be a genius, but I think Musk greatly underestimates the time it will take to get his human brain implant into clinical trails, especially if the US Food and Drug Administration shares his concern that a rogue computer with AGI might take over the patient's brain. (But it certainly should make the patient's Informed Consent form an interesting read.) So I don't think we'll be seeing patients sporting Neuralink implants for a few years yet. I hope I'm wrong.

In summary, my bottom-line assessment is that I think Neuralink's plan to treat human brain diseases "with a chip" is a worthwhile and laudable endeavor. It may very well fail in its ultimate goal of curing brain diseases, but regardless of success or failure, there will be a considerable number of useful scientific and technological innovations spun off along the way. So, why not go for it? I wish Neuralink the best of luck.

But now I am concerned. Musk has been correct about battery-powered cars and reusable rockets. And now he may be right about treating brain diseases with computer chips. I wonder if it's wise to bet against a person with such a track record for winning. Maybe Musk is also correct that AGI is a real and present danger to humanity. I need to suppress that thought. My human intelligence, frail as it is, can't deal with a problem that big right now. I'll leave it to Musk to deal with AGI. Maybe his "symbiosis" plan will fix that problem too.

I am certainly unwilling, however, to let my angst about AGI intimidate me into abandoning ordinary AI and its wonderful artificial neural networks. I like having the bank's automatic teller machine read my checks, and who knows when I'll need to find a picture of a cat? No, I'm not ready to ditch ordinary AI simply because it may lead to AGI.

Speaking of pictures and AI, I've recently seen an advertisement for AI software that can organize your personal collection of family and vacation photos for you.[14] Having accumulated thousands of such photos that I've never had the time to sort into albums, this product appeals to me. So I'm seriously considering putting this artificial neural network to work on the huge stash of photos I've collected over the years.

Another thing that appeals to me about this particular photographic software package is its name. As I've said a couple of times already, I'm not fond of using cheesy acronyms as product names, so I'm glad the AI software company executives didn't stoop to that level. Rather, they gave their product a prestigious moniker that's associated with both a distinguished past and a promising future. They call their photo-sorting software *Amber*.

ACKNOWLEDGMENTS

I want to thank all of my friends and colleagues who read drafts of chapters and provided critical input. These individuals represent a wide range of scientific expertise, but also those with no scientific background. The scientists helped me ensure that the information in this book is scientifically accurate, while the nonscientists helped safeguard that the technical content was understandable and never became so dense as to overwhelm the narrative. Among the scientists, William Rebeck was particularly astute in his critique, and the book is significantly better for his input. Among the nonscientists, Paul Laurenza and Matt Estes provided incisive literary criticisms, which greatly improved the book's prose.

Others who contributed include Kim Annis, Robert Armiger, Mary Katherine Atkins, Jeff Behary, Augusto Beléndez, Richard Blanchard, Autumn Bullard, Kenneth Catania, Gregory Clark, Patrick Cooney, Nancy Cowdin, Anders Damgaard, Jan Dean, Patrick Forcelli, Jacob George, John Gookin, John Jensenius, Anna Jorgensen, Helen Jorgensen, Matthew Jorgensen, Melissa Loomis, Richard McDonald, Birgitte Niclasen, Jake Ponce, Irving Reti, James Reynolds, Carole Sargent, Ajay Seth, Carlos Suárez-Quian, Punit Vaidya, and Mark Watkins.

I am especially indebted to Jessica Papin, my literary agent at Dystel, Goderich & Bourret LLC, and Ingrid Gnerlich, my editor at Princeton University Press. They both provided me with advice and support at critical junctures in the publication process. There would have been no book without them.

Science and scientists have taken a severe political beating over the last few years. I, therefore, thank my readers who, like me, still believe science is real and can be a force for good in the world, and who are committed to the idea that all people can be enriched through a better understanding of science. Science is not just for the scientists; it is for everyone. I hope this book contributes, if only in a small way, to communicating that important message.

NOTES

Some of the citations in the endnotes, as well as some references listed in the bibliography, include web pages with URLs that may expire after this book has been published. Such web pages may be recoverable using the Internet Archive Wayback Machine (http://www.archive.org/web/web.php).

Chapter 1: Sparks Will Fly

1. http://www.houseofamber.com

2. Chemically, amber is a very heterogeneous substance. The bulk of it is composed of very large organic molecules produced by polymerization of different precursors of labdanes, which are major chemical constituents of natural resins. Elementally, amber is mostly carbon, hydrogen, oxygen, and sulfur, just as are all substances derived from living things.

3. Dahlström and Brost, *The Amber Book*, 113.

4. Hopp, *Amber: Jewelry, Art, and Science*, 6–7.

5. Niclasen warns, however, that counterfeiters have taken to making large pieces of amber out of smaller ones by pressing multiple smaller pieces together. Such "pressed amber" is a type of fake that can be electrified by rubbing, so the electricity test itself isn't definitive in authenticating large amber pieces. Amber experts, however, can detect the fusion lines of pressed amber through examination under a microscope.

6. The Pleistocene epoch (Ice Age) ended about 11,700 years ago.

7. A short video showing amber's attractive properties can be viewed here: https://www.youtube.com/watch?v=kfq3pQj3vgw

8. The man commonly known as Paracelsus (1493–1541) was a prominent physician and scientist of the German Renaissance. His birth name was Theophrastus von Hohenheim.

9. Marchant, *Cure*, 1–40.

10. Michael Faraday (1791–1867) was a self-educated British scientist, most famous for inventing the electric motor.

11. Fisher, H. J., *Faraday's Experimental Researches in Electricity*, 28.

12. Franklin published a description of how one would make and use a lightning rod even before he had his kite evidence that lightning was actually an electrical discharge (Cohen, *Benjamin Franklin's Science*, 6).

13. "European papers on electricity frequently speak of rubbing the tube as a fatiguing exercise. Our [American] spheres are fixed on iron axes, which pass through them. At one axis there

is a small handle with which we turn the sphere like a common grindstone" (Benjamin Franklin, in a letter to his British friend Peter Collinson, May 25, 1747).

14. Heilbron, *Electricity in the 17th and 18th Centuries*, 324.

15. Gizmodo, "The flying boy experiment entertained audiences by electrifying a kid."

16. Other variations of the Flying Boy demonstration involved levitating feathers or other light objects, rather than turning book pages.

17. Although static electric forces aren't powerful enough to make a boy actually fly, it has recently been shown they are strong enough to make spiders fly. Researchers at the University of Bristol, England, have shown spiders can use electrostatic forces to become airborne and then travel great distances on air currents using a technique known as *ballooning*. (Morley and Robert, "Electric fields elicit ballooning in spiders").

18. Scientists had been seriously studying electrical phenomena since the seventeenth century, but they encountered problems resolving issues such as whether "electrics" like amber, which gain attractive forces via rubbing, were the same as or different from "magnetics" such as loadstone, which attract metals without rubbing. They also got bogged down trying to reconcile electricity's attractive forces, which worked at a distance, with their mechanical concepts of force, which required some intervening material to transmit the force (Heilbron, *Electricity in the 17th and 18th Centuries*, 11–323).

19. Pieter van Musschenbroek (1692–1761) was a Dutch scientist and academic.

20. The modern electronic component that serves the same function as a Leyden jar is called a *capacitor* (or sometimes a *condenser*). Both Leyden jars and capacitors work on the principle of separating two electrically charged surfaces using some intervening electrical insulator. The unit of capacitance is the *farad* (F), named after Michael Faraday.

21. Jean-Antoine Nollet (1700–1770) was both a Catholic priest and a physicist.

22. Americans say "ground," while the British say "earth."

23. If you would like to try to make your own Leyden jar, this video shows how to do so in just a few minutes, using plastic drinking cups and other household items: https://www.youtube.com/watch?v=sUXfMwZBxmU

24. Smyth, *The Writings of Benjamin Franklin*, 255.

25. Franklin had theorized that lightning rods must be pointed to be maximally effective because his experience with static electricity showed him that electric charge tended to be drawn to points. He, therefore, always specified that lightning rods be constructed with a pointed end. This turned out to be somewhat of an international controversy when British electrical scientist Benjamin Wilson asserted that rounded or blunt-end rods work better than points (Heilbron, *Electricity in the 17th and 18th Centuries*, 381). The two Bens feuded for years about this. The end result was that most American-made lightning rods were constructed with pointed ends, while most British-made lightning rods were blunted. The modern-day scientific consensus is that the shape of the lightning rod's end doesn't make much of a difference toward its performance.

26. Truly pure water actually does not conduct electricity, but most water in the environment contains some small level of conductive dissolved solids.

27. Cohen, *Benjamin Franklin's Science*, 66–109; Isaacson, *Benjamin Franklin*, 142–143; Heathcote, "Franklin's introduction to electricity."

28. Franklin was slow to publish his kite experiment, perhaps precisely because the French had already published their findings with the Sentry Box. But his tardiness in publishing would haunt his historical legacy, leading critics to speculate that the kite experiment was an apocryphal story that Franklin invented to bolster his image. Fortunately for Franklin's legacy, I. Bernard Cohen, a respected science historian, has taken on each criticism and shown it to be without merit (Cohen, *Benjamin Franklin's Science*, 66–109).

29. Georg Wilhelm Richmann (1711–1753) was born in Pernau, Swedish Livonia (now Pärnu, Estonia), which became part of the Russian Empire in 1721.

30. In fact, the idea of electricity representing some kind of fluid predates the concepts of current, voltage, and Leyden jars. Puzzled by how amber could attract objects through apparently free space, William Gilbert (1544–1603), a famous sixteenth-century English physician and scientist, proposed that rubbing amber caused it to release some type of internal liquid as a vapor, and it was this vapor that stuck to small objects and drew them back toward the amber (Heilbron, *Electricity in the 17th and 18th Centuries*, 169–179).

31. Heilbron, *Electricity in the 17th and 18th Centuries*, 431–448.

32. Franklin, *Experiments and Observations on Electricity*, 14.

33. The scientist Franz Aepinus (1724–1802), an admirer of Franklin's theory, pointed out that, in order for the theory to account for the repulsion of two negative bodies, it must also include the assumption that matter stripped of its fluid is self-repellent (Millikan, *The Electron, Its Isolation and Measurement and the Determination of Some of Its Properties*, 12).

34. Franklin coined the following electrical terms, which are still in use today, although our understanding of what they actually mean physically has evolved greatly since he first described them: positive and negative charge; electrical battery; conductor; and condenser.

35. *National Geographic*, "Fossil daddy longlegs sports a 99-million-year erection."

Chapter 2: Shocking Developments

1. This weight estimate is based on a visual comparison to a stuffed specimen that stood in our lodge's lobby. I did not actually weigh the running bear.

2. On the infamous night of August 12, 1967, dubbed "Night of the Grizzlies" by the locals, two women were killed and a man mauled in two separate incidents on opposite sides of Glacier National Park. For more information about these attacks, see http://www.montanapbs.org/GlacierParksNightoftheGrizzlies/

3. British scientist Henry Cavendish (1731–1810) was one of the first to argue that there were two parameters associated with electricity that he called "intensity" (voltage) and "quantity" (amperage) (Finger and Piccolino, *The Shocking History of the Electric Fishes*, 252–256).

4. Luigi Galvani (1737–1798) was a prominent Italian scientist. His wife, Lucia Galeazzi Galvani, worked very closely with him, but she wasn't credited for any scientific work because of the conventions of the time against women scientists.

5. Glickstein, *Neuroscience: A Historical Introduction*, 55–58.

6. Piccolino and Bresadola, *Shocking Frogs*, 141–214.

7. Alessandro Giuseppe Volta (1747–1827) was a brilliant Italian scientist who is credited with both the discovery of methane and the invention of the electric battery.

8. A short (and humorous) video dramatizing Faraday's experiments with electricity can be seen at this website: http://www.rigb.org/christmas-lectures/supercharged-fuelling-the-future/thermodynamics-2016-advent-calendar/15--faradays-frogs

9. Hermann von Helmholtz (1821–1894) was a German scientific polymath. Germany's largest association of research institutions, the Helmholtz Association, bears his name.

10. Glickstein, *Neuroscience: A Historical Introduction*, 60.

11. Turkel, *Spark from the Deep*, 29.

12. Turkel, 45.

13. Turkel, 44.

14. Turkel, 55.

15. Jan Ingenhouz (1730–1799) was a scientific colleague of Benjamin Franklin. He is most famous for discovering photosynthesis in plants.

16. Finger and Piccolino, *The Shocking History of the Electric Fishes*, 285.

17. Walsh's success may have been more related to his superior experimental technique than to his use of *Gymnotus*. *Gymnotus* species are only weakly electric compared to electric eels and torpedo fish.

18. Wulf, *The Invention of Nature*, 9.

19. From Humboldt's own travel reports, first published in 1807 (Humboldt, "[Hunt and fight of electric eels with horses]").

20. Carlson, B. A., "Animal behavior: Electric eels amp up for an easy meal"; Catania, "Electric eels concentrate their electric field to induce involuntary fatigue in struggling prey."

21. Gizmodo, "This electric eel kills its prey with a sophisticated coiling maneuver."

22. Finger and Piccolino, *The Shocking History of the Electric Fishes*, 406–407.

23. I am using the term *potential* colloquially here, simply to indicate a charge differential between two different areas. Technically, an electric potential is defined as the amount of work needed to move a unit of positive charge from a reference point to a specific point inside an electrical field without producing any acceleration.

24. Another apt analogy would be a Leyden jar, which uses glass as its insulator and segregates electrical charge between the inside and outside of the jar.

25. Physics World, "Electric eel inspires new power source."

26. A short video addressing the question of why electric eels don't electrocute themselves can be viewed at this website: https://www.popsci.com/why-dont-electric-eels-electrocute-themselves

27. Klein, "Testing the electric eel's shock powers with his own arm."

28. Klein, "Testing."

Chapter 3: Bolt from the Blue

1. Lang et al., "WMO world record lightning extremes: Longest reported flash distance and longest reported flash duration."

2. *Independent*, "Two of the longest and biggest lighting strikes on earth recorded."

3. True to Cerveny's words, improved satellite technology has allowed these records to be broken yet again. On June 26, 2020, the WMO announced that a lightning strike across southern Brazil on October 31, 2018, was certified as the longest distance record at 440 miles, and a flash

in northern Argentina on March 19, 2019, was the new titleholder for longest duration at 16.73 seconds. WMO officials predict that even these latest records might soon be broken because it is likely that even greater "megaflashes" exist but thus far have eluded measurement.

4. Gookin, *Lightning*.

5. Gookin, *Lightning*, 8–10.

6. Elson, "Striking reduction in annual number of lightning fatalities in the United Kingdom since the 1850s."

7. Verge, "How exactly did lightning kill 323 reindeer in Norway?"

8. *ScienceBlogs*, "Death by lightning for giraffes, elephants, sheep and cows."

9. *Orlando Sentinel*, "Lightning kills giraffe at Disney Park."

10. WPEC CBS12 Television News, June 11, 2019.

11. Andrews, C. J. et al., "Lightning injury—a review of clinical aspects, pathophysiology, and treatment."

12. This little leg maneuver may give you some protection from a ground current, but you're no better protected from a direct strike, which likely will enter through your head on its way to the ground.

13. John Jensenius retired from NOAA in 2019.

14. Ritenour et al., "Lightning injury: A review."

15. Account from the *American Weekly Mercury*, August 12, 1732 (Marrin, *A Glance Back in Time*, 274).

16. More information about the Franklin lightning rod on Maryland's State House can be found on this website: https://msa.maryland.gov/msa/mdstatehouse/html/lightrod.html

17. VOA News, "Franklin-designed lightning rod saves historic Maryland building."

18. Uman, *All About Lightning*, 8.

19. Scientists call this the *charge-neutrality principle*. That is, all the positive and negative charges of the Earth add to zero. Strictly speaking, it is true for the entire Earth, but there can be localized differences in charge on the Earth's surface due to the inhomogeneity of the soil. Also, the highly negatively charged thunderclouds can induce patches of positively charged ground beneath them by repelling all the local negative charge in the ground.

20. Viemeister, *The Lightning Book*, 104.

21. While they are of significantly different sizes, the differences in density are also important. The smaller ice particles are low-density small crystals; the larger particles are high-density small pellets.

22. For a more sophisticated treatment of the underlying science of lightning, see Uman, *The Lightning Discharge*.

23. Uman, *All About Lightning*, 65–70; Dwyer and Uman, "The physics of lightning."

24. Cohen, *Benjamin Franklin's Science*, 7 (Franklin's emphasis).

25. It isn't clear how Franklin came to know the charge of the clouds, but he most likely compared the charge he collected in Leyden jars from the clouds with the charge he collected from glass by rubbing it with silk—Franklin's standard for bona fide positive charge—and found them to be different.

26. Franklin's "positive" and "negative" designations for static electricity were actually replacements for the older terms *vitreous* (like glass) and *resinous* (like resin) static electricity, respectively, which were first coined in 1733 by Charles François de Cisternay du Fay. The French

chemist had noticed that rubbing objects made of glass (*vitreous* in Latin) with silk and rubbing objects made of copal (a type of resin, similar to amber) with wool resulted in two different types of static electricity, each type repelling itself but attracting the other type. Rubbing objects made of other materials produced static electricity that behaved like either the vitreous or the resinous types. This suggested to many people that static electricity represented two different kinds of fluid, but it suggested nothing about the direction of each fluid's flow. Franklin's model differed from the two-fluid model in that it claimed there was just one fluid moving from one material to another during the process of rubbing. In some cases, the fluid was moving onto the rubbed item and in other cases it was moving off of it. He said the fluid was moving onto the glass, but off of the copal, putting the glass in a fluid surplus (i.e., positive) and the copal in a fluid deficit (i.e., negative). Franklin never explained, however, his rationale for the positive and negative charge allocations to the different materials, nor how he determined the alleged direction of the single fluid's movement. So it appears his assignments of positive and negative charge to vitreous and resinous static electricity, respectively, were quite arbitrary.

27. Typically, the enormous amount of negative charge in a passing thundercloud has the effect of pushing aside the negative charge in the ground immediately below it (since like charges repel each other), leaving the local ground area relatively positively charged. This can result in the formation of upward positively charged leaders, sometimes called *streamers*, that emanate from the ground and reach upward toward the sky. A step leader (negatively charged) coming down from the cloud encounters a streamer (positively charged) coming up from the ground to complete the electrical connection a few yards (or a few meters) above the ground surface, so that the step leader never actually touches the ground. (This is why victims of direct strikes by lightning sometimes report they felt electrified with static electricity just before they were struck. The positive ground charge of the streamer was moving through their bodies and attracting the down-coming step leader.) Streamers that emanate from tall objects, like flagpoles, lightning rods, and ship masts, sometimes give off a visible glow in the sky. Sailors of tall ships called such a glow coming from a ship's mast *St. Elmo's fire*. St. Elmo is the patron saint of sailors, and the sailors believed the glow was a sign that St. Elmo was protecting the ship from the storm, when in actuality the glow was foretelling the possibility of an imminent lightning strike to the mast.

28. An animated video of the direction of charge versus light movement within a lightning bolt can be viewed at this US National Weather Service site: https://www.weather.gov/safety/lightning-science-return-stroke

29. To make things a little more complicated, a lightning strike can, on rare occasions, come from the ground and move up toward the sky. It happens when weather conditions are such that negative ground charge can step all the way up to the positively charged regions of the cloud. When this happens, you can end up with an upside-down lightning bolt. But in this case, it actually looks like an upside-down lightning bolt—the main bolt splits into multiple forks facing upward rather than downward (Uman, *All About Lightning*, 51). I told you lightning was a mercurial beast.

30. Uman, *All About Lightning*, 49.

31. Cohen, *Benjamin Franklin's Science*, 120.

32. Ramón y Cajal, *Recollections of My Life*, 22–23.

33. Ramón y Cajal, 24.

34. Ritenour et al., "Lightning injury: A review."

35. Andrews, C., "Electrical aspects of lightning strikes to humans."

36. "Electrocution death" is actually redundant. The term *electrocution* really means to die by electricity. Its origins come from the combination of "electro" and "execution." But the word has now come to include any severe electrical shock resulting in bodily injury, whether fatal or nonfatal.

37. Peng and Shikui, "Study on electrocution death by low-voltage."

38. Bikson, "A review of hazards associated with exposure to low voltages." See also https://www.allaboutcircuits.com/textbook/direct-current/chpt-3/ohms-law-again/

39. Cohen, *Benjamin Franklin's Science*, 122.

40. Timeline, "One unlucky cat was the only victim when the ancient St. Mark's bell tower collapsed in Venice."

41. Gensel, "The medical world of Benjamin Franklin."

42. Quote taken from a letter sent by Franklin to his parents, posted from Philadelphia on September 6, 1744.

43. Letter to Pringle from Franklin, December 21, 1757 (Gensel, "The medical world of Benjamin Franklin").

Chapter 4: For All That Ails You

1. Influence machines slowly replaced machines that generated electricity purely through friction. They operated under the principle of *electrostatic induction* to convert mechanical work into electrical energy by the aid of a small initial charge that was continually replenished. For a historical account of the development of influence machines and a comprehensive description of the influence machines commercially available in 1890, see Gray, *Electrical Influence Machines*.

2. King, *Electricity in Medicine and Surgery*, part II, 27.

3. Licht, "History of electrotherapy."

4. In his book, Wesley demonstrates a strong command of electrical science, and he includes a detailed and accurate description of Franklin's kite experiment and his lightning rod (Wesley, *Desideratum*, 28–29).

5. Wesley, *Desideratum*, vi.

6. Wesley, vi (Wesley's emphasis).

7. Wesley was not the first to suggest electrical fluid and nervous fluid may be one and the same. It is an idea that can be traced as far back as Isaac Newton.

8. Wesley, *Desideratum*, 9.

9. Wesley was also among the first to suggest that electricity might be composed of charge particles (Wesley, *Desideratum*, 14).

10. Cleaves, "Franklinization as a therapeutic measure in neurasthenia."

11. King, *Electricity in Medicine and Surgery*, part I, 31–36.

12. King, part I, 43.

13. Adolphe Gaiffe (1832–1887) was an inventor and founder of a Parisian electrical instrumentation firm.

14. King, *Electricity in Medicine and Surgery*, part I, 61.

15. King, part II, 13.

16. Wootton, *Bad Medicine*, 67–70.

17. Joseph Lister (1827–1912) was a British surgeon who promoted cleanliness in the operating room. He introduced phenol (also known as carbolic acid) as the first clinical antiseptic. The modern germ-killing mouthwash Listerine is named after him.

18. Penicillin, the first antibiotic, was discovered in 1928 by Alexander Fleming (1881–1955).

19. Pitzer, *Electricity in Medicine and Surgery*, 55.

20. King, *Electricity in Medicine and Surgery*, 269–270.

21. Wootton, *Bad Medicine*, 148.

22. Wootton, 144.

23. Guillaume-Benjamin-Amand Duchenne (1806–1875) is sometimes referred to as Duchenne de Boulogne.

24. Jean-Martin Charlot (1825–1893) is often credited as being the "father of neurology," but Charlot himself acknowledged he owed much to his predecessor, Duchenne.

25. Poore, *Selections from the Clinical Works of Dr. Duchenne.*

26. Poore, 369.

27. Poore, 375–376.

28. Poore, xv.

29. De La Pena, *The Body Electric*, 143–161.

30. Pulvermacher Galvanic Company, *The Best Known Curative Agent*, 2.

31. Pitzer, *Electricity in Medicine and Surgery*, 82.

32. Wexler, "The medical battery in the United States (1870–1920)."

33. King, *Electricity in Medicine and Surgery.*

34. King, *Electricity in Medicine and Surgery.*

35. Parent, "Giovanni Aldini: From animal electricity to human brain stimulation."

36. *Inside Science*, "The Science That Made Frankenstein." (The quote taken from this online article originates from the *Newgate Calendar*, a monthly bulletin of prison executions, published by the Keeper of Newgate Prison, London.)

37. It is easy to see why people would have thought Aldini was using electricity to bring people back to life. At the time, both the public and scientists were struggling to define exactly what life is, and the renowned French anatomist and pathologist Marie François Xavier Bichat (1771–1802) had famously promoted his working definition: "The set of functions which resist death." In light of this definition, Aldini's demonstration that electricity could animate a corpse seemed to suggest electricity was one of these death-resisting functions. Unfortunately, Bichat himself had a very low resistance to death. He died from a fever at age 30 (Strauss, *Human Remains*, 60–62, 111–112).

38. Seymour, *Mary Shelley*, 44.

39. Flexner, *Medical Education in the United States and Canada.*

40. For a comprehensive biography chronicling Flexner's achievements in educational reform, see Bonner, *Iconoclast.*

41. The AMA currently has over 30,000 members and publishes weekly the *Journal of the American Medical Association* (*JAMA*), a highly influential medical journal.

42. Flexner, *Medical Education in the United States and Canada*, 157 (emphasis added).

43. Flexner, 156–166.

44. Homeopathy and allopathy were directly opposing treatment theories. Homeopathy was based on the notion that a disease could be successfully treated with low doses of the agent that caused the disease. Allopathy was based on the notion that the best treatments were high doses of agents other than the substance that caused the disease.

45. The report also praised the good schools by name. Flexner hailed the Johns Hopkins Medical School as "unsurpassed" in laboratory facilities and said the school provided "ideal opportunities" in clinical teaching. Consequently, many reform-minded schools soon adopted the Johns Hopkins curriculum as their own. In fact, the educational model of two years of classroom study followed by two years of clinical training that is used in most American medicals schools today finds its origins in the Johns Hopkins curriculum.

46. X-rays appeared on the medical scene at precisely the time that electrotherapy began to fade away. In fact, X-rays, which were discovered in 1895, may have actually hastened electrotherapy's demise. Immediately after their discovery, X-rays were used both diagnostically and therapeutically, and their effectiveness for shrinking cancerous tumors was obvious to all who used them. Ironically, electrotherapy physicians were the best prepared of any physicians to make the switch to X-rays. Their electricity-generating machines could easily be used to power X-ray tubes, and many of them likely saw great opportunities in shifting medical careers from being "electricians" to being "radiologists." See Jorgensen, *Strange Glow*, 116–140.

47. This state of affairs mirrors the situation with radium that was happening at about the same time. The medical demand for radium to treat cancer drove costly radium mining and refining activities. Once the mining and refining methods were in place, radium prices fell low enough that radium could be used in consumer products, such as watches with luminous radium-painted dials. See Jorgensen, *Strange Glow*, 81–140.

Chapter 5: A Circuitous Route

1. For a biography of Volta, see Pancaldi, *Volta*.

2. Johnson, *The Ten Most Beautiful Experiments*, 60–74.

3. How Galvani reconciled this experiment with his claim that frog legs severed from the body retained their electricity-producing capabilities is not clear.

4. Finger and Piccolino, *The Shocking History of Electric Fishes*, 334.

5. Although William Nicholson (1753–1815) is often described as a professional chemist, he was also a writer, publisher, inventor, patent agent, and engineer.

6. Pancaldi, *Volta*, 183.

7. Finger and Piccolino, *The Shocking History of Electric Fishes*, 330.

8. Pancaldi, *Volta*, 191–192.

9. Nicholson, "Observations on the *Electrophorus*, tending to explain the means by which torpedo and other fish communicate electric shocks."

10. Volta was so convinced he had discovered the mechanism of the fish electrical organ that his first choice for a name for his new device was *organe electrique artificiel*—French for "artificial electric organ"—but that name did not catch on.

11. More accurately, the term was first used in electrical science by Franklin. He likened the discharge of multiply linked Leyden jars to a battery of artillery firing together. The term was later adopted to describe multicell electrical batteries, but we now use the term to describe even batteries with only a single cell.

12. Pancaldi, *Volta*, 185.

13. If you reduce the diameter of the garden hose by constricting its end, the force of the exiting water increases, but the rate of water flow actually will have been reduced. Some people find this counterintuitive because the higher-force water exiting the constricted hose shoots out further, which they misinterpret as an indication of a higher flow rate. But, in reality, pinching the end of a hose will not make a bucket fill any faster; rather, it will increase the time needed to fill the bucket.

14. This is easier said than done. To double the water pressure at the tap, we would have to increase the height of the water tower that's providing the pressure.

15. Schlesinger, *The Battery*, 50. (There are few accessible, nontechnical, and credible sources of information about the electrical battery. Henry Schlesinger's book is a notable exception.)

16. It is, unfortunately, quite common to use Michael Faraday's terminology and call the positive electrode of a battery an "anode" and the negative one a "cathode." The names refer to the Greek words related to the rising and setting sun, and were introduced by Faraday (at the suggestion of William Whewell) because, like the sun, electrical current figuratively rises in the east (*anodos*) and sets in the west (*kathodos*). Pretty obscure and confusing, isn't it? Because I believe they confuse more than enlighten, I don't use these terms in this book. Rather, I call a battery's terminals the positive and negative electrodes.

17. The development of the lithium-ion battery has been transformative to modern electronics, so it seems appropriate that the 2019 Nobel Prize in Chemistry was awarded to three scientists who significantly contributed to the development of the lithium-ion battery: John B. Goodenough, M. Stanley Whittingham, and Akira Yoshino.

18. The 18650 battery has a diameter of 18 mm and a height of 65.0 mm (hence the name), while a typical AA-size battery is about 14 mm by 50 mm.

19. DO NOT try to disassemble an 18650 or any other lithium battery! It will explode into flames due to the tremendous amount of energy it stores.

Chapter 6: Jolted Back to Reality

1. Inspector General, "Review of electrocution deaths in Iraq: Part I," 3.

2. Inspector General, "Review of electrocution deaths in Iraq: Part II."

3. Wick and Byard, "Electrocution and the autopsy."

4. Price and Cooper, "Electrical and lightning injuries."

5. Rådman et al., "Electrical injury in relation to voltage, 'no-let-go' phenomenon, symptoms and perceived safety culture."

6. Cooper, "Emergent care of lightning and electrical injuries."

7. Hyldgaard et al., "Autopsies of fatal electrocutions in Jutland."

8. It is not typically necessary for every outlet in a house to be a GFCI outlet; one outlet within any electrical circuit will suffice. Replacing only one outlet in the circuit will keep down costs. But consult with a licensed electrician before replacing any old-style electrical outlets.

9. Wick and Byard, "Electrocution and autopsy."

10. *New York Public Library Blog.* "Ben Franklin on cooking turkey . . . with electricity."

11. Note that this current route—one hand to the other hand—would have passed through Franklin's heart, the route of highest risk to Franklin's life.

12. Quote taken from Benjamin Franklin's letter to his brother John Franklin, December 25, 1750. Original letter housed in Winthrop Family Papers, Massachusetts Historical Society. (Spelling, capitalization, and punctuation of the quoted section were modernized for reading clarity.)

13. This is *Joule's first law,* named after physicist James Prescott Joule (1818–1889). It states the heat output from an electrical conductor is proportional to the product of the square of the current times the resistance.

14. *New York Public Library Blog.* "Ben Franklin on cooking turkey . . . with electricity."

15. The Humane Slaughter Association is a UK-based nonprofit organization that works to improve the welfare of food animals during transport, marketing, and slaughter. They provide scientific research grants to help identify better ways to humanely slaughter all food animals from fish to cows: https://www.hsa.org.uk

16. There recently has been some controversy surrounding the electrified water bath immersion method as it is performed by the US poultry industry. Critics claim it paralyzes the birds without making them insensible: https://www.huffingtonpost.com/entry/chickens-slaughtered-conscious_us_580e3d35e4b000d0b157bf98

17. Berg and Raj, "A review of different stunning methods for poultry—animal welfare aspects."

18. *Modern Farmer,* "Here's why a chicken can live without its head."

19. Lest you think brainlessness is just a phenomenon of birds, the same thing has been demonstrated in house cats: https://www.regenexx.com/blog/can-learn-walking-cat-no-brain/

20. Gezgin and Karakaya, "The effects of electrical water bath stunning on meat quality of broiler produced in accordance with Turkish slaughter procedures."

21. Lethal injection was not known yet and is not among the 34 methods the commission reviewed. Lethal injection with toxic doses of morphine was first proposed in 1888 by Julius Mount Beyer, a New York physician, after the commission had issued its final report. But the medical profession was largely opposed to lethal injection. The modern hypodermic needle, a medical instrument invented by Francis Rynd in 1844, was seen as a medical breakthrough. Physicians didn't want to taint the image of the hypodermic needle, which was frightening enough to some patients, by having it associated with killing people.

22. Galvin, *Old Sparky,* 15–41.

23. Moran, R., *Executioner's Current,* 81.

24. *Popular Mechanics,* "The most gruesome government report ever written evaluates 34 ways to execute a man."

25. Capital punishment gets its name from the Latin word *capitalis,* meaning "of the head." So the term *capital punishment* literally refers to a punishment of the head, because the death sentence was typically delivered by decapitation.

26. Galvin, *Old Sparky,* 63.

27. "Brain-dead" means the irreversible loss of the functions of the brain, including the brain stem. A patient determined to be brain-dead is legally and clinically dead even if her heart has not yet stopped beating.

28. *Buffalo Evening News*, "Never mind the intentions."

29. *Auburn Daily Advertiser*, "Kemmler."

30. Moran, R., *Executioner's Current*, 34.

31. Moran, 17–18.

32. The events surrounding McElvaine's electrocution as described here are summarized from a description in Moran, R., *Executioner's Current*, 215–217. The original source of that information is the *New York Times* newspaper article "Two shocks were needed."

33. *New York Times*, "Two shocks were needed."

34. Elliott, *Agent of Death*.

35. Elliott, 14–15.

36. Elliott did not immediately succeed Davis as state executioner. Davis retired in 1914 and was succeeded by John Hulbert, and then Elliott took over when Hulbert retired in 1926.

37. Wick and Byard, "Electrocution and autopsy."

38. Krider, "Benjamin Franklin and lightning rods." https://physicstoday.scitation.org/doi/10.1063/1.2180176

39. Elliott, *Agent of Death*, 150.

40. Elliott, 211.

41. Elliott, 149–150.

42. *New York Times*, "G. M. Ogle, authority on the death chair, dies."

43. Elliott, *Agent of Death*, 148.

44. Elliott, 147.

45. This was before lethal injection was introduced in the United States; the first execution by lethal injection occurred in Texas in 1982.

46. Elliott, *Agent of Death*, 141.

47. Elliott, 150.

48. Elliott, 309.

49. There were 4,374 electric chair executions from 1890 to 2010 in the United States. For comparison, there were about 21,000 reported lightning deaths over the same period, although there were probably many more since not all lightning deaths are reported to authorities (López and Holle, "Changes in the number of lightning deaths in the United States during the twentieth century").

50. The Eighth Amendment to the US Constitution prohibits "cruel and unusual punishment."

51. Choosing the method of execution is an option only for Tennessee prisoners who received their death sentence prior to 1999. For later death sentences, lethal injection is the prescribed execution method.

52. *Tennessean*, "Death row inmates weigh the choice between the electric chair and lethal injection."

53. *Tennessean*, "Tennessee executes David Earl Miller by electric chair."

54. Dorman et al., "Tennessee execution."

55. See McNichol, *AC/DC*.

Chapter 7: A Field Day

1. To be precise, direction changes every 6 hours and 12.5 minutes. It's slightly longer than 6 hours because the moon keeps advancing its position in the sky as it does its 27.32-day revolution around the Earth.

2. Carlson, W. B., *Tesla*.

3. The transistor was invented at Bell Laboratories in the 1940s. A transistor allows electrical signals to be amplified by means of a semiconductor material, such as silicon. I speak more about amplification of electrical signals in chapter 11.

4. A *diode* is a semiconductor device that allows current to flow in one direction only.

5. Faraday went on to be trained by Humphry Davy, Britain's leading chemist at the time.

6. Electric fields and magnetic fields are not exactly the same thing. They are different but related phenomena. If you are having trouble understanding the distinction between the two, consider yourself normal; virtually everyone has trouble with it. Let's just say that a moving magnetic field will create an electric field, and a moving electric field will create a magnetic field. And that "movement" is relative to some particular observer, which gets into relativistic theory (think Albert Einstein)—a place where we don't necessarily want to go for our purposes here. Fortunately, the physical properties of such force fields are comparable enough that we often conflate the two terms and simply call them *electromagnetic fields* when talking about field theory. This is the approach I take in this book. Henceforth, I will exclusively talk about the general properties of electromagnetic fields, unless a distinction needs to be made.

7. Hans Christian Ørsted (1777–1851) was a highly prominent physicist and chemist, credited with both discovering the link between electricity and magnetism and being the first to isolate the element aluminum.

8. *Magnetic field* was a term Faraday coined in 1845.

9. Although the concept is a direct outgrowth of Ørsted's work, the electromagnet was developed for practical purposes by William Sturgeon, who is usually credited with its invention.

10. Naturally magnetized lodestones were allegedly first discovered in ancient Greece in the region then known as Magnesia, hence the origin of the term *magnet*. But the story is likely apocryphal.

11. Faraday, "Experimental researches in electricity—fifteenth series," Part I, 11.

12. Duchenne, *A Treatise on Localized Electrization*, 288–291.

13. Licht, "History of electrotherapy."

14. Duchenne, *A Treatise on Localized Electrization*, 192.

15. Forbes and Mahon, *Faraday, Maxwell, and the Electromagnetic Field*, 101–103.

16. Faraday, "Thoughts on ray-vibrations."

17. Forbes and Mahon, *Faraday, Maxwell, and the Electromagnetic Field*, 117.

18. Maxwell, "On Faraday's lines of force."

19. Maxwell actually used the more precise term *incompressible fluids*. Water and most other liquids are considered incompressible fluids, while gases, which are considered compressible fluids, behave differently.

20. Maxwell, "On Faraday's lines of force."

21. Forbes and Mahon, *Faraday, Maxwell, and the Electromagnetic Field*, 158.

22. The modern standard international unit for field intensity is newtons per coulomb.

23. Forbes and Mahon, *Faraday, Maxwell, and the Electromagnetic Field*, 160–161.

24. Forbes and Mahon, 107.

25. One unit of horsepower is equal to 745.7 watts.

26. James Watt (1736–1819) was a Scottish inventor whose steam engines contributed significantly to the Industrial Revolution.

27. Sebastian Ziani de Ferranti, along with Lord Kelvin, designed one of the earliest AC power systems, which included a basic AC transformer. Lucien Gaulard and John Dixon Gibbs later designed a very similar transformer and marketed it. They ultimately lost to Ferranti in a patent suit in the British courts.

28. There are other methods to modulate DC voltage, such as step-up converters or alternators coupled with transformers and rectifiers, but these approaches are far more complex than a relatively simple AC transformer.

29. Brown, "Death in the wires."

30. Moran, R., *Executioner's Current*, 101.

31. Guttman, "A review of Mr. Harold P. Brown's experiments."

32. A full description of Harold Brown's campaign against AC and its consequences can be found in Moran, R., *Executioner's Current*, 92–118.

33. To be more precise, the nominal "110 volts" of alternating current delivered to American homes is not the "peak" (i.e., top of the cycle) voltage (V_{peak}). Rather, the 110 volts refers to the root-mean-square voltage (V_{rms}). The root-mean-square voltage is the average voltage over time. The peak home voltage is actually about 155 volts ($V_{rms} = 0.71\ V_{peak}$).

34. *IEEE Spectrum*, "DC microgrids and the virtues of local electricity."

35. Fairley, "Edison's revenge: The rise of DC power."

36. James, *The Correspondence of Michael Faraday* (vol. 2).

37. Faraday, "Experimental researches in electricity—fifteenth series."

Chapter 8: Zombie Fish

1. "Electrofishing" is the term commonly used in North America. Elsewhere, it is often called "electric fishing."

2. For a contemporary and comprehensive description of the uses of electricity in fishery management and research, see Beaumont, *Electricity in Fish Research and Management*.

3. Reynolds, "Development and status of electric fishing as a scientific sampling technique."

4. Up to 60 hertz is allowed for hardier fish species.

5. Electrofishing has been around for a very long time. The first patent on electrofishing equipment was issued in London in 1863.

6. You can also use AC to shock fish. It works quite well at stunning. However, since AC doesn't have fixed positive and negative electrodes (because the current is constantly switching back and forth many times per second), you don't get any electrotaxis effect, so it makes the fish harder to catch. For this and other reasons, virtually all electrofishing is done with DC.

7. Pure water is actually a very poor conductor of electricity. But most water in the environment has some dissolved solids in it. They act as electrolytes to increase water's conductivity.

8. Actually, there is a good way to see an electrical field using crystals of potassium permanganate—a purple-colored salt—and wet filter paper. An example of the visualization of various electric field geometries with potassium permanganate can be seen in this video: https://www.youtube.com/watch?v=63FnToW-Hxc

9. Electricity moving through wires can produce electromagnetic fields (EMFs) and nonionizing electromagnetic waves—a type of radiation. Some people have alleged that there are health effects of EMFs and low-energy nonionizing radiation, but scientific evidence supporting that contention is weak to nonexistent. For more information on the known biological effects and health consequences of various types of radiation, see Jorgensen, *Strange Glow*.

10. The largescale sucker (*Catostomus macrocheilus*) is native to the Pacific Northwest, from British Columbia to Oregon.

11. Stream electrofishers trail behind them a long, thick wire that they call a "rattail," which serves as the negative electrode. It comes out of the bottom of the battery pack and is insulated with a rubber coating for the first 4 feet or so, and then becomes a bare, uninsulated wire for another 4 feet. The long, linear wire is dragged downstream behind the electrofishing team and serves to place the end of the electric field well behind them, such that the electrofishers' legs are actually within the electric field, hence the need for protective rubberized waders.

12. Van Harreveld, "On the galvanotropism and oxcillotaxis of fish."

13. Stimulus-response theory doesn't require the stimulus to be pain. You probably have heard of Pavlov's famous dog, an animal that began to salivate on hearing the bell, a behavior produced by preconditioning the dog to associate a ringing bell with feeding time.

14. Van Harreveld, "On the galvanotropism and oxcillotaxis of fish."

15. For a complete history of research on reflexes, see Liddell, *The Discovery of Reflexes*.

16. To spare the squeamish, I left out the details of how the sensory neurons were eliminated. But for those who are curious, it turns out that the sensory neurons in a fish are primarily in its skin and fins. So a skinned fish with its fins cut off is left with only the spinal cord and the motor neurons, which extend from the spinal cord and infiltrate the lateral muscles. The additional destruction of the spinal cord, however, abolishes the response; so the response cannot be attributed to the motor neurons alone. (I should also mention that these experiments predate animal rights protection regulations for vertebrates.)

17. The dependency of fish electrotaxis on an intact spinal cord despite nondependency on a reflex arc probably is because the cell bodies of motor neurons in fish actually penetrate into the spinal cord, so that destroying the spinal cord also damages the motor neurons.

18. Since the head of an electric eel is positively charged, it's possible the eel uses its electric field to actually attract prey fish to its head. But so far, there is no evidence electric eels exploit electrotaxis in attracting their victims (Kenneth Catania, personal communication).

19. Vibert, "Neurophysiology of electric fishing"; Blancheteau et al., "Etude neurophysiologique de la pêche électrique"; Lamarque, "Electrophysiology of fish subject to the action of an electric field."

20. People use the noun *polarity* and the verb *polarize* to mean different things in different contexts, and this can be confusing. I don't want to add to this confusion. In its most general definition, an object is polarized when it has distinct ends that give it a defined functional orientation. A pencil is polarized; it only writes when its point is facing in a specific direction. But a piece of blackboard chalk is not polarized; flip it around and it writes just the same. In this

book, I use the term *polarized* in this general sense, to mean something that has a defined functional orientation. Thus, magnets, battery terminals, and cells such as motor neurons are all polar entities. In contrast, red blood cells, which look like little round doughnuts and function the same regardless of which direction they happen to be facing, are not polarized.

21. Rolls and Jegla, "Neuronal polarity: An evolutionary perspective."

22. Sharber and Black, "Epilepsy as a unifying principle in electrofishing theory."

23. The term *shock therapy* is actually a misnomer for *electroconvulsive therapy*, as I explain in chapter 15.

24. In an electric field, the voltage drops in roughly linear patterns along field lines. So a fish's head is at a different voltage than its tail when oriented along a field line. The difference in voltage from head to tail is the shocking voltage the fish's body experiences. In contrast, a fish oriented perpendicular to a field line experiences a much lower voltage difference across its body.

25. Peimani et al., "A microfluidic device to study electrotaxis and dopaminergic system of zebrafish larvae."

26. Zebrafish eggs typically hatch two days after fertilization. So a larva at 5 to 7 days postfertilization is about 3 to 5 days posthatching.

27. You would expect that some control larvae would exhibit an apparent electrotaxis response, because they must swim in just one direction or the other in the channel. And the probability they would randomly swim in the direction of the positive anode, twice in a row, is not zero. So, a zero result for the control group would be highly suspect.

Chapter 9: Crookes of the Matter

1. Though Crookes's tube was entertaining, it was not simply a toy. Crookes's work was in a tradition of research into the strange effects that took place inside evacuated glass tubes under high voltages that stretched back over a quarter of a century. Crooke himself thought that his experiments demonstrated the existence of a fourth state of matter.

2. You may think you've never seen a Crookes tube before, but if you're over 30, you probably have. Remember those old monochrome computer monitors, where the letters on the screen glowed green? They were actually glorified Crookes tubes. The insides of the screens were coated with fluorescent paint that glowed when hit by beams of "cathode rays" coming from the back of the monitor.

3. Amdahl, *There Are No Electrons*, 1.

4. For a collection of various papers that try to define the electron, see Simulik, *What Is the Electron?*

5. For the earliest articulation of the importance of the electron to chemistry, see Thomson, *The Electron in Chemistry*.

6. For a concise biography of Isaac Newton and a description of his work in alchemy, see Gleick, *Isaac Newton*.

7. Thomson, "The connection between chemical combination and the discharge of electricity through gases," 493.

8. We tend to think of Faraday as a physicist rather than a chemist, but his work actually spanned both disciplines.

9. Forbes and Mahon, *Faraday, Maxwell, and the Electromagnetic Field*, 83 (emphasis added).

10. As we've seen, when current flows through a wire, it produces a magnetic field. If two parallel wires have currents running through them, they each have their own magnetic field. And those two fields either attract each other or repel each other, depending upon whether the currents are running in the same (repel) or opposite (attract) directions. If we set the current level in each wire at exactly 1 amp, there should be a specific amount of charge flow, and the forces between the wires should produce a specific pulling (or pushing) against each other.

11. For a biography of Coulomb and a description of his work with electrical charge, see Gillmor, *Coulomb and the Evolution of Physics and Engineering in Eighteenth-Century France.*

12. Thomson won the 1906 Nobel Prize in Physics for his work on conduction of electricity by gases. This seminal work ultimately resulted in the discovery of the electron.

13. At Cambridge University, undergraduate coursework was assessed via examinations divided into parts covering a set of related subject areas called a tripos. Thomson's Mathematical Tripos included training in mathematical physics.

14. Actually, Maxwell was also notorious for having annoying little mathematical mistakes that peppered his publications, making them even harder to understand, until Thomson came along and corrected the mathematical errors.

15. Navarro, *A History of the Electron*, 21.

16. Navarro, *A History of the Electron*, 56. It is interesting to note that some modern theoretical particle physicists have criticized the current state of their own profession, saying that the mathematics has taken on a life of its own, entirely divorced from experimental validation; in this case, not because experimentation isn't valued but because it has become too difficult and costly to perform. For a full discussion of this controversy, see Hossenfelder, *Lost in Math.*

17. Thomson, *Notes on Recent Researches in Electricity and Magnetism*, 189.

18. *Momentum* is the force that a physical object has when it is moving. It is the product of the object's mass and velocity, and it is a *vector* quantity, meaning it has a defined direction. As the momentum of an object increases, a greater external force must be applied to divert it from its path.

19. For Millikan's firsthand account of the sequence of scientific events contributing to the discovery of the electron, see Millikan, *The Electron.*

20. Millikan won the 1923 Nobel Prize in Physics for this work.

21. We now know the electron is a type of subatomic particle known as an *elementary fermion.* Protons and neutrons are *composite fermions*, made up of odd numbers of elementary fermions. Thus, every atom can be subdivided into even smaller fermionic particles, and the electron is their prototype.

22. To be perfectly accurate, the term *electron* actually predates the discovery of the electron particle. By the late 1800s, it had become evident that electrical charge must have some fundamental unit that cannot be further subdivided. As such, it represented the electrical counterpart of the atom—the component of matter thought at the time to be indivisible. In 1891, Johnstone Stoney gave the name *electron* to this indivisible electricity unit, and the name was later applied to the particle Thomson discovered since the newly discovered particle itself was the apparent carrier of a single unit of electrical charge.

23. The "solar system" model is known among physicists as the *Rutherford-Bohr model.*

24. The intrinsic angular momentum of an electron (spin) or other elementary particle is strictly a quantum mechanical property. It has no relationship to what is meant by angular momentum in classical physics.

25. I am calling such atoms "electrically" unstable to distinguish atoms with an imbalance in their electron spin from atoms that have an unstable nucleus due to imbalances between the numbers of protons and neutrons. Atoms with an unstable nucleus are prone to spontaneous decay—a nuclear physics phenomenon commonly referred to as *radioactivity*. See Jorgensen, *Strange Glow*, 43–44.

26. Lewis is considered the discoverer of the covalent bond. He was nominated 41 times for the Nobel Prize in Chemistry for this tremendous achievement. He never won.

27. Thomson initially called them "corpuscles"—a corpuscle is a minute particle of matter—rather than electrons. The *electron* term was not adopted until later.

28. Navarro, *A History of the Electron*, 84.

29. Very recent experimental evidence suggests "flexing" may be more important than "scraping" when it comes to producing static electricity by rubbing materials together. See Mizzi et al., "Does flexoelectricity drive triboelectricity?"

30. The *law of conservation of charge* says charge cannot be created or destroyed; it can just be moved around. This "law" was first proposed by William Watson and Benjamin Franklin in the eighteenth century, but not experimentally proved until Michael Faraday did so in the nineteenth century.

31. Millikan, *The Electron*, 5.

32. The Einstein quote is from the lecture of Hans G. Dehmelt, inventor of the ion trap technique, on the occasion of his Nobel Prize in Physics award ceremony in 1989.

Chapter 10: Redbud

1. This was known as the *reticulum theory* of nerve signal transmission.

2. Ramón y Cajal, *Recollections of My Life*, 295.

3. Ramón y Cajal, 294.

4. Ramón y Cajal, 324.

5. Alturkistani et al., "Histological stains."

6. The mechanism underlying the apparently capricious and random staining of neurons using Golgi's method remains unknown to this day.

7. In fact, Golgi was in the habit of naming certain classes of neurons by their alleged function (e.g., sensory neurons). Cajal frowned on that because the true functions of the individual neurons hadn't yet been established. Cajal preferred to name cell types that he discovered by a simple descriptor of their morphology. For example, he named the brain cells that had pyramid-shaped cell bodies *pyramidal neurons*—a purely descriptive name with no allusion to their alleged function.

8. Some of Cajal's most beautiful and important drawings of neurons are shown in color and very high resolution in Swanson et al. (eds.), *The Beautiful Brain*.

9. Ramón y Cajal, *Recollections of My Life*, 322.

10. Ramón y Cajal, 155.

11. Ramón y Cajal, 156.

12. Ramón y Cajal, 155.

13. For a complete description of the value of the squid axon to neuroscience, see Kleinzeller and Baker, *The Squid Axon.*

14. This discovery has been credited to John Zachary Young, who confirmed that the tubular structure seen on the squid mantle is the very large axon of a motor neuron (Keynes, "J.Z. and the discovery of squid giant nerve fibres"; Young, "The function of the giant nerve fibres of the squid").

15. Electric fields can also be measured in terms of newtons (the unit of force) per coulomb (the unit of charge): 1 volt/meter equals 1 newton/coulomb.

16. A similar electric field value can be obtained using the *Nernst equation,* which is an electrochemical approach to the same electromagnetic field question; but employing the Nernst equation requires knowledge of the relative concentrations of ions inside and outside the membrane.

17. Electrical signals are transmitted from one neuron to the next at the synapse. The transmission of the electrical signal can be direct, as in the case of an *electrical synapse.* But, more commonly, the signal transmission occurs through a *chemical synapse.* At a chemical synapse, the axon side of the synapse releases a chemical neurotransmitter in response to the arrival of an action potential. The released neurotransmitter then binds to a receptor on the dendritic side of the synapse. The binding of the neurotransmitter to the receptor causes the membrane in the area of the receptor to depolarize, thereby causing the initiation of a new action potential, which moves from the cell body down the axon to the next downstream neuron. Electrical synapses transmit signals faster than chemical synapses. But electrical synapses aren't as versatile and are not amenable to outside modulation. Electrical synapses can also transmit signals in either direction, while chemical synapses ensure that the signal moves only in the intended direction. Thus, for reasons of both control and directionality, chemical synapses predominate over electrical synapses throughout the nervous system, and even at the junctions between neurons and muscle cells.

18. Sometimes scientists attempt to improve a poor model by adding mathematical parameters. The added complexity of the model can sometimes give the illusion of a better fit (a problem known by statisticians as *overfitting*). But the best models—the ones that reveal true insight—are often the simplest. So the most useful models are ones that the data fit well, while at the same time having the fewest number of parameters. Albert Einstein was able to find such a model for the relationship between energy and mass: $E = mc^2$. Can't get much simpler than that mathematical model, and time has shown the data fit his model perfectly.

19. Actually, their data fit best when a third parallel branch was added to the model. This minor third branch of the circuit model is thought to be from a small amount of current flow across the membrane due to nonspecific ion leakage through the membrane (Hodgkin and Huxley, "A quantitative description of membrane current and its application to conduction and excitation in nerve").

20. The relatively simple mathematical model is defined by the equation $I_m = C_m\, dV/dt + I_{Na} + I_K + I_{leak}$, where I_m is the total current across the membrane, and I_{Na}, I_K, and I_{leak} are additive contributions of the transmembrane sodium and potassium ion currents, plus a small

minor current contribution from nonspecific ion leakage. C_m is a membrane capacitance constant, and dV/dt is the change in voltage with change in time (Kleinzeller and Baker, *The Squid Axon*, 308).

21. We will use "pulses" here rather than "waves," so as not to confuse action potentials with electromagnetic waves.

22. Occam's razor is believed to have originated with the fourteenth-century Franciscan friar William of Occam (or Ockham), who studied what we would now call the philosophy of science. The meaning of "razor" is lost to history, but some speculate that it refers to shaving away unnecessary things.

Chapter 11: Zombie People

1. Del Guercio, "From the archives: The secret of Haiti's living dead."

2. Nugent et al., "The undead in culture and science"; Littlewood and Douyon, "Clinical findings in three cases of zombification."

3. The notion that Haitian zombies are created by means of some type of poisonous narcotic was not new. Similar versions on the hypothesis had been postulated by different people since at least 1860, when the French ambassador to Haiti, Marquis de Forbin Janson, reported: "Two days after my arrival in Port-au-Prince a woman sent to sleep by means of a narcotic and buried the same evening in a cemetery of the town was disinterred during the night. She still breathed" (Davis, *Passage of Darkness*, 66–71).

4. If you are interested in the long version of the story, Davis wrote a book for the lay public in the style of an Indiana Jones novel, in which he describes in detail his quest for the zombie potion and the discovery of its key ingredients. See Davis, *The Serpent and the Rainbow*. The book was subsequently made into a tawdry horror film by Universal Studios. Davis has also written a scholarly book on the ethnobiology of Haitian zombies. See Davis, *Passage of Darkness*.

5. Of course, the real discoverers of the Hawaiian Islands were the Polynesians who sailed up from Polynesia centuries before, and from whom the native Hawaiians are descendants.

6. There are several dozen sodium and potassium ion gates per square micron in the membrane of a squid's axon.

7. There are also calcium and chloride ion channels in axon membranes, but they do not participate in action potentials.

8. Curiously enough, lithium—sodium's next-door neighbor on the periodic table of elements—acts like a wolf in sheep's clothing. Lithium's chemistry is so similar to sodium that the sodium voltage-gated membrane channel cannot distinguish a lithium ion from a sodium ion. Thus, the wolf sneaks inside the axon through the same gate as the sheep. Once inside the neuron, it disrupts ionic balances and suppresses neuronal activity. This mechanism is thought to contribute, in part, to the efficacy of lithium as a drug treatment for the manic phase of bipolar disorder (Jakobsson et al., "Towards a unified understanding of lithium action in basic biology and its significance for applied biology").

9. Faraday, *Chemical Manipulations*, 1.

10. Warshofsky, *The Chip War*, 21.

11. Such approaches to amplify electrical signals often rely on *relays*, which are electromagnetic switches that use small currents to turn on large ones.

12. It is worth noting that patch clamp amplification technology has been improved and miniaturized since then to the point that it is now possible to put an amplifier on a 3 × 3 mm chip. Remarkably, this makes such a chip-based patch clamp amplifier nearly as small as the diameter of the squid axon itself (0.5 mm) (Weerakoon et al., "Patch-clamp amplifiers on a chip").

13. The positively charged amino acids are lysine, arginine, and histidine; the negatively charged amino acids are aspartate and glutamate.

14. Horn, "How ion channels sense membrane potential."

15. Of course, if sodium ion entry into the axon is simultaneously and instantaneously matched by potassium ion exit, a charge balance would be constantly maintained and there would be no upward spike in voltage. But voltage-gated potassium channels trigger more slowly than the voltage-gated sodium channels, and this lag in response between the two types of channels causes a momentary imbalance in charge, resulting in a transient voltage spike.

16. The more astute will notice that the charge equilibrium between inside and outside may have been restored, but the chemical equilibrium still has not returned to the original state. Specifically, the sodium ions are now inside the membrane and the potassium is out—the exact opposite of the original distribution of ions and an obstacle for transmission of the next electrical signal. But the axon has developed another type of membrane channel to deal with that. To restore the chemical equilibrium, the membrane also contains a *sodium/potassium "pump,"* which is a protein channel that acts similarly to a revolving door, pushing sodium ions out while at the same time pushing potassium ions in.

17. Sheumack et al., "Maculotoxin: A neurotoxin from the venom glands of the octopus *Hapalochlaena maculosa* identified as tetrodotoxin."

18. Poisons differ from venom in that poisons are toxins that are ingested and venoms are toxins that are injected, for example, through snake bites and bee stings. It is possible for toxins to be present in both poisons and venoms. See Wilcox, *Venomous*.

19. Mouhat et al., "Animal toxins acting on voltage-gated potassium channels."

20. *MythBusters* is a science entertainment television series created by Peter Rees. It is produced by Beyond Television Productions of Australia.

21. Anderson, "Tetrodotoxin and the zombie phenomenon."

Chapter 12: The Wanderer

1. *Vagus* means "wandering" in Latin.

2. Galen of Pergamon was born in Greece in 129 AD. He became the greatest physician of the Roman Empire and was personal physician to multiple Roman emperors. His teachings dominated the practice of medicine in European and Arab countries for 1,500 years.

3. Nerves that transmit signals toward the brain or spinal cord are collectively called *afferent* nerves, while nerves that transmit signals away from the brain or spinal cord are called *efferent* nerves.

4. To be completely accurate, cranial nerves III, VII, IX, and X also carry parasympathetic outflow to specific muscles and glands, and all cranial nerves (except the cranial nerves I and II) probably carry *proprioception*, the sense of self-movement and body position.

5. Because of its influence on heart rate, there are ongoing investigations into the potential for treating heart failure with electrical stimulation of the vagus nerve, but recent clinical trials have reported mixed results. See Premchand et al., "Extended follow-up of patients with heart failure receiving autonomic regulation therapy in the ANTHEM-HF study"; Gold et al., "Vagus nerve stimulation for the treatment of heart failure."

6. Unfortunately, this never worked in practice. Today, a pacemaker is implanted when a heart transplant is performed (see chapter 14).

7. Rosenberg, *Assessing the Healing Powers of the Vagus Nerve*, 29–55; Habib, *Activate Your Vagus Nerve*, 122–169.

8. Kim et al., "Transneuronal propagation of pathologic alpha-synuclein from the gut to the brain models Parkinson's disease."

9. Heiko et al., "Staging of brain pathology related to sporadic Parkinson's disease."

10. Panebianco et al., "Vagus nerve stimulation for partial seizures."

11. The human bodies studied at Georgetown University School of Medicine have been donated by individuals who wish to benefit the living after their death by assisting in medical education. The University appreciates their generosity and treats all donated bodies with the utmost respect. Bodies are ultimately buried or cremated, depending upon the family's wishes. Donors are remembered each year in a memorial Mass conducted in the spring on the Georgetown University campus. Families, friends, and all medical students whose training has benefited from the donation frequently attend the memorial services.

12. The *flexor carpi ulnaris* flexes the wrist; the *flexor digitorum profundus* flexes the ring and pinky fingers.

13. More specifically, it emerges at the C7-T1 spinal segment.

14. This is why damage to the ulnar nerve results in a "clawed hand" and makes the hand useless.

15. The name of the electronic device that can do this is commonly called a *TENS*, an acronym for *transcutaneous electrical nerve stimulation*. It is used to direct electrical current toward specific body parts underlying the skin.

16. G. D. Dawson and J. W. Scott did some of the pioneering work on measuring ulnar nerve action potentials through skin in 1949 (Dawson and Scott, "The recording of nerve action potentials through skin in man").

17. See chapter 2 for a description of the frog experiment.

18. Nerve axons connect with muscles at *neuromuscular junctions*. These highly specialized structures allow motor neurons to transmit their action potentials to muscle fibers. Transmission occurs by means of a neurotransmitter (acetylcholine) that is released by the neuron's cell membrane and then binds to a receptor on the muscle fiber's cell membrane. The binding causes the muscle cell to electrically depolarize, which precipitates a sequence of electrical and molecular events that spread throughout the muscle and cause a coordinated contraction.

19. Georgetown Visitation Preparatory School is not affiliated with Georgetown University.

20. A short presentation about Backyard Brain, delivered by cofounder Greg Gage, can be seen in this video: https://www.youtube.com/watch?v=c5u4k8Spxrk

21. Marzullo and Gage, "The SpikerBox."

22. National Academy of Sciences, "Laboratory experiences and student learning."

23. Feighan, "DIY-ers with a mind-boggling medium."

24. For a comprehensive collection of possible projects, see Baichtal et al., *Make: Lego and Arduino Projects.*

25. Scherz and Monk, *Practical Electronics for Inventors,* 843–895.

26. Yoder, "An Arduino-based alternative to the traditional electronics laboratory."

27. For a video demonstration of the Inebriator dispensing a Voodoo cocktail, see https://www.youtube.com/watch?v=9rSgAu4qYaU

28. They will feel a shock because the ulnar nerve carries both motor and sensory nerve signals—motor signals moving in one direction (from brain to forearm) and sensory signals moving in the opposite direction (forearm to brain).

29. You can see for yourself in this video of a demonstration by cofounder Gregory Gage: https://www.youtube.com/watch?v=c5u4k8Spxrk

30. As far as the cost goes, let's run the numbers. Backyard Brains' "Brain-Machine Interface Classroom Bundle" currently (2020) retails for just $1,000 and contains enough classroom materials to accommodate a class of 30 students. And much of the kit is reusable, so you only need to purchase refill packs for the consumable components of the kit. That means teaching the second class of 30 students will cost only half of the original $1,000 investment. This puts the cost for an ongoing program at less than $25 per student. Will this stretch high school science budgets? Perhaps, but I don't think by much. A kit for frog dissection—the traditional high school lab exercise—costs about $12.50 per student (frog included), and you cannot reuse the dissected frog. But the advent of the synthetic dissection frog may change that too. See https://www.cbsnews.com/news/florida-high-school-introduces-synthetic-frogs-for-dissection-in-science-class/

31. Formerly known as the Intel Science Talent Search (and, before that, the Westinghouse Science Talent Search).

32. See Backyard Brains' artificial hand video demonstration here: https://www.youtube.com/watch?time_continue=8&v=oeoGGj9SDeE&feature=emb_logo

Chapter 13: Crossed Circuits

1. The details of Melissa Loomis's ordeal are chronicled by her surgeon in Seth, *Rewired.*

2. For amputation statistics, see https://advancedamputees.com/amputee-statistics-you-ought-know

3. The leg has since been lost to science. It was held at the Royal College of Surgeons in London but was destroyed by a German air raid during World War II.

4. The reason that the Civil War had a disproportionately greater number of amputees than previous wars can be attributed to the widespread use of the Minié ball, invented by Claude-Étienne Minié. It wasn't really a musket ball at all. Rather, it was a pointed bullet with a hollow base. Unlike a solid ball, which typically passed through the body intact, the Minié ball flattened

and deformed upon impact. It would shatter bones rather than just break them. See https://opinionator.blogs.nytimes.com/2012/08/31/the-bullet-that-changed-history/

5. Figg and Farrell-Beck, "Amputation in the Civil War: Physical and social dimensions."

6. Figg and Farrell-Beck, "Amputation."

7. The medical field is called neuroprosthetics. But, increasingly, the term *neuroprosthetics* is also being used as the terminology for the artificial limbs themselves, rather than using the word *neuroprostheses*. Either word is considered acceptable for describing the limbs.

8. "Carpal" comes from the Latin word *carpus*, which means "wrist."

9. A degree of freedom (DOF) of motion is a direction of displacement that a physical object can move. The number of degrees of freedom of an object is the total number of independent displacements an object can make. There are about 22 DOFs in an arm and hand (depending on whether the wrist is included). For example, the index finger has 4 DOFs: flexion/extension of the three joints from the base to the fingertip, plus the ability to move sideways (abduction-adduction).

10. Sometimes the term *brain-computer interface* (BCI) is used, but BMI is more encompassing because the electronic device need not be a computer.

11. Moran, D., "Brain computer interfaces."

12. Skin electrodes and other surface interfaces act as *electrical transducers*—devices able to convert some type of physical or chemical quantity into a proportional amount of electricity. The transduction actually occurs at the very thin physical interface where the electrode and the electrolyte meet. The flow of ions in the electrolyte produces a flow of electrons in the electrode because of chemical oxidation reactions with the metal atoms at the electrode's surface. The result is that negative ions in the electrolyte flow toward the boundary line of the interface. Positive ions, in contrast, flow away from the interface boundary. To counterbalance this movement of charges, electrons in the electrode flow away from the boundary surface, creating a current within the electrode.

13. The brain also can recruit more muscle cells within the muscle—metaphorically adding people "knocking" on the door. But each of those additional people knock with the exact same force.

14. It is also possible to digitalize an analog electrical signal through a technique called *sampling*. See Sanchez, *Neuroprosthetics*, 34–36.

15. Electrical signals detected at the muscles are called *electromyographic signals*.

16. With amputations below the elbow, some "hand" muscles may remain. This is because the *extrinsic hand muscles* are actually located in the forearm. These hand muscles can remain even after the hand itself is completely gone.

17. Mioton and Dumanian, "Targeted muscle reinnervation and prosthetic rehabilitation after limb loss."

18. Kuiken, "Targeted reinnervation for improved prosthetic function."

19. As of 2016, approximately 300 amputees had undergone successful TMR surgeries (Meek, "Prosthetic limbs").

20. A version of the LUKE arm is now commercially manufactured by Mobius Bionics, although the commercial version does not have all the sensors of the arm described later in this chapter.

21. Delgado-Martinez et al., "Fascicular topography of the human median nerve for neuroprosthetic surgery."

22. The scalp electrode is usually placed over the contralateral primary somatosensory cortex of the brain. When you strap on one of those silly cone-shaped birthday hats, the hat is sitting just above your somatosensory cortex.

23. Some fascicles have a mixture of motor and sensory neurons. They are identified by an attenuated sensory signal on SSEP.

24. The disks are actually little electric motors that have an off-center axis, causing them to wobble (vibrate) when running.

25. Unfortunately, osteointegration surgery for purposes of an advanced prosthetic arm is still in the research phase and not FDA-approved for the general public as of this writing. However, it is approved for leg prostheses.

26. This wireless technology was named Bluetooth by the Danish engineers who developed it. They named it after King Bluetooth, a tenth-century Danish king who is famous for uniting all of Scandinavia. (Legend has it that the king actually had a blue tooth.) The engineers felt that their new technology united computers with their various peripheral devices just as King Bluetooth had united Scandinavia.

27. Seth, *Rewired*, 221.

28. Saal and Bensmala, "Touch as a team effort."

29. The slanted design is considered an improvement over an earlier Utah array design that had flat (i.e., level) electrodes.

30. Sanchez, *Neuroprosthetics*, 26–27; Tyler, "Peripheral nerve stimulation," 335–336.

31. George et al., "Biomimetic sensory feedback through peripheral nerve stimulation improves dexterous use of a bionic hand."

32. Dhar, "Touching moments in prosthetics: New bionic limbs that can feel."

33. Loomis's commercial advanced prosthetic cost $236,000, which was covered by her health insurance, although most American health insurance policies don't cover advanced prosthetics.

Chapter 14: Sounds of Silence

1. Mark Twain and Helen Keller were good friends. She said of Twain, "He treated me not as a freak, but as a handicapped woman seeking a way to circumvent extraordinary difficulties."

2. For statistics on deafness, see https://www.gatecommunications.org/statistics

3. The mutated gene most commonly responsible for congenital deafness is *GJB2*, which codes for a protein called Gap junction beta-2.

4. Eshraghi et al., "The cochlear implant."

5. This question on the nature of sound was first proposed by philosopher George Berkeley in 1710.

6. Lim et al., "Restoring hearing with neuroprostheses."

7. For a simulation of the sound improvement in a 120-channel cochlear implant see this video: https://youtu.be/eo-HNAoCzjw

8. This approach works even for people with damaged hair cells because the next neuron in the chain that transmits the action potentials to the brain is usually still intact. That neuron has been waiting around hoping to relay any signals it might receive from the hair cell up to the brain, but it never had anything to do before because its hair cell wasn't working. But, provided with artificial stimulation via electricity, these neurons are able to relay electrical signals upstairs where the brain can sort them out.

9. Aquilina, "A brief history of cardiac pacing."

10. Heart pacemakers should not be confused with heart defibrillators. *Ventricular fibrillation* (VF) is a potentially fatal arrhythmia of the bottom chambers of the heart. VF is an emergency condition that requires immediate electrical defibrillation with external electrodes in the form of handheld chest paddles. To prevent recurrence, smaller implantable defibrillators are often used. These devices can detect a VF and automatically deliver a strong shock to the heart to reset the normal heart rhythm. Implantable defibrillators look similar to heart pacemakers, but they aren't the same thing. You may wonder how electricity delivered to the heart can be life saving when we know the heart is the target for the lethality of electricity. This turns out to be another serendipitous finding put to good clinical use. Two electrical engineering professors at the Johns Hopkins University during the 1940s, William Kouwenhoven and Guy Knickerbocker, were studying the mechanism of electrically induced cardiac death by shocking the hearts of dogs. They observed that shocking the dog's heart one time would often produce a potentially fatal VF, but a second shock would often "reboot" the heart and restore a normal beat rhythm. But they were simply rediscovering the same phenomenon that had been reported in chickens two centuries earlier. In 1775, Peter Christian Abildgaard, a Danish physician, conducted experiments in which he would first render a chicken lifeless with electricity and then revive it with a "countershock" applied to the chest. Although Abildgaard's ideas about how the countershock worked were wrong, his experiments were likely the first demonstration of electrical defibrillation of the heart (Driscol et al., "The remarkable Dr. Abildgaard and countershock").

11. Inspire is manufactured by Inspire Medical Systems, based in Minneapolis, MN.

12. Wilson and Dorman, "Cochlear implants."

13. *Clarion Ledger*, "Toddler among youngest cochlear implant patients in world."

14. Mitchell et al., "Auditory comprehension outcomes in children who receive a cochlear implant before 12 months of age."

15. The auditory nerve has a secondary function in maintaining balance.

16. If you are among the few people who can actually wiggle their ears, that's because you have good control of your *auricular muscles*, which are controlled by the temporal branch of the facial nerve—the seventh (or VIIth) cranial nerve. Your auditory nerve has nothing to do with it.

17. Kaplan et al., "Auditory brainstem implant candidacy in the United States in children 0–17 years old."

18. For detailed descriptions of the latest advancements in artificial vision technology, see Humayun et al. (eds.), *Artificial Sight*.

19. Scientists call the eye an *immune privileged* site, meaning that it has a number of physical barriers to entry by the body's immune system, which results in the tolerance of certain

materials within the eye that otherwise would be subject to immune rejection if elsewhere in the body.

20. *Occipital* comes from the Latin word *occiput*, which literally means "back of the head."

21. Fernandez, "Development of visual neuroprostheses."

22. See list of approved systems in Fernandez, "Development of visual neuroprostheses."

23. Nirenberg and Pandarinath, "Retinal prosthetic strategy with the capacity to restore normal vision."

24. An interesting irony is that the computer programs used to decode nervous system signals often rely on an *artificial neural network*, which is a machine learning algorithm modeled upon the way the human brain teaches itself to learn new information. I describe the learning strategy of artificial neural networks further in chapter 16. See Schwemmer et al., "Meeting brain-computer interface user performance expectations using a deep neural network decoding framework."

25. Yan and Nirenberg, "An embedded real-time processing platform for optogenetic neuroprosthetic applications."

26. A description of the Orion feasibility study can be found at https://clinicaltrials.gov/ct2/show/NCT03344848

27. To learn more about McDonald's experience with his Orion implant, see his autobiographical account: McDonald, *My Brain Implant for Bionic Vision*.

28. Bushdid et al., "Humans can discriminate more than 1 trillion olfactory stimuli."

29. Studies have also shown that loss of the sense of smell can produce lasting depression (Seo et al., "Influences of olfactory impairment on depression, cognitive performance, and quality of life in Korean elderly").

30. Macpherson, "Sensory substitution and augmentation."

31. See a BrainPort video demonstration at this website: https://www.youtube.com/watch?v=CNR2gLKndog

32. Twilley, "Sight unseen."

Chapter 15: Inner Sanctum

1. Sengupta and Stemmler, "Power consumption during neuronal computation."

2. Recent work with cochlear implants supports this contention. When the electronic voltage pulses from cochlear implants mimicked the shape of neuronal pulses, energy requirements went down. This change in pulse shape could lead to more battery-efficient cochlear implants. See Navntoft et al., "Ramped pulse shapes are more efficient for cochlear implant stimulation in an animal model."

3. Sengupta et al., "The effect of cell size and channel density on neuronal information encoding and energy efficiency."

4. For other examples of how nervous systems optimize their energy efficiency, see Sengupta and Stemmler, "Power consumption during neuronal computation."

5. Cobb, "Why your brain is not a computer."

6. For a review of current state-of-the-art cellular techniques used to study the cognitive properties of the brain, see Fried et al., *Single Neuron Studies of the Brain*.

7. The drop in electrical potential is a function of the distance between the neural tissue and the electrode. The voltage drops by one over the square root of the distance between the two, or $1/d^2$. See Sanchez, *Neuroprosthetics*, 20.

8. Some scientists estimate an area of 6 cm^2 (about 1 square inch, or the size of a typical postage stamp) of cortical tissue must fire in synchrony in order to be detected by a scalp electrode (Schwartz et al., "Brain-controlled interfaces").

9. Schwartz et al., "Brain-controlled interfaces."

10. Berger, "[On the electroencephalogram of man]."

11. Fields, *Electric Brain*, 159–160.

12. Between seizures, the EEGs of patients with epilepsy can show a pattern of activity designated as an *interictal epileptiform discharge* (IED). But the sensitivity of IED detection for diagnosing epilepsy is less than 50%.

13. Smith, "EEG in the diagnosis, classification, and management of patients with epilepsy."

14. EKG is commonly used rather than ECG (*C* for *cardio*) because, in spoken medical orders, ECG and EEG sound too similar and would, therefore, cause confusion in a clinical setting. The *K* comes from the German word *kardio*.

15. Although EEG hypersynchrony has long been associated with epilepsy, recent studies suggest the development of the hypersynchronous state in epileptics is a more dynamic and complicated process than previously believed. See Jiruska et al., "Synchronization and desynchronization in epilepsy."

16. For an accessible science book about brain waves that portrays their potential value to both fundamental brain research and the therapy of mental disorders in a more positive light than I have, see Fields, *Electric Brain*.

17. Shorter and Healy, *Shock Therapy*.

18. With medications (namely, camphor and pentylenetetrazol), the seizures were not immediate. There was typically a period of many minutes during which the patient felt increasingly anxious until the seizure occurred. Also, medically induced seizures were often prolonged and the convulsions violent. Furthermore, there were frequent cardiac complications. Many patients developed an intense fear of medically induced seizures and would refuse to have more. ECT has none of these issues.

19. UK ECT Review Group, "Efficacy and safety of electroconvulsive therapy in depressive disorders."

20. Gazdag and Ungvari, "Electroconvulsive therapy"; Leiknes et al., "Contemporary use and practice of electroconvulsive therapy worldwide."

21. Hermida et al., "Electroconvulsive therapy in depression."

22. The typical patient starts with a mean threshold dose of about 170 millicoulombs, and that value more than doubles during the course of ECT treatment (Duthie et al., "Anticonvulsive mechanisms of electroconvulsive therapy and relation to therapeutic effect").

23. Bouckaert et al., "ECT."

24. Chang et al., "Narp mediates antidepressant-like effects of electroconvulsive seizures."

25. You may be wondering how one knows when a mouse is depressed. Since it is difficult to say whether a rodent is "depressed," protocols like the forced-swim test, sucrose preference

test, tail suspension test, and others are generally used in animal models of depression. Terms like "depression-like behavior" or "antidepressant-like effect" tend to be preferred as well. The Johns Hopkins researchers used the standardized mouse "swim test," an accepted metric to assess the depression level of a mouse. When placed in water, a depressed mouse will spend more time just floating than actively swimming, compared to a nondepressed mouse. During a 6-minute test, healthy mice spend about 50 seconds floating, whereas the mice without Narp spent about 80 seconds floating.

26. Johns Hopkins Medicine. "How electroconvulsive therapy relieves depression per animal experiments."

27. As we discussed in chapter 6, amperage, not voltage, is the major driver of biological effects. The coulomb unit captures the amperage component of the treatment because a coulomb (as defined in chapter 9) is the amperage of the current multiplied by the time the current is on (coulombs = amps × seconds). In short, a coulomb is a measure of the duration that a patient was exposed to a specific amperage during the treatment.

28. Rajagopal et al., "Satisfaction with electroconvulsive therapy among patients and their relatives."

29. These criteria are fundamental to all randomized clinical trials. But the gold standard is the "double-blind" randomized clinical trial, in which neither the patient nor the physician knows which patient is getting which treatment. The double-blind trials are seen as superior because neither patient nor physician can intentionally or unintentionally influence the outcomes of the different treatments since neither knows what treatment group a particular patient is in. In practice, it is easier to blind the patients (single-blind) than it is the physicians. So achieving double-blind is not always possible.

30. For a fuller discussion of the power of randomization to reveal the truth in a variety of questions even beyond assessing medical treatments, I recommend this entertaining and accessible book: Leigh, *Randomistas: How Radical Researchers Are Changing the World.*

31. Based on the case descriptions, it's possible that some of Cerletti's early schizophrenia patients treated with ECT actually were suffering from psychotic depressions or catatonic episodes. For example, the first ECT patient (the man wandering in a train station in Rome) might be considered catatonic by today's criteria. ECT is thought to be an effective treatment for catatonia (a syndrome that can occur in schizophrenia, depression, mania, or medical illnesses).

32. Sinclair et al., "Electroconvulsive therapy for treatment-resistant schizophrenia."

33. Ray, "Does electroconvulsive therapy cause epilepsy?"

34. Zanchetti et al., "The effect of vagal afferent stimulation on the EEG pattern of the cat."

35. Binnie, "Vagus nerve stimulation."

36. The historical importance of cortical stimulation mapping is reviewed in Catani, "A little man of some importance."

37. The term *homunculus* originates with the sixteenth-century physician Paracelsus, who famously claimed that a deformed little man, whom he called a homunculus, could be produced by incubating putrefied human sperm in the womb of a horse. It's unfortunate that Paracelsus is remembered for this nonsense. He was otherwise far ahead of his time in his medical ideas, particularly in the field of toxicology.

38. Kogan et al., "Deep brain stimulation for Parkinson disease."

39. This abbreviated story of the birth of deep brain stimulation doesn't do justice to the rich history of both the technology and neurosurgery advances that came into play and the many players who were involved. For a comprehensive review of all aspects of deep brain stimulation, see Perlmutter and Mink, "Deep brain stimulation."

40. Schuurman et al., "A comparison of continuous thalamic stimulation and thalamotomy for suppression of severe tremor."

41. For a history of the use of electricity to stimulate the pleasure centers of the brain, see Frank, *The Pleasure Shock*.

42. An exception to this is repetitive transcranial magnetic stimulation (rTMS), which is an FDA-approved treatment for depression, headaches, and obsessive-compulsive disorder. It is believed that rTMS works by producing a localized electric current in a specific area of the brain through an electromagnetic induction mechanism. A changing magnetic field applied to the brain causes current to flow within the brain, just as Michael Faraday was able to get current to flow in a wire just by changing the magnetic field around the wire (see chapter 7).

Chapter 16: Future Shock

1. Although different versions of this quote have been attributed variously to Mark Twain, Yogi Berra, and Niels Bohr, among others, linguistic evidence suggests that they are all English translations of an old Danish proverb that says the same thing: "Det er vanskeligt at spaa, især naar det gælder Fremtiden."

2. A video recording of the Neuralink Launch Event can be viewed here: https://youtu.be/r-vbh3t7WVI

3. For a biography of Musk and a description of his various business ventures, see Vance, *Elon Musk*.

4. Cross, "The novelist who inspired Elon Musk."

5. If you are an artificial intelligence programmer and you feel that my greatly simplified description of artificial neural networks does your field a great disservice, I apologize. If you are not a programmer but would like to learn more about artificial intelligence, I recommend this excellent book by Melanie Mitchell, which is written in a thorough and yet very accessible style: *Artificial Intelligence: A Guide for Thinking Humans*.

6. Musk and Neuralink, "An integrated brain-machine interface platform with thousands of channels."

7. Russell, *Human Compatible*, 164.

8. Russell, 165.

9. Russell, 165.

10. Hires's quotes come from a Business Insider interview, "Inside the science behind Elon Musk's crazy plan to put chips in people's brains and create human-AI hybrids."

11. Bullard et al., "Estimating risk for future intracranial, fully implanted, modular neuroprosthetic systems."

12. Healey et al., "Complications in surgical patients."

13. On August 28, 2020, Musk provided an update on Neuralink's progress. He reported that the company's device had been implanted in the brains of three pigs, claiming, "Pigs are quite similar to people. If we're going to figure out things for people, then pigs are a good choice." He also said they had scrapped the behind-the-ear receiver idea in favor of a fully implanted receiving device, a change in receiver technology that seems much more invasive than what he had previously described. He further reported that Neuralink had recently secured an FDA "breakthrough device" designation, paving the way for the first implant in a human, but he backed away from an updated projection as to exactly when that human milestone would be achieved. In January of 2021, Musk predicted Neuralink's human clinical trials would begin by the end of 2021. On April 8, 2021, Neuralink reported that it had also implanted its device in the brain of a Macaque monkey and had trained that monkey to play the video game Pong using only its brain to control the joystick. The company even released a video of the monkey playing Pong that is currently posted on YouTube: https://www.youtube.com/watch?v=rsCul1sp4hQ&t=6s.

14. For the company's description of its Amber software, see https://www.myamberlife.com/learn/how-ambers-ai-technology-works/

BIBLIOGRAPHY

Alturkistani, H. A., F. M. Tashkandi, and Z. M. Mohammedsaleh. "Histological stains: A literature review and case study." *Glob J Health Sci* 8(3): 72–79, 2015.

Amdahl, K. *There Are No Electrons*. Broomfield, CO: Clearwater Publishing, 1991.

Anderson, W. H. "Tetrodotoxin and the zombie phenomenon." *Journal of Ethnopharmacology* 23: 121–126, 1988.

Andrews, C. "Electrical aspects of lightning strikes to humans." Pages 701–723 (chapter 16) in Cooray, V. (ed.), *The Lightning Flash*. London: Institution of Engineering and Technology, 2014.

Andrews, C. J., M. Darveniza, and D. Mackerras. "Lightning injury: A review of clinical aspects, pathophysiology, and treatment." *Adv Trauma* 4: 241, 1989.

Aquilina, O. "A brief history of cardiac pacing." *Images Paediatr Cardiol* 8(2): 17–81, 2006.

Arabatzis, T. *Representing Electrons: A Biographical Approach to Theoretical Entities*. Chicago: University of Chicago Press, 2006.

Ashcroft, F. *The Spark of Life: Electricity and the Human Body*. New York: W. W. Norton, 2012.

Assis, A.K.T. *The Experimental and Historical Foundations of Electricity*. Montreal: C. Roys Keys, 2010.

Auburn Daily Advertiser. "Kemmler." August 12, 1890.

Baichtal, J., M. Beckler, and A. Wolf. *Make: Lego and Arduino Projects*. Sebastopol, CA: Maker Media, 2013.

Beaumont, W.R.C. *Electricity in Fish Research and Management: Theory and Practice*. 2nd ed. Chichester, UK: Wiley & Sons, 2016.

Berg, C., and M. Raj. "A review of different stunning methods for poultry—animal welfare aspects (stunning methods for poultry)." *Animals* 5: 1207–1219, 2015.

Berger, H. "[On the electroencephalogram of man]." *Archiv für Psychiatrie* 87: 527–570, 1929.

Bikson, M. "A review of hazards associated with exposure to low voltages." Department of Biomedical Engineering, Graduate School and University Center of the City University of New York, 2004. http://bme.ccny.cuny.edu/faculty/mbikson/BiksonMSafeVoltageReview.pdf

Binnie, C. D. "Vagus nerve stimulation: A review." *Seizure* 9: 161–169, 2000.

Blancheteau, M., P. Lamarque, G. Mousset, and R. Vibert. "Etude neurophysiologique de la pêche électrique." *Bull Cent Etud Rech Sci Biarritz* 3(3): 277–382, 1961. [French with English abstract.]

Bonner, T. N. *Iconoclast: Abraham Flexner and a Life in Learning*. Baltimore, MD: Johns Hopkins University Press, 2002.

Bouckaert, F., P. Sienaert, J. Obbels, A. Dols, M. Vandenbulcke, M. Stek, and T. Bolwig. "ECT: Its brain enabling effects." *Journal of ECT* 30(2): 143–151, 2014.

Brown, H. P. "Death in the wires." *New York Evening Post*, June 5, 1888.

Buffalo Evening News. "Never mind the intentions." August 8, 1890.

Bullard, A. J., B. C. Hutchison, J. Lee, C. A. Chestek, and P. G. Patil. "Estimating risk for future intracranial, fully implanted, modular neuroprosthetic systems: A systematic review of hardware complications in clinical deep brain stimulation and experimental human intracortical arrays." *Neuromodulation: Technology and the Neural Interface* 23(4): 411–426, 2020.

Bushdid, C., M. O. Magnasco, L. B. Vosshall, and A. Keller. "Humans can discriminate more than 1 trillion olfactory stimuli." *Science* 343(6177): 1370–1372, 2014.

Business Insider (website). "Inside the science behind Elon Musk's crazy plan to put chips in people's brains and create human-AI hybrids." October 6, 2019. https://www.businessinsider.in/tech/news/inside-the-science-behind-elon-musks-crazy-plan-to-put-chips-in-peoples-brains-and-create-human-ai-hybrids/articleshow/71463463.cms

Carlson, B. A. "Animal behavior: Electric eels amp up for an easy meal." *Current Biology* 25(22): R1070–1072, 2015.

Carlson, W. B. *Tesla: Inventor of the Electrical Age*. Princeton, NJ: Princeton University Press, 2013.

Catani, M. "A little man of some importance." *Brain* 140(11): 3055–3061, 2017.

Catania, K. C. "Electric eels concentrate their electric field to induce involuntary fatigue in struggling prey." *Current Biology* 25(22): 2889–2898, 2015.

Chang, A. D., P. V. Vaidya, E. P. Retzbach, S. J. Chung, U. Kim, K. Baselice, K. Mynard, et al. "Narp mediates antidepressant-like effects of electroconvulsive seizures." *Neuropsychopharmacology* 43: 1088–1098, 2018.

Charlier, P. *Zombies: An Anthropological Investigation of the Living Dead*. Gainesville, FL: University Press of Florida, 2015.

Clarion Ledger. "Toddler among youngest cochlear implant patients in world." April 28, 2018.

Cleaves, M. A. "Franklinization as a therapeutic measure in neurasthenia." *JAMA* XXVII(20): 1043–1052, 1896.

Cobb, M. "Why your brain is not a computer." *The Guardian*, February 27, 2020.

Cohen, I. B. *Benjamin Franklin's Science*. Cambridge, MA: Harvard University Press, 1990.

Cooper, M. A. "Emergent care of lightning and electrical injuries." *Semin Neurol* 15(3): 268–278, 1995.

Cowx, I. G. (ed.). *Developments in Electric Fishing*. Oxford, UK: Fishing News Books, 1990.

Cowx, I. G., and P. Lamarque (eds.). *Fishing with Electricity: Applications in Freshwater Fisheries and Management*. Oxford, UK: Fishing News Books, 1990.

Cross, T. "The novelist who inspired Elon Musk." *The Economist*, March 31, 2017.

Dahlström, Å., and L. Brost. *The Amber Book*. Tucson, AZ: Geoscience Press, 1996.

Davis, E. W. "The ethnobiology of the Haitian zombi." *Journal of Ethnopharmacology* 9: 85–104, 1983.

Davis, W. *Passage of Darkness: The Ethnobiology of the Haitian Zombie*. Chapel Hill: University of North Carolina Press, 1988.

———. *The Serpent and the Rainbow: A Harvard Scientist's Astonishing Journey in the Secret Societies of Haitian Voodoo, Zombis, and Magic*. New York: Simon & Schuster, 1985.

Dawson, G. D., and J. W. Scott. "The recording of nerve action potentials through skin in man." *J Neurol Neurosurg Psychiat* 12: 259267, 1949.

De La Pena, C. T. *The Body Electric: How Strange Machines Built the Modern American*. New York: New York University Press, 2003.

Delgado-Martinez, I., J. Badia, A. Pascual-Font, A. Rodriguez-Baeva, and X. Navarro. "Fascicular topography of the human median nerve for neuroprosthetic surgery." *Frontiers in Neuroscience* 10(286): 1–13, 2016.

Del Guercio, G. "From the archives: The secret of Haiti's living dead." *Harvard Magazine*, October 31, 2017.

Dhar, P. "Touching moments in prosthetics: New bionic limbs that can feel." *Washington Post*, December 17, 2019.

Dorman, T., N. Allison, and A. Tamburin. "Tennessee execution: Nicholas Todd Sutton executed by electric chair." *Tennessean*, February 20, 2020.

Driscol, T. E., O. D. Ratnoff, and O. F. Nygaard. "The remarkable Dr. Abildgaard and countershock: The bicentennial of his electrical experiments on animals." *Ann Intern Med* 83(6): 878–882, 1975.

Duchenne, G. B. *A Treatise on Localized Electrization, and Its Applications to Pathology and Therapeutics*. Miami, FL: HardPress Publishing, 2013. [Originally published in 1871 by Lindsay & Blakiston, Philadelphia.]

Duthie, A. C., J. S. Perrin, D. M. Bennett, J. Currie, and I. C. Reid. "Anticonvulsive mechanisms of electroconvulsive therapy and relation to therapeutic effect." *Journal of ECT* 31(3): 173–178, 2015.

Dwyer, J. R., and U. A. Uman. "The physics of lightning." *Physics Reports* 534: 147–241, 2014.

Elliott, R. G. *Agent of Death: Memoirs of an Executioner*. New York: E. P. Dutton, 1940.

Elson, D. M. "Striking reduction in annual number of lightning fatalities in the United Kingdom since the 1850s." *Weather* 70(9): 251–257, 2015.

Eshraghi, A. A., R. Nazarian, F. F. Telischi, S. M. Rajguru, E. Truy, and C. Gupta. "The cochlear implant: Historical aspects and future prospects." *Anat Rev* (Hoboken) 295(11): 1967–1980, 2012.

Fairley, P. "Edison's revenge: The rise of DC power." *MIT Technology Review*, April 24, 2012.

Faraday, M. *Chemical Manipulations: Instructions to Students in Chemistry*. London: John Murray, 1842.

———. "Experimental researches in electricity—fifteenth series." *Philosophical Transactions of the Royal Society of London* 129: 1–12, 1839.

———. "Thoughts on ray-vibrations." *Philosophical Magazine* S.3, vol. XXVIII, N188, May 1846.

Feighan, M. "DIY-ers with a mind-boggling medium: The brain." *Detroit News*, September 20, 2018.

Fernandez, E. "Development of visual neuroprostheses: Trends and challenges." *Bioelectronic Medicine* 4(1): 12, 2018.

Fields, R. D. *Electric Brain: How the New Science of Brainwaves Reads Minds, Tells Us How We Learn, and Helps Us Change for the Better*. Dallas, TX: BenBella Books, 2020.

Figg, L., and J. Farrell-Beck. "Amputation in the Civil War: Physical and social dimensions." *Journal of the History of Medicine and Allied Sciences* 48: 456–463, 1993.

Finger, S., and M. Piccolino. *The Shocking History of the Electric Fishes*. Oxford, UK: Oxford University Press, 2011.

Fisher, H. J. *Faraday's Experimental Researches in Electricity: Guide to First Reading*. Sante Fe, NM: Green Lion Press, 2001.

Fisher, R. A. *The Design of Experiments*. Edinburgh: Oliver and Boyd, 1935.

Flexner, A. *The American College: A Criticism*. New York: Century Co., 1908.

———. *Medical Education in the United States and Canada: A Report to the Carnegie Foundation for the Advancement of Science*. Boston: D. B. Updike, Merrymount Press, 1910.

Forbes, N., and B. Mahon. *Faraday, Maxwell, and the Electromagnetic Field: How Two Men Revolutionized Physics*. Amherst, NY: Prometheus Books, 2014.

Frank, L. *The Pleasure Shock: The Rise of Deep Brain Stimulation and Its Forgotten Inventor*. New York: Dutton, 2018.

Franklin, B. *Experiments and Observations on Electricity, Made at Philadelphia in America, by Mr. Benjamin Franklin, and Communicated in Several Letters to Mr. P. Collinson, of London, F.R.S.* London: E. Cave at St. John's Gate, 1751.

Fried, I., U. Rutishauser, M. Cerf, and G. Kreiman (eds.). *Single Neuron Studies of the Brain: Probing Cognition*. Cambridge, MA: MIT Press, 2014.

Galvin, A. *Old Sparky: The Electric Chair and the History of the Death Penalty*. New York: Carrel Books, 2015.

Garratt, A. C. *Electro-Physiology and Electro-Therapeutics: Showing the Best Methods for the Medical Uses of Electricity*. 2nd ed. Boston: Ticknor & Fields, 1861.

Gazdag, G., and G. S. Ungvari. "Electroconvulsive therapy: 80 years old and still going strong." *World J Psychiatr* 9(1): 1–6, 2019.

Gensel, L. "The medical world of Benjamin Franklin." *J R Soc Med* 98(12): 534–538, 2005.

George, J. A., D. T. Kluger, T. S. Davis, S. M. Wendelken, E. V. Okorokova, Q. He, C. C. Duncan, et al. "Biomimetic sensory feedback through peripheral nerve stimulation improves dexterous use of a bionic hand." *Science Robotics* 4(32): eaax2352, 2019.

Gezgin, T., and M. Karakaya. "The effects of electrical water bath stunning on meat quality of broiler produced in accordance with Turkish slaughter procedures." *Journal of Poultry Research* 13(1): 22–26, 2016.

Gillmor, C. S. *Coulomb and the Evolution of Physics and Engineering in Eighteenth-Century France*. Princeton, NJ: Princeton University Press, 1971.

Gizmodo (website). "The flying boy experiment entertained audiences by electrifying a kid." January 15, 2015. https://io9.gizmodo.com/the-flying-boy-experiment-entertained-audiences-by-elec-1679627835

———. "This electric eel kills its prey with a sophisticated coiling maneuver." October 30, 2015. https://gizmodo.com/this-electric-eel-kills-its-prey-with-a-sophisticated-c-1739230210

Gleick, J. *Isaac Newton*. New York: Vintage, 2004.

Glickstein, M. *Neuroscience: A Historical Introduction*. Cambridge, MA: MIT Press, 2014.

Gold, M. R., D. J. Van Veldhuisen, P. J. Hauptman, M. Borggrefe, S. H. Kubo, R. A. Lieberman, G. Milasinovic, et al. "Vagus nerve stimulation for the treatment of heart failure: The INOVATE-HF trial." *J Am Coll Cardiol* 68(2): 149–158, 2016.

Gookin, J. *Lightning*. Mechanicsburg, PA: Stackpole Books, 2014.

Gray, J. *Electrical Influence Machines: A Full Account of Their Historical Development, and Modern Forms, with Instructions for Making Them*. London: Chiswick Press, 1890.

Guttman, L. "A review of Mr. Harold P. Brown's experiments," *Electrical World* 14: 25–26, 1889.

Habib, N. *Activate Your Vagus Nerve: Unleash Your Body's Natural Ability to Heal*. Berkeley, CA: Ulysses Press, 2019.

Healey, M. A., S. R. Shackford, T. M. Osler, F. B. Rogers, and E. Burns. "Complications in surgical patients." *Arch Surg* 137(5): 611–617, 2002.

Heathcote, N. H. de V. "Franklin's introduction to electricity." *Isis* 46(1): 29–35, 1955.

Heiko, B., K. Del Tredici, U. Rüba, R.A.I deVos, E.N.H. Jansen Steur, and E. Braaka. "Staging of brain pathology related to sporadic Parkinson's disease." *Neurobiology of Aging* 24(2): 197–211, 2003.

Heilbron, J. L. *Electricity in the 17th and 18th Centuries: A Study of Early Modern Physics*. Berkeley: University of California Press, 1979.

Hermida, A. P., O. M. Glass, H. Shafi, and W. M. McDonald. "Electroconvulsive therapy in depression: Current practice and future directions." *Psychiatr Clin N Am* 41: 341–353, 2018.

Hines, T. *Pseudoscience and the Paranormal*. Amherst, NY: Prometheus Books, 2003.

Hodgkin, A. L., and A. F. Huxley. "A quantitative description of membrane current and its application to conduction and excitation in nerve." *J Physiol* 117(4): 500–544, 1952.

Hopp, N.P.S. *Amber: Jewelry, Art, and Science*. Atglen, PA: Schiffer Publishing, 2009.

Horn, R. "How ion channels sense membrane potential." *Proceedings of the National Academy of Sciences, USA* 102(14): 4929–4930, 2005.

Hossenfelder, S. *Lost in Math*. New York: Basic Books, 2018.

Humayun, M. S., J. D. Weiland, G. Chader, and E. Greenbaum (eds.). *Artificial Sight: Basic Research, Biomedical Engineering, and Clinical Advances*. New York: Springer, 2007.

Humboldt, A. "[Hunt and fight of electric eels with horses]." *Annalen der Physik* 2(1): 34–43, 1807. https://doi.org/10.1002/andp.18070250103

Hyldgaard, L., E. Søndergaard, and P. M. Leth. "Autopsies of fatal electrocutions in Jutland." *Scandinavian Journal of Forensic Science*. 10(1): s8–12, 2004.

IEEE Spectrum. "DC microgrids and the virtues of local electricity." February 6, 2014. https://spectrum.ieee.org/green-tech/buildings/dc-microgrids-and-the-virtues-of-local-electricity

Independent. "Two of the longest and biggest lightning strikes on earth recorded." September 22, 2016. https://www.independent.co.uk/news/science/lightning-strikes-world-record-longest-biggest-recorded-a7323251.html

Inside Science. "The Science That Made Frankenstein." October 27, 2010. https://www.insidescience.org/news/science-made-frankenstein

Inspector General, Department of Defense. Report No. IE-2009–006. "Review of electrocution deaths in Iraq: Part I—electrocution of Staff Sergeant Ryan D. Maseth, U.S. Army," July 24, 2009.

———. Report No. IPO-2009-E001. "Review of electrocution deaths in Iraq: Part II—seventeen incidents apart from Staff Sergeant Ryan D. Maseth, U.S. Army," July 24, 2009.

Isaacson, W. *Benjamin Franklin: An American Life*. New York: Simon & Schuster, 2003.

Jakobsson, E., O. Argüello-Miranda, S.W. Chiu, Z. Fazal, J. Kruczek, S. Nunez-Corrales, S. Pandit, and L. Pritchet. "Towards a unified understanding of lithium action in basic biology and its significance for applied biology." *Journal of Membrane Biology* 250(6): 587–604, 2017. https://doi.org/10.1007/s00232-017-9998-2

James, F.A.J.L. *Michael Faraday: A Very Short Introduction*. Oxford, UK: Oxford University Press, 2010.

James, F.A.J.L. (ed.). *The Correspondence of Michael Faraday* (vol. 2). London: Institution of Electrical Engineers, 1993.

Jiruska, P., M. de Curtis, J.G.R. Jefferys, C. A. Schevon, S. J. Schiff, and K. Schindler. "Synchronization and desynchronization in epilepsy: Controversies and hypotheses." *J Physiol* 591(4): 787–797, 2013.

Johns Hopkins Medicine (website). "How electroconvulsive therapy relieves depression per animal experiments." December 18, 2017. https://www.hopkinsmedicine.org/news/media/releases/how-electroconvulsive-therapy-relieves-depression-per-animal-experiments

Johnson, G. *The Ten Most Beautiful Experiments*. New York: Vintage Books, 2009.

Jorgensen, T. J. *Strange Glow: The Story of Radiation*. Princeton, NJ: Princeton University Press, 2016.

Kaplan, A. B., E. D. Kozin, S. V. Puram, M. S. Owoc, P. V. Shah, A. E. Hight, K. V. Rosh, R.K.V. Sethi, A. K. Remenschneider, and D. J. Lee. "Auditory brainstem implant candidacy in the United States in children 0–17 years old." *Int J Pediatr Otorhinolaryngol* 79(3): 310–315, 2015.

Keynes, R. "J.Z. and the discovery of squid giant nerve fibres." *Journal of Experimental Biology* 208: 179–180, 2005.

Kim, S., S. H. Kwon, T. I. Kam, N. Panicker, S. S. Karuppagounder, S. Lee, J. H. Lee, et al. "Transneuronal propagation of pathologic alpha-synuclein from the gut to the brain models Parkinson's disease." *Neuron* 103: 627–641, 2019.

King, W. H. *Electricity in Medicine and Surgery, Including the X Ray*. New York: Boericke & Runyon, 1901.

Klein, J. "Testing the electric eel's shock powers with his own arm." *New York Times*, September 14, 2017.

Kleinzeller, A., and P. F. Baker (eds.). *The Squid Axon (Vol. 22 of Current Topics in Membranes and Transport)*. New York: Academic Press, 1984.

Kogan, M., M. McGuire, and J. Riley. "Deep brain stimulation for Parkinson disease." *Neurosurg Clin N Am* 30: 137–146, 2019.

Krider, P. "Benjamin Franklin and lightning rods." *Physics Today* 59(1): 42, 2006.

Kuiken, T. "Targeted reinnervation for improved prosthetic function." *Phys Med Rehabil Clin N Am* 17(1): 1–13, 2006.

Lamarque, P. "Electrophysiology of fish subject to the action of an electric field." Pages 65–92 in Vibert, R. (ed.), *Fishing with Electricity, Its Application to Biology and Management*. London: Fishing News Books, 1967.

Lang, T. J., S. Pédeboy, W. Rison, R. S. Cerveny, J. Montanyà., S. Chauzy, D. R. MacGorman, et al. "WMO world record lightning extremes: Longest reported flash distance and longest reported flash duration." *Bull Am Meteorol Soc* 98(6): 1153–1168, 2017.

Leigh, A. *Randomistas: How Radical Researchers Are Changing the World*. New Haven, CT: Yale University Press, 2018.

Leiknes, K. A., L. Jarosh-von Schweder, and B. Høie. "Contemporary use and practice of electroconvulsive therapy worldwide." *Brain and Behavior* 2(3): 283–344, 2012.

Licht, S. "History of electrotherapy." Chapter I in Licht, S. (ed.), *Therapeutic Electricity and Ultraviolet Radiation*. 2nd ed. Baltimore, MD: Waverly Press, 1967.

Liddell, E.G.T. *The Discovery of Reflexes*. Oxford, UK: Oxford University Press, 1960.

Lim, H. H., M. E. Adams, P. B. Nelson, and A. J. Oxenham. "Restoring hearing with neuroprostheses: Current status and future directions" Chapter 3.6 in Horch, K., and Kipke, D. (eds.), *Neuroprosthetics: Theory and Practice*. 2nd ed. Series on Bioengineering and Biomedical Engineering—vol. 8. Singapore: World Scientific Publishing, 2017.

Littlewood, R., and Douyon, C. "Clinical findings in three cases of zombification." *Lancet* 350: 1094–1096, 1997.

López, R. E., and R. L. Holle. "Changes in the number of lightning deaths in the United States during the twentieth century." *Journal of Climate* 11: 2070–2077, 1998.

Macpherson, F. "Sensory substitution and augmentation: An introduction." Chapter 1 in Macpherson, F. (ed.), *Sensory Substitutions and Augmentation*. Oxford, UK: Oxford University Press, 2018.

Marchant, J. *Cure: A Journey into the Science of Mind Over Body*. New York: Broadway Books, 2016.

Marrin, R. B. *A Glance Back in Time: Life in Colonial New Jersey (1704–1770) as Depicted in News Accounts of the Day*. Bowie, MD: Heritage Books, 1994.

Marzullo, T. C., and G. J. Gage. "The SpikerBox: A low cost, open-source bioamplifier for increasing public participation in neuroscience inquiry." *PLoS ONE* 7(3): e30837, 2012.

Maxwell, J. C. "On Faraday's lines of force." *Trans Camb Phil Soc* X: 27–83, 1856.

———. *A Treatise on Electricity and Magnetism*. Oxford, UK: Clarendon Press, 1873.

McDonald, R. B. *My Brain Implant for Bionic Vision: The First Trial for Artificial Sight for the Blind*. Middletown, DE: Self-published, 2019.

McNichol, T. *AC/DC: The Savage Tale of the First Standards War*. San Francisco: Jossey-Bass Wiley, 2008.

Meek, S. G. "Prosthetic limbs." Chapter 4.1 in Horch, K., and Kipke, D. (eds.), *Neuroprosthetics: Theory and Practice*. 2nd ed. Series on Bioengineering and Biomedical Engineering—vol. 8. Singapore: World Scientific Publishing, 2017.

Millikan, R. A. *The Electron, Its Isolation and Measurement and the Determination of Some of Its Properties*. Miami, FL: HardPress Publishing, 2012. [Originally published in 1917 by University of Chicago Press, Chicago.]

Mioton, L. M., and G. A. Dumanian. "Targeted muscle reinnervation and prosthetic rehabilitation after limb loss." *Journal of Surgical Oncology* 118(5): 807–814, 2018.

Mitchell, M. *Artificial Intelligence: A Guide for Thinking Humans*. New York: Farrar, Straus and Giroux, 2019.

Mitchell, R. M., E. Christianson, R. Ramirez, F. M. Onchiri, D. L. Horn, L. Pontis, C. Miller, S. Norton, and K.C.Y. Sie. "Auditory comprehension outcomes in children who receive a cochlear implant before 12 months of age." *Laryngoscope* 130(3): 776–781, 2020.

Mizzi, C. A., A.Y.W. Lin, and L. D. Marks. "Does flexoelectricity drive triboelectricity?" *Physical Review Letters* 123: 116103, 2019.

Modern Farmer. "Here's why a chicken can live without its head." August 14, 2014. https://modernfarmer.com/2014/08/heres-chicken-can-live-without-head/

Moran, D. "Brain computer interfaces." Chapter 4.3 in Horch, K., and Kipke, D. (eds.), *Neuroprosthetics: Theory and Practice*. 2nd ed. Series on Bioengineering and Biomedical Engineering—vol. 8. Singapore: World Scientific Publishing, 2017.

Moran, R. *Executioner's Current: Thomas Edison, George Westinghouse, and the Invention of the Electric Chair*. New York: Vintage Books, 2002.

Morley, E. L., and D. Robert. "Electric fields elicit ballooning in spiders." *Current Biology* 28(14): 2324–2330.e2, 2018.

Mouhat, S., N. Andreotti, B. Jouirou, and J. M. Sabatier. "Animal toxins acting on voltage-gated potassium channels." *Current Pharmaceutical Design* 14: 2503–2518, 2008.

Musk, E., and Neuralink. "An integrated brain-machine interface platform with thousands of channels." *J Med Internet Res* 21(10): e16194, 2019.

National Academy of Sciences. "Laboratory experiences and student learning." Pages 75–115 (chapter 3) in *America's Lab Report: Investigations in High School Science*. Washington, DC: National Academies Press, 2006. https://doi.org/10.17226/11311

National Geographic. "Fossil daddy longlegs sports a 99-million-year erection." February 1, 2016. https://news.nationalgeographic.com/2016/02/160201-arachnids-harvestman-penis-amber-fossil-animals-science

Navarro, J. *A History of the Electron*. Cambridge, UK: Cambridge University Press, 2012.

Navntoft, C. A., J. Marozeau, and T. R. Barkat. "Ramped pulse shapes are more efficient for cochlear implant stimulation in an animal model." *Sci Rep* 10: 3288, 2020. https://doi.org/10.1038/s41598-020-60181-5

New York Public Library Blog. "Ben Franklin on cooking turkey . . . with electricity." November 24, 2014. https://www.nypl.org/blog/2014/11/24/ben-franklin-turkey-electricity

New York Times. "G. M. Ogle, authority on the death chair, dies." October 15, 1931.

———. "Two shocks were needed." February 9, 1892.

Nicholson, W. "Observations on the *Electrophorus*, tending to explain the means by which the torpedo and other fish communicate the electric shock." *Journal of Natural Philosophy, Chemistry and the Arts* 1: 355–359, 1797.

Nirenberg, S., and C. Pandarinath. "Retinal prosthetic strategy with the capacity to restore normal vision." *Proceedings of the National Academy of Sciences, USA* 109(37): 15012–15017, 2012.

Nugent, C., G. Berdine, and K. Nugent. "The undead in culture and science." *Proceedings of the Baylor University Medical Center* 31(2): 244–249, 2018.

Orlando Sentinel. "Lightning kills giraffe in Disney Park." July 24, 2003. http://articles.orlandosentinel.com/2003-07-24/news/0307240055_1_giraffe-disney-world-animal-animal-kingdom

Pancaldi, G. *Volta: Science and Culture in the Age of Enlightenment*. Princeton, NJ: Princeton University Press, 2003.

Panebianco, M., A. Rigby, J. Weston, and A. G. Marson. "Vagus nerve stimulation for partial seizures." *Cochrane Database of Systematic Reviews* 4: CD002896, 2015. DOI: 10.1002/14651858.CD002896.pub2.

Parent, A. "Giovanni Aldini: From animal electricity to human brain stimulation." *Canadian Journal of Neurological Sciences* 31(4):576–584, 2004.

Peimani, A. R., G. Zoidil, and P. Rezai. "A microfluidic device to study electrotaxis and dopaminergic system of zebrafish larvae." *Microfluidics* 12(1): 014113, 2018.

Peng, Z., and C. Shikui. "Study on electrocution death by low-voltage." *Forensic Science International* 76(2): 115–119, 1995.

Perlmutter, J. S., and J. W. Mink. "Deep brain stimulation." *Annu Rev Neurosci* 29: 229–257, 2006.

Physics World (website). "Electric eel inspires new power source," December 13, 2017. https://physicsworld.com/a/electric-eel-inspires-new-power-source/

Piccolino, M., and M. Bresadola. *Shocking Frogs: Galvani, Volta, and the Electrical Origins of Neuroscience.* Oxford, UK: Oxford University Press, 2013.

Pitzer, G. C. *Electricity in Medicine and Surgery*. St. Louis, MO: Self-published, 1883.

Poore, G. V. *Selections from the Clinical Works of Dr. Duchenne (de Boulogne)*. London: New Sydenham Society, 1883.

Popular Mechanics. "The most gruesome government report ever written evaluates 34 ways to execute a man." March 16, 2017. https://www.popularmechanics.com/technology/a25689/gerry-commission-report-methods-of-execution/

Premchand R. K., K. Sharma, S. Mittal, R. Monteiro, S. Dixit, I. Libbus, L. A. DiCarlo, et al. "Extended follow-up of patients with heart failure receiving autonomic regulation therapy in the ANTHEM-HF study." *Journal of Cardiac Failure* 22(8): 639–642, 2016. https://doi.org/10.1016/j.cardfail.2015.11.002

Price, T. G., and M. A. Cooper. "Electrical and lightning injuries." Pages 1906–1914 in Marx, J., Walls, R., and Hockberger, R., (eds.), *Rosen's Emergency Medicine-Concepts and Clinical Practice*. 8th ed. Philadelphia, PA: Elsevier Health Sciences, 2013.

Pulvermacher Galvanic Company. *The Best Known Curative Agent: Pulvermacher's Electric Belts and Bands for Self-Application*. Cincinnati, OH: Pulvermacher, 1879. [Company's sales brochure.]

Rådman, L., Y. Nilsagård, K. Jakobsson, Å. Ek, and L. G. Gunnarsson. "Electrical injury in relation to voltage, 'no-let-go' phenomenon, symptoms and perceived safety culture: A survey of Swedish male electricians." *Int Arch Occup Environ Health* 89(2): 261–270, 2016.

Rajagopal, R., S. Chakrabarti, and S. Graver. "Satisfaction with electroconvulsive therapy among patients and their relatives." *J ECT* 29: 283–290, 2013.

Ramón y Cajal, S. *Recollections of My Life* (Memoirs of the American Philosophical Society, vol. 8, 1937). Cambridge, MA: MIT Press, 1966 (reprint ed.).

Ray, A. K. "Does electroconvulsive therapy cause epilepsy?" *J ECT* 29(3): 201–205, 2013.

Reynolds, J. B. "Development and status of electric fishing as a scientific sampling technique." Pages 49–61 in Sakagawa, G. T. (ed.), *Assessment Methodologies and Management. Proceedings of the World Fisheries Congress, Theme 5*. New Delhi: Oxford and IBH Publishing, 1995.

Ritenour, A. E., M. J. Morton, J. G. McManus, D. J. Barillo, and L. C. Cancio. "Lightning injury: A review." *Burns* 34(5): 585–594, 2008.

Rolls, M. M, and T. J. Jegla. "Neuronal polarity: An evolutionary perspective." *J Exp Biol* 218(4): 572–580, 2015.

Rosenberg, S. *Assessing the Healing Powers of the Vagus Nerve: Self-Help Exercises for Anxiety, Depression, Trauma, and Autism*. Berkeley, CA: North Atlantic Books, 2017.

Russell, S. *Human Compatible: Artificial Intelligence and the Problem of Control*. New York: Viking Press, 2019.

Saal, H. P., and S. J. Bensmala. "Touch as a team effort: Interplay of submodalities in cutaneous sensibility." *Trends Neurosci* 37: 689–697, 2014.

Sanchez, J. C. *Neuroprosthetics: Principles and Applications* (Rehabilitation Science in Practice Series). New York: CRC Press, 2016.

Sanchez, J. C., and J. C. Principe. *Brain-Machine Interface Engineering*. New York: Morgan & Claypool, 2007.

Scherz, P., and S. Monk. *Practical Electronics for Inventors*. 4th ed. New York: McGraw-Hill, 2016.

Schlesinger, H. *The Battery: How Portable Power Sparked a Technological Revolution*. New York: HarperCollins, 2010.

Schuurman, P. R., D. A. Bosch, P. M. Bossuyt, G. J. Bonsel, E. J. van Someren, R. M. de Bie, M. P. Merkus, and J. D. Speelman. "A comparison of continuous thalamic stimulation and thalamotomy for suppression of severe tremor." *N Engl J Med* 342(7): 461–468, 2000.

Schwartz, A. B., X. T. Cui, D. J. Weber, and D. W. Moran. "Brain-controlled interfaces: Movement restoration with neural prosthetics." *Neuron* 52: 205–220, 2006.

Schwemmer, M. A., N. D. Shomrock, P. B. Sederberg, J. E. Ting, G. Sharma, M. A. Bockbrader, and D. A. Friedenberg. "Meeting brain-computer interface user performance expectations using a deep neural network decoding framework." *Nature Medicine* 24: 1669–1676, 2018.

ScienceBlogs. "Death by lightning for giraffes, elephants, sheep and cows." July 14, 2009. http://scienceblogs.com/tetrapodzoology/2009/07/15/mammal-deaths-by-lightning

Sengupta, B., A. A. Faisal, S. B. Laughlin, and J. E. Niven. "The effect of cell size and channel density on neuronal information encoding and energy efficiency." *J Cereb Blood Flow Metab* 33(9): 1465–1473, 2013.

Sengupta, B., and M. B. Stemmler. "Power consumption during neuronal computation." *Proceedings of the IEEE* 102(5): 738–750, 2014.

Seo, H. S., K. J. Jeon, T. Hummel, and B. C. Min. "Influences of olfactory impairment on depression, cognitive performance, and quality of life in Korean elderly." *European Archives of Otorhinolarygology* 266(11): 1739–1745, 2009.

Seth, A. K. *Rewired: An Unlikely Doctor, a Brave Amputee, and a Medical Miracle That Made History*. Nashville, TN: Thomas Nelson, 2019.

Seymour, M. *Mary Shelley*. New York: Grove Press, 2000.

Sharber, N. G., and J. S. Black. "Epilepsy as a unifying principle in electrofishing theory: A proposal." *Transactions of the American Fisheries Society* 128: 666–671, 1999.

Sheumack, D. D., M. E. Howden, I. Spence, and R. J. Quinn. "Maculotoxin: A neurotoxin from the venom glands of the octopus *Hapalochlaena maculosa* identified as tetrodotoxin." *Science* 199(4325): 188–189, 1978.

Shorter, E., and D. Healy. *Shock Therapy: A History of Electroconvulsive Treatment in Mental Illness*. New Brunswick, NJ: Rutgers University Press, 2007.

Simulik, V. (ed.). *What Is the Electron?* Montreal: C. Roys Keys, 2005.

Sinclair, D.J.M., S. Zhao, F. Qi., K. Nyakyoma, J.S.W. Kwong, and C. E. Adams. "Electroconvulsive therapy for treatment-resistant schizophrenia." *Cochrane Database Syst Rev* 3(3): CD011847, 2019.

Smith, S.J.M. "EEG in the diagnosis, classification, and management of patients with epilepsy." *J Neurol Neurosurg Psychiatry* 76(Suppl II): ii2–ii7, 2005.

Smyth, A. H. *The Writings of Benjamin Franklin*, vol. 3. New York: Macmillan, 1905.

Stone, A. D. *Einstein and the Quantum: The Quest of the Valiant Swabian*. Princeton, NJ: Princeton University Press, 2013.

Strauss, J. *Human Remains: Medicine, Death, and Desire in Nineteenth Century Paris*. New York: Fordham University Press, 2012.

Swanson, L. W., E. Newman, A. Araque, and J. M. Dubinsky (eds.). *The Beautiful Brain: The Drawings of Santiago Ramón y Cajal*. New York: Abrams, 2017.

Talan, J. *Deep Brain Stimulation: A New Treatment Shows Promise in the Most Difficult Cases*. New York: Dana Press, 2009.

Tennessean. "Death row inmates weigh the choice between the electric chair and lethal injection." November 3, 2018. https://www.tennessean.com/story/news/2018/11/03/tennessee-death-penalty-electric-chair-lethal-injection/1855440002/

———. "Tennessee executes David Earl Miller by electric chair." December 7, 2018. https://www.tennessean.com/story/news/2018/12/06/david-earl-miller-execution-electric-chair-tennessee/2158239002/

Thomson, J. J. "The connection between chemical combination and the discharge of electricity through gases." In *Report of the British Association for the Advancement of Science*. London: John Murray, 1894.

———. *The Electron in Chemistry: Being Five Lectures Delivered at the Franklin Institute, Philadelphia*. London: Chapman & Hall, 1923.

———. *Notes on Recent Researches in Electricity and Magnetism*. Oxford, UK: Clarendon Press, 1893.

Timeline (website). "One unlucky cat was the only victim when the ancient St. Mark's bell tower collapsed in Venice." August 4, 2017. https://timeline.com/venice-bell-tower-collapse-5789fb4434b9

Turkel, W. J. *Spark from the Deep: How Shocking Experiments with Strongly Electric Fish Powered Scientific Discovery*. Baltimore, MD: Johns Hopkins University Press, 2013.

Twilley, N. "Sight unseen" (print ed.); "Seeing with your tongue" (online ed.). *New Yorker*, May 15, 2017.

Tyler, D. J. "Peripheral nerve stimulation." Chapter 2.4 in Horch, K., and Kipke, D. (eds.), *Neuroprosthetics: Theory and Practice*. 2nd ed. Series on Bioengineering and Biomedical Engineering—vol. 8. Singapore: World Scientific Publishing, 2017.

UK ECT Review Group. "Efficacy and safety of electroconvulsive therapy in depressive disorders: A systematic review and meta-analysis." *Lancet* 361(9360): 799–808, 2003.

Uman, M. A. *All About Lightning*. 2nd ed. Mineola, NY: Dover Publications, 1986.

———. *The Lightning Discharge*. Orlando, FL: Academic Press, 1987.

US Environmental Protection Agency. *New Perspectives in Electrofishing*. Office of Research and Development, EPA/600/R-99/108, February 2000.

Vance, A. *Elon Musk: Tesla, SpaceX, and the Quest for a Fantastic Future*. New York: HarperCollins, 2015.

Van Harreveld, A. "On the galvanotropism and oscillotaxis of fish." *Journal of Experimental Biology* 15: 197–208, 1938.

Verge (website). "How exactly did lightning kill 323 reindeer in Norway?" August 29, 2016. https://www.theverge.com/2016/8/29/12690402/lightning-strike-kills-norway-reindeer-death-why-science

Vibert, R. "Neurophysiology of electric fishing." *Transactions of the American Fisheries Society* 92: 265–275, 1963.

Vibert, R. (ed.). *Fishing with Electricity, Its Application to Biology and Management*. London: Fishing News Books, 1967.

Viemeister, P. E. *The Lightning Book*. Cambridge, MA: MIT Press, 1972.

VOA News (website). "Franklin-designed lightning rod saves historic Maryland building." July 2, 2016. https://www.voanews.com/a/franklin-designed-lightning-rod-saves-historic-maryland-building/3401691.html

Warshofsky, F. *The Chip War: The Battle for the World of Tomorrow*. (1st ed.). New York: Scribner, 1989.

Weerakoon, P., E. Culurciello, Y. Yang, J. Santos-Sacchi, P. J. Kindlmann, and F. J. Sigworth. "Patch-clamp amplifiers on a chip." *Journal of Neuroscience Methods* 192(2): 187–192, 2010.

Wesley, J. *Desideratum: Or, Electricity Made Plain and Useful*. London: W. Flexney, 1760.

Wexler, A. "The medical battery in the United States (1870–1920): Electrotherapy at home and in the clinic." *Journal of the History of Medicine and Allied Sciences* 72(2): 166–192, 2017.

Wick, R., and R. W. Byard. "Electrocution and the autopsy." Pages 53–66 in Tsokos, M. (ed.), *Forensic Pathology Reviews*, vol 5. Totowa, NJ: Humana Press, 2008.

Wilcox, C. *Venomous: How Earth's Deadliest Creatures Mastered Biochemistry*. New York: Scientific American/Farrar, Straus and Giroux, 2016.

Wilson, B. S., and M. F. Dorman. "Cochlear implants: A remarkable past and a brilliant future." *Hear Res* 242(1–2): 3–21, 2008.

Wootton, D. *Bad Medicine: Doctors Doing Harm since Hippocrates*. Oxford, UK: Oxford University Press, 2007.

Wulf, A. *The Invention of Nature: Alexander von Humboldt's New World*. New York: Vantage Books, 2015.

Yan, B., and S. Nirenberg. "An embedded real-time processing platform for optogenetic neuroprosthetic applications." *IEEE Transactions on Neural Systems and Rehabilitation Engineering* 26(1): 233–243, 2018.

Yoder, R. B. "An Arduino-based alternative to the traditional electronics laboratory." Paper presented at the Conference on Laboratory Instruction beyond the First Year of College, College Park, MD, July 22–24, 2015.

Young, J. Z. "The function of the giant nerve fibres of the squid." *Journal of Experimental Biology* 15: 170–185, 1938.

Zanchetti, A., S. C. Wang, and G. Moruzzi. "The effect of vagal afferent stimulation on the EEG pattern of the cat." *Electroencephalography and Clinical Neurophysiology* 4(3): 357–361, 1952.

INDEX

Page numbers in italics refer to figures.